T0265120

DAIRY SCIENCE HANDBOOK
VOLUME 15

International Stockmen's School Handbooks

Dairy Science Handbook
Volume 15
edited by Frank H. Baker

The 1983 *International Stockmen's School Handbooks* include more than 200 technical papers presented at this year's Stockmen's School—sponsored by Winrock International—by outstanding animal scientists, agribusiness leaders, and livestock producers expert in animal technology, animal management, and general fields relevant to animal agriculture.

The *Handbooks* represent advanced technology in a problem-oriented form readily accessible to livestock producers, operators of family farms, managers of agribusinesses, scholars, and students of animal agriculture. The *Beef Cattle Science Handbook*, the *Dairy Science Handbook*, the *Sheep and Goat Handbook*, and the *Stud Managers' Handbook* each include papers on such general topics as genetics and selection; general anatomy and physiology; reproduction; behavior and animal welfare; feeds and nutrition; pastures, ranges, and forests; health, diseases, and parasites; buildings, equipment, and environment; animal management; marketing and economics (including product processing, when relevant); farm and ranch business management and economics; computer use in animal enterprises; and production systems. The four *Handbooks* also contain papers specifically related to the type of animal considered.

Frank H. Baker is director of the International Stockmen's School at Winrock International, where he is also program officer of the National Program. An animal production and nutrition specialist, Dr. Baker has served as dean of the School of Agriculture at Oklahoma State University, president of the American Society of Animal Science, president of the Council on Agricultural Science and Technology, and executive secretary of the National Beef Improvement Federation.

A Winrock International Project

Serving People Through Animal Agriculture

This handbook is composed of papers presented at the
International Stockmen's School
January 2-6, 1983, San Antonio, Texas
sponsored by Winrock International

A worldwide need exists to more productively exploit animal
agriculture in the efficient utilization of natural and human
resources. It is in filling this need and carrying out the public
service aspirations of the late Winthrop Rockefeller, Governor
of Arkansas, that Winrock International bases its mission to
advance agriculture for the benefit of people. Winrock's focus
is to help generate income, supply employment, and provide
food through the use of animals.

DAIRY SCIENCE HANDBOOK VOLUME 15

edited by Frank H. Baker

Routledge

Taylor & Francis Group

LONDON AND NEW YORK

First published 1983 by Westview Press

Published 2018 by Routledge
52 Vanderbilt Avenue, New York, NY 10017
2 Park Square, Milton Park, Abingdon, Oxon OX14 4RN

Routledge is an imprint of the Taylor & Francis Group, an informa business

Copyright © 1983 by Winrock International

All rights reserved. No part of this book may be reprinted or reproduced or utilised in any form or by any electronic, mechanical, or other means, now known or hereafter invented, including photocopying and recording, or in any information storage or retrieval system, without permission in writing from the publishers.

Notice:
Product or corporate names may be trademarks or registered trademarks, and are used only for identification and explanation without intent to infringe.

ISBN 13: 978-0-367-01950-1 (hbk)
ISBN 13: 978-0-367-16937-4 (pbk)

CONTENTS

PREFACE

The Dairy Science Handbook includes presentations made at the International Stockmen's School, January 2-6, 1983. The faculty members of the School who authored this fifteenth volume of the Handbook, along with books on Beef Cattle, Horses, and Sheep and Goats, are scholars, stockmen, and agribusiness leaders with national and international reputations. The papers are a mixture of tried and true technology and practices with new concepts from the latest research results of experiments in all parts of the world. Relevant information and concepts from many related disciplines are included.

The School has been held annually since 1963 under Agriservices Foundation sponsorship; before that it was held for 20 years at Washington State University. Dr. M. E. Ensminger, the School's founder, is now Chairman Emeritus. Transfer of the School to sponsorship by Winrock International with Dr. Frank H. Baker as Director occurred late in 1981. The 1983 School is the first under Winrock International's sponsorship after a one-year hiatus to transfer sponsorship from one organization to the other.

The five basic aims of the School are to:
1. address needs identified by commercial livestock producers and industries of the United States and other countries,
2. serve as an educational bridge between the livestock industry and its technical base in the universities,
3. mobilize and interact with the livestock industry's best minds and most experienced workers,
4. incorporate new livestock industry audiences into the technology transfer process on a continuing basis, and
5. improve the teaching of animal science technology.

Wide dissemination of the technology to livestock producers throughout the world is an important purpose of the Handbooks and the School. Improvement of animal production and management is vital to the ultimate solution of hunger problems of many nations. The subject matter, the style of presentation, and opinions expressed in the papers are those of the authors and do not necessarily reflect the opinions of Winrock International.

This handbook is copyrighted in the name of Winrock International and Winrock International encourages its use and the use of its contents. Permission of Winrock International and the authors should be requested for reproduction of the material. In the case of papers with individual copyrights or of illustrations reproduced by permission of other publishers (indicated on the copyright page), contact those authors or publishers.

ACKNOWLEDGMENTS

Winrock International expresses special appreciation to the individual authors, staff members, and all others who contributed to the preparation of the Dairy Science Handbook. Each of the papers (lectures) was prepared by the individual authors. The following editorial, word processing, and secretarial staff of Winrock International assisted the School Director in reading and editing the papers for delivery to the publishers.

Editorial Assistance

Jim Bemis, Production Editor
Essie Raun
Betty Stonaker

Word Processing and Secretarial Assistance

Patty Allison, General Coordinator
Shirley Zimmerman, Coordinator of Word Processing
Darlene Galloway
Tammy Chism
Jamie Whittington
Venetta Vaughn
Kerri Alexander, Computing Specialist Assistant
Ramona Jolly, Assistant to the School Director
Natalie Young, Secretary for the School

Part 1

A GLOBAL VIEW
OF ANIMAL AGRICULTURE

us in this day and time to find that more pounds of salted or pickled meat were exported from the U.S. to Europe and the West Indies over a hundred years ago than are exported fresh today.

Then we made great progress--refrigeration was developed. This was an amazing advance--fresh meat was available year round! But refrigeration also became a shackle and a chain. In recent years, especially in the U.S., we have become chained to our refrigerators. We can go only so far as we can take our refrigerators!

On the other hand, refrigeration has also brought to the U.S. and Europe unparalleled adventures in good taste, variety, and quality of meat. Refrigeration has allowed us in the U.S. to enter the roast, steak, and hamburger era in a way never thought possible. But it has also allowed U.S. beef producers to go wandering off into the insignificant. We began looking inward and forgot the rest of the world, thinking, "Those dumb people overseas wouldn't know good beef if they saw it."

Instead of seeking wider markets, we have spent our time over the last 70 to 80 years arguing over such insignificant things as horns or lack of horns; or whether a few hairs of a different color are wrongly placed; or trying to make bulls and cows look like steers. Now we are rushing about and exchanging thousands of dollars on the difference of a centimeter or two in testicle size, trying to put legs on cattle that might look better on thoroughbred horses. And, oh, the hours we spend washing, puffing, combing tail hair so as to make a nice round-looking ball at the end of a show animal's tail! Insignificant fadism in the extreme!

We in the U.S.--the greatest food-animal producers in the world--have perhaps been so busy arguing over minute details of grade standards, which no one but a few experts understand, and over other inward-looking insignificant details, that we have missed the greatest world market for meat that has ever existed since the earliest Homo sapiens killed their first Bos primigenius (auroch or wild ox). In short, "we have been so busy fighting gnats that we let the herd get away."

There is a big world all around us--some four billion people--all hungry for meat! There are in this populous world of ours today approximately one billion people who exist at a malnourished level. Most of the people in the world today do not get the health benefit that we have proved comes from a diet in which food animal protein is the major ingredient. If, indeed, it is true that a good diet containing major elements of food animal protein makes for more intelligent, more ambitious, harder working people, doesn't it follow that U.S. livestock producers should seek to contribute that food animal protein to mankind--at a profit? Would not this contribution be greater than wars or multibillion dollar giveaway programs?

Think about it! What if the approximately 800 million Chinese could be induced to consume only one (1) pound of food animal protein per year above their present one pound? What would demand for 800 million pounds of meat do for beef, pork, sheep, and goat--food-animal--producers in the U.S.? Add to the Chinese the Russians, the Indians, and the Southeast Asians, and you have a need for food-animal protein that boggles the mind!

Thus, U.S. livestock producers must move from their present "cowboy" mentality to one in which they seek to become world marketeers of food-animal protein. We must move from being little more than herders to thinking of ourselves as food-animal producers whose goal is to sell meat protein to a hungry world--at a profit!

And this shift to world food-animal-protein suppliers can be made while we continue to supply the American consumer with desired beef, milk, pork, and mutton.

How?

Think big.

Think systems.

Let's be more specific. Here are what I believe to be the essential elements of developing a Global Food Animal Protein System.

Phase I. Search out a small or medium-sized country that has unfulfilled nutrition and food problems, heavy population density, and a reasonable amount of foreign exchange.

Study the food habits of the population in detail. Find out exactly how the people prefer to eat meat. Learn all there is to know about their overall food habits, preferences. Study their tariff laws, their customs, their religion--everything that could make an impact on their meat consumption. Learn to think as the natives think and react.

All the while your goal is a modest one: Increase per capita meat consumption, on the average, by one (1) pound.

The key to this phase is to drop the American habit of turning up our noses at the way other people like their meat. Just because we like roasts, there's no reason that all the rest of the world must eat roast the way we do. If the natives want barbecued tail bone, let's not call them stupid and say there is no market, and retreat to the nearest McDonald's. Let's sell them tail bones! The goal is a good old American custom: Study a market, decide what can be sold at a profit, then produce it well enough to earn a profit.

Phase II. Devise a production and processing system to deliver the product without waste of time or effort from conception to consumption. We in U.S. agriculture now do just exactly the opposite: We produce something and then try to peddle it in the form we want--or "to hell with 'em."

Phase III. Take the basic lessons learned in one country with modest goals, and apply them to other countries one by one--at a profit. Soon, we will have made a world contribution.

Let's explore some other dimensions of the idea of supplying food-animal protein to a malnourished world.

Let's look at Russia. Russia desperately needs meat. At present, they get their meager supplies from their own limited meat production system. They import some meat from Australia and some from their satellite European countries, but not nearly as much as the Russian population needs and wants. Instead of importing meat, the Russians are trying to do it the long, hard way: Import grains from all over the world, then process and feed them through their herds and flocks.

Has the U.S. ever seriously studied with the Russians (or the Hungarians, or the Saudi Arabians) a policy of importing ready-to-consume meat products (mutton and chicken) instead of raw grains and the resulting timelag to consumers? Wouldn't it be possible, technologically, for us to ship lean grass beef to Russia at a profit? Russians prefer beef that tastes much different from our own grain-fed beef. Mostly they get cull cow or bull beef--if they get any beef at all. Let's sell them what they are used to eating.

We can load forage-produced boneless beef on planes at Atlanta or Dallas, go up to 35,000 feet and quick-freeze it at no cost, and land it in Moscow 16 hours later. Too expensive? We don't know. So far as I know, we have never even tried it. Flying too much moisture? We can dehydrate beef fibers, reconstitute them in Istanbul or Pakistan. Remember our hunting, frontier-busting forebears? How did they transport beef supplies when traveling alone? Jerky beef. It needed no refrigeration. Wouldn't a malnourished native of Gambia be glad to have some jerky beef to mix with his root foods and maize?

American technological expertise is envied the world over. But we haven't used our skill to try to supply food-animal protein to a food-deficient world. Can we not preserve meat and milk through irradiation and/or treatment so that it can be held on a pantry shelf? Good U.S. agricultural policy, it seems to me, should be one that uses U.S. technical advantages to the benefit of the U.S.--and then for other citizens of the world. If so, we must begin to shift from a policy of shipping raw, unprocessed grains to a policy of exporting value-added, more nearly ready-to-eat food-animal products. The U.S. has the livestock. We have more feed and feedgrains than anyone. We have the nutrition knowledge. We have financing. Shouldn't we ship food-animal protein products and short-circuit the long food production chain that most countries now are trying to establish by buying our grain and soybeans?

What about the strategic military considerations of such a policy? Ship food protein ready to eat instead of supplying grain? That should be the U.S. goal. We must do this instead of supplying grains so the countries can build their own livestock infrastructure for future self-sufficiency.

The U.S. food-animal production industry has seen some quantum jumps in the last few years in development of all-tender, flavorful meat products. Meat protein cubes, dehydrated meat fibers, ready-cooked meals are just a few of the new forms and products pouring out of labs. More are about to emerge. However, we cannot hope to supply all the world with edible animal protein. Perhaps we should adopt this policy: Sell edible animal food protein to those countries whose needs are greatest and most immediate. To those countries with developed livestock industries, sell genetic and germ plasm materials to use in upgrading their production. This whole field of genetic engineering now in its infancy in the U.S. may shatter much of our previous livestock thinking. Livestock producer leaders must be equally bold in their thinking!

I have no doubt that if U.S. food animal producers--at one time called cowmen, sheepmen, dairymen--would set out to develop a Global Food Animal Protein System, they could do so rather easily. But to do so they must throw off the shackles of the past, think aggressively American, think profit, think systems, and above all think BIG.

* * *

"Populations of the lower- and middle-income countries still are increasing rapidly, and in many countries there is a growing number of affluent people whose diets are being upgraded to include more red meat, dairy products, and eggs. As more countries are unable to meet their national requirements for staple foods, their governments are looking for outside sources of supply, often on an urgent basis. It is dangerous politically for national leaders to let food shortages occur, driving prices up; they run the risk of unrest, violence, and even overthrow of governments."

Beyond the Bottom Line
The Rockefeller Foundation

REFERENCES

Fagan, B. M. Men of the Earth: An Introduction to World Prehistory. Little, Brown and Co.

2

FOUNDATION OF CIVILIZATION: FOOD

Allen D. Tillman

> "And he gave it for his opinion that whoever
> could make two ears of corn or two blades of
> grass grow on a spot of ground where only one
> grew before would deserve better of mankind and
> do more service to his country than the whole
> race of politicians put together."
> —Johnathan Swift
> The Voyage to Brobdingnag
> in Gulliver's Travels

I chose this quotation because it is apparent to me that man now has the knowledge and power to make two ears of corn or two blades of grass grow on a spot of ground that formerly would grow one or less.

Civilization is defined as "an advanced state of human society in which there is a high level of culture, science, industry, and government." The high level of civilization that we enjoy today has resulted from many technological developments in agriculture that increased the amount of food produced and the efficiency of human labor in producing it. Each innovation freed more people for the development of human society and of the culture, science, industry, and government found in it.

In discussing some of the developments, this paper is divided into major sections as follows: (1) a brief history of agricultural development worldwide; (2) the close relationship of agricultural and industrial developments in modern societies; (3) some characteristics of a successful agriculture at national levels, and (4) reasons for optimism about the world food problem.

HISTORY OF AGRICULTURAL DEVELOPMENTS IN THE WORLD

> "History celebrates the battlefield whereon we
> meet our death, but scorns to speak of the plow-
> ed fields whereby we thrive; it knows the names
> of the king's bastards, but cannot tell us the

origin of wheat. This is the way of human folly."

--Jean Henri Fabre

For convenience, I have divided this section into three parts, as follows: the gathering and hunting stage (2,000,000 - 7000 B.C.), the low-technology stage (7000 B.C. - 1750 A.D.); and the scientific stage (1750 A.D. - the present).

The Gathering/Hunting Stage

"Cultural man has been on earth for some 2 million years and for 99% of this period he has lived as a hunter/gatherer. Only in the last 10,000 years has man begun to domesticate plants and animals, to use metals, and to harness energy sources other than the human body. Of the estimated 90 billion people who have lived out a life span on earth, over 90% have lived as hunters/gatherers, about 6% by agriculture, and 4% have lived in industrial societies. To date, the hunting/gathering way of life has been the most successful and persistent adaptation man has ever achieved."

--Lee and Devore (1968)

The first ancestor of man, Australopithecus, appeared on earth about two million years ago. His main invention was the knife, which was made by putting an edge on a pebble, an invention that permitted him to kill animals, skin, and cut meat, thereby changing him from an herbivora to an omnivora. This change was dramatic, because the addition of dietary meat reduced the bulkiness of his diet by about two-thirds, permitting him to leave the trees and to become more mobile to better utilize the rapidly developing savannas. Also, meat required less time to gather, thus he had more time for social activities - improvement in communication skills and in his tools. And so man for the first time released the brake that environment imposes on his fellow creatures. It is significant that this basic tool was not changed very much for the next million years, attesting to the strength of the invention.

Homo erectus came onto the scene about one million years ago. His ability to walk upright freed his arms, which improved his ability to hunt, and his greater ability to adapt to many ecosystems permitted him to spread out from his place of origin, Africa. In fact, the classical discovery of Homo erectus was the Peking man, who lived in China about 400,000 years ago. He was the discoverer of fire, which was used for warmth and cooking. The Neanderthal man, who was discovered in Europe, appears to have led directly to us, Homo sapiens.

The test of the ability of Homo erectus to adapt came about 500,000 years ago when the Pleistocene Ice Age covered much of the earth. Clans of 40 or more moved to caves for protection and work, a move that required a new organization - the young and stronger men (usually 10 per clan - Willham, 1980) were the hunters, while the remainder were assigned duties in keeping with their abilities. Some of the dwellers even had time to paint on the cave walls. Bronowski (1973) felt that these are saying to all - "This is my mark, this is man." Man is now saying that he has the ability to shape the world and is not a mere creature to be shaped by the environment.

The great glaciers began to retreat about 30,000 years ago. Left in their wakes were great savannas that were soon filled with grasses of all kinds and with cloven-hoofed ruminants to consume these. In response, the clans came out of the caves and spread out over the plains following the animals, going north in the summer and returning south in the winter. This was the beginning of a transhumant way of life, which later became dominant in many areas of Eurasia. In fact, there are cases of this way of life even today-- East Africa, North Africa, Finland.

The dog was domesticated in about 20,000 B.C., and this greatly increased man's ability to hunt. Also , there developed a symbiotic relationship between man and animals: animals furnished meat, skins, and other products for man, while man furnished some protection and salt to the animals. Urine of meat-eating man contains salt, a valuable commodity to ruminants in many salt-deficient areas. Man developed oars in about 20,000 B.C. and the bow and arrow in about 15,000 B.C.; both inventions increased his efficiency as a hunter. Man domesticated the gregarious sheep and goats during the latter part of this stage, and benefited greatly by the increased food supply from these animals.

At the end of this stage, about 10,000 to 7,000 B.C., and for a long time after man had already established village agriculture, animals continued to be an important source of food for man. The live animal represents, until it is sacrificed, a reserve food supply. This fact is often forgotten or overlooked by planning economists, who plan for the aid programs that are given to the developing countries. Early man recognized the importance of animals: so much so that the root word used for money in many languages reflects the importance of animals to man (Leeds and Vayda, 1965).

Low Technology Agriculture (7000 B.C. - 1750 A.D.)

> "The greatest single step in the ascent of man is the change from nomad to village agriculture."
>
> --Bronowski (1973)

With the receding of the Pleistocene ice, there came great environmental changes in the Old World. The hot and dry winds off the Eurasian Steppes eliminated all but the hardier grasses in much of North Africa and some of Asia, thereby turning some of the lands into semideserts or deserts. As a result, wild animals migrated to the river valleys of the Euphrates, Tigris, Indus, Nile, and Yellow rivers. Agriculture began when the nomads decided to stay put to exploit plants. Whether the nomads planned to develop agriculture or were the benefactors of two genetic accidents is not clear. What is clear is the fact that the early ones came to hunt and to gather wild wheat. By a genetic accident (Harlan, 1975), wild wheat containing 14 chromosomes, crossed with wild oat grass, also containing 14 chromosomes, to produce a fertile hybrid, called Emmer. It contains 28 chromosomes, thus the grain was much larger than wild wheat. Therefore, man began to cultivate it. Emmer's grain is so tightly bound to the husk and chaff that it is easily dispersed by the wind. As a result, it spread over wide areas. It appears that by another genetic acident, bread wheat came to the settled agriculturalists: Emmer crossed with another wild oat grass to produce still another fertile hybrid, bread wheat, containing 42 chromosomes, which has a large grain. In contrast to Emmer, when bread wheat is broken, the chaff flies off, leaving the grain in place, thus it is not spread by the wind. With the advent of bread wheat, which man has to plant and cultivate, man and wheat developed a symbiotic relationship that remains up to this day (Heiser, 1978).

Farming and husbandry in a settled agriculture creates an atmosphere from which technology and science take off (Bronowski, 1973). The first tools used by village agriculturalists were the digging stick, which was invented about 7000 B.C. This later evolved to a crude plow, the footplow that used human labor, which appeared in about 6000 B.C. The cow was domesticated in about 6000 B.C., and when man yoked the ox to the plow he for the first time began to utilize a power source greater than the human muscle. This was undoubtedly the most powerful invention of this early period, making it possible for man to wrest a great surplus of agricultural products from nature. The surplus food released more men to create, invent, innovate, and to build great civilizations--something for which the nomads had never had time. Civilizations, with their specific cultures, developed on the flood plains on the Nile, the Tigris-Euphrates, the Indus, and the Yellow Rivers. Many of their activities, such as irrigation, required cooperation by many men; therefore there developed administrative systems that led to the building of empires. Law and government developed. The great cities of that period thrived on a cereal-based agriculture. Trade between cities developed in order for them to acquire necessities - salt, spice, metals, etc. (Thomas, 1979).

The animals found in the settled villages, up to about 3000 B.C., were goats, sheep, the oxen and the onager, a kind of wild ass. As long as the animals were the servants of agriculture, all went well. But some time after 4000

B.C., the horse was domesticated and the nomads learned to ride in about 2000 B.C. Thus, the nomad was transformed from a poor wanderer to a threat to the settled villages. Warfare of that period was intensified by the discovery of how to ride the horse, and warfare became a nomad activity. The nomads battered on doors of the settled villages from about 2000 B.C. until the early part of the 14th century A.D. Sometimes the nomads were successful and took over villages, but in all instances, the nomads were absorbed into the villages. Historians have made all of us aware of the great wars waged by the nomads, recording the names of the famous nomads--the Huns, the Phrygians, and the Mongols. The Mongols were defeated in about 1300 A.D., thereby ending the threat of their making nomad life supreme throughout sections of the Old World.

When the horse collar was discovered, the horse became an important draft animal, especially in northern Europe were great teams of horses turned the heavy sod. Without these teams of horses, which permitted the Vikings to produce great surpluses of grains, they could not have been such a military power and threat to much of northern Europe.

The low-technology stage continues right on through the European Renaissance (Thomas, 1979), during which time there were many contributions to agricultural innovations, each with its consequent improvements in food production and the efficiency by which man produced it. As there were many such improvements over time, for sake of time and space, let us summarize the advances:

- Man developed methods for the systematic exploitation of plants.
- Man developed methods for the cultivation of plants for the production of grain--wheat, barley, millet, and rice (Heiser, 1978).
- Man domesticated animals--dog, cow, sheep, goat, and the horse.
- Man developed systems of irrigation.
- Man developed some degrees of mechanization--the digging stick, the plow, the ox-drawn plow, the wheel, and others.

The developments in limited technology were not continuous but came in ebbs and flows throughout the period up to about 1750. There appeared to be a de facto technological ceiling upon agricultural production throughout the entire stage. At the core was the simple Malthusian element--population expansion ultimately pressed against the land, thereby producing malnutrition, famine, disease and, finally, a decrease in population. In China, Hung Liang Chi (Rostow, 1978) the predecessor of Malthus, wrote "during a long reign of peace--the government could not prevent the people from multiplying themselves." Rostow said it well--"During this period of limited technology, if war did not get you, peace did." And so ended the second era.

The Scientific State (1750 - present)

The scientific state began in about 1750 and continues until the present. In this period, the western nations for the first time broke the ceilings on agricultural and industrial technology so that invention and innovation came at a regular flow. The key to these, I feel, was the advent of the "scientific revolution which brought with it experimental science." Experimental science, for the first time, permitted and motivated the formulation of scientific laws to describe general and natural scientific phenomena. This led scientists to design experiments that would lead to the manipulation of nature to man's advantage. This exciting time saw the advent of scientific agricultural societies in which agriculturalists met together for the discussion of scientific discoveries and how these could be put to use by farmers. Innovations came at a faster pace, and as in the past, each invention or innovation increased the level of food production and its efficiency of production. The rates increased rapidly and we now have the development of high-technology agriculture. Some of the basic advancements that are characteristic of this stage are as follows:
- Classification of soils, along with estimates of their fertilities.
- Improved plants by gene manipulations (genetic engineering).
- Improved animals by gene manipulations.
- Scientific utilization of fertilizers.
- Proper use of irrigation.
- Proper use of fermentation and other means of food preservation (Tannahill, 1973).
- Use of insecticides, fungicides, herbicides, vaccines, etc.
- Use of modern techniques in farm mangement.

Many names stand out during this third era; however, it is significant to mention that King George III (better known to Americans for other reasons) gave much support to the newly developing agricultural research in England. Some feel that the innovations resulting from this agricultural research greatly increased agricultural production in Great Britain. In fact, some suggest that it was the English agricultural revolution that permitted that nation to defeat Napoleon at a time when agricultural imports were essentially cut off by France.

Agricultural research is so important that every modern nation has now developed a national agricultural production plan in which agricultural research is a powerful component. Those countries that are lagging in agricultural production are the ones that have been slow in developing good agricultural research, teaching, and extension programs.

How about the United States? Our country had its beginning in 1776, or sixteen years after the third stage began. When our Declaration of Independence was signed, the Revolutionary War was fought, and our people set out to build a nation, fully 90% of our people were directly engaged in agricultural production. If time and space permitted, it would be useful to point out the significant inventions, such as the first cotton gin, the first wheat thresher, the first steam or gasoline-propelled tractor and many others, all American inventions or innovations, and to note the effect of each upon food production and efficiency. However, I will end this section by saying that during the 200 years since Independence, our farm population has decreased from 90% of the total population to only 5%. However, these fewer people are producing food of improved variety and quality, providing nourishment for a vigorous population. Our national agricultural program has been one of the modern success stories.

THE CLOSE RELATIONSHIP OF AGRICULTURAL AND INDUSTRIAL DEVELOPMENT IN MODERN SOCIETIES

In reviewing agricultural developments worldwide, Rostow (1978) noted that the modernization of agriculture and industry have gone hand-in-hand in successful national development programs in the past. In fact, one finds that the modernization of agriculture must precede industrialization in most countries. There are many reasons for the close interrelationship; successful agriculture (Foster, 1978) provides:

- An ever-expanding supply of high quality foods to nourish its people, increasing their vigor.
- An adequate supply of high quality food, available at a reasonable price, to combat inflation which, if left uncontrolled, hampers national development.
- Capital for the expansion of the nonagricultural sector of the economy. This is very critical in the early stages of industrial development because at least 90% of the population in every country studied were farmers at this critical period.
- More food for the nonagricultural population. Therefore, all through the modernization process, more and more people are freed for work in the nonfarm sector.
- Educated and motivated people for work in the nonagricultural sector in the developed countries.
- Land for the nonfarm sector--for highways, railroad, airports, shopping areas, etc.
- A market for the nonfarm sector--tools, machines, medicines, insecticides, clothing, gasoline, etc. Again, this market is most critical in the early stages of industrialization.

SOME CHARACTERISTICS OF A MODERN AND PRODUCTIVE AGRICULTURE AT THE NATIONAL LEVEL

In general, a successful national agriculture results from a good agricultural plan or programs that provide farmers with relevant production information and assures an adequate infrastructure that is needed for both production and marketing of agricultural products. In addition, farmers must receive a fair return from their investment of land, labor, and capital. Otherwise, production will be sporadic rather than continuous. Some specific requirements are:
- Adequate government financial support for research, teaching, and extension (public service) for agriculture.
- Infrastructure: The national government also has to provide certain components of the infrastructure needed by farmers--such as roads, harbors, railroads (in some cases). (Many of the infrastructures in the U.S. are now furnished by industry.)

Some inputs needed by modern farmers are as follows:
- Farm machinery and spare parts
- Farm tools and spare parts
- Power source--gasoline, electricity, diesel fuel
- Fertilizers
- Insecticides, fungicides, herbicides, vaccines, etc.
- Irrigation tools--pumps, pipe, valves, etc.
- Credit
- Others

Those who are familiar with the ready availability of inputs on the American scene might well question why private industry hasn't made these available in many developing countries and in developing agricultural industries. In many cases, private industry will not take the risk of production and distribution of many necessary inputs unless the volume and price justifies the risk. In such situations, the government has to subsidize these inputs until the industry is well along toward development. The writer has spent over 10 years in four developing countries and found that the unavailability of certain inputs, such as vaccines, insecticides, and fertilizers, has served as a severe constraint on production. In the same vein, lack of marketing services results in severe losses of the harvested produce.

Marketing services needed to assist production increases are:
- Farm produce collection and storage
- Processing and subsequent storage of the processed produce
- Wholesale distribution of produce
- Retail distribution of produce
- Marketing--farm markets, small stores, supermarkets, and others.

Even with all of the above components in place, farmers will not produce continuously unless the price for agricultural products pays for the costs of inputs and allows them a fair return on their investments.

REASONS FOR OPTIMISM ABOUT THE WORLD FOOD PROBLEM

The world food problem has two components--the demand side (population) and the supply side (production). Many reports today emphasize that the demand side is growing faster than the supply side. Let us analyze the population problem first.

Population

The demographic facts at first glance are frightening! If we plot population against time, we see that the population growth from about 8000 B.C. until about 1650 A.D. was almost a straight line function, a very slow growth rate. In 1650 A.D., there were about 0.5 billion people, and the rate of increase was only 0.3% per year. At that rate, the population would require 250 years to double. Instead we have had an exploding population since about 1700 A.D., and now some authorities estimate that we will have 6 to 7 billion people in the year 2000.

It is my belief that the population estimates have all been wrong, from Malthus down to the United Nations' recent estimates. For example, the U.N. estimated in 1976 that the world population by the year 2000 would be 7 billion. However, in 1979 that estimate was reduced to 6 billion. Therefore, in 3 years, we "lost" 1 billion people. There are many other examples of inaccuracies. Why is this so? Population projections require assumptions about the choices people have and make in regard to the size of their family. Therefore, it is easy to see why all past assumptions have been wrong.

Simon (1981) maintains that there is a built-in, self-reinforcing logic that forces the rate of population growth to respond to resource conditions. It does so by reducing population growth and size when food resources are limited, and expands it when resources are plentiful. Others have studied population changes in Europe for 1400 A.D. until 1800 A.D., and found that the population did not grow at a constant rate, and that it did not always grow. Instead, there were advances and reverses, provoked by many forces--famine and disease, however, were not the major forces. We know that increased incomes, associated with economic developments, reduce birthrate and population growth. For example, the populations in Singapore, Hong-kong, Japan, and other places have tended to stabilize within the last 20 years.

In summary, these facts lead me to suggest that population size tends to adjust to the production conditions.

Following a technological advance in agriculture, there is an increase in production followed by an increase in population size. However, the rate of population increase levels off as the new technology is "used up".

Food Production

If I am optimistic about the demand side of population problem, I have even more reasons to be optimistic about the supply side--food production. As pointed out by Schultz (1964), the world food problem exists because of low productivity in the poor countries. It is estimated that the amount of crops produced per ha of land in the poor countries was only about one-fourth of that in the industrial countries, and the production of animal products/animal unit was even lower. Therefore, the potential for greatly increasing production lies in the poor countries where it is needed.

Starting with the success of the Green Revolution in India during the 1965-70 period, there has been an awareness worldwide that a nation can increase its food supply if it has the will to do so.

Since the World Food Congress in Rome in the early 1970s, the FAO, the World Bank, the bilateral aid programs and other organizations have given priority in their aid programs to agricultural development in the poor countries. Up to 1975, the proportion of foreign aid dedicated to increasing food production was less than 10% of the total program of aid and the level has now more than doubled.

Some of the reasons for my optimism about future world food production are:
- The nature of the food/population problem as a whole is now becoming understood.
- The complexity of technology transfer from the industrial countries to the developing countries is also becoming understood; the international research and development centers have made excellent progress in these endeavors.
- The potential for increasing yields per land unit in the poor countries, most of which are located in the tropics, is enormous. We call this "payoff on research." I have made some estimates of the annual rates of return on monetary investments in agricultural research, both by commodities and by countries. If we consider rice research in the tropics, the figures run from 46% to 71%; however, if only Asia is considered the figures vary from 74% to 102%. the International Rice Research Institute, in the Philippines, has done an excellent job.
- Fertilizers and the knowledge of how to use them are now available to farmers in the poor countries. Many national governments are now subsidizing the price of this input.

- New and better adapted plant varieties are now available to most poor countries; many of these plants are resistant to certain diseases.
- Many national governments in the poor countries are now using aid and loans from the FAO, World Bank, bilateral sources and philanthropic organizations to support agricultural production by improving some or all of the following: (a) The infrastructure (such as roads, communication, and harbors), (b) The teaching, research, and extension structure facilities and activities. The research is now directed at solving their own production problems. (c) The credit structure. (d) Water resource utilization (e) Active intervention programs to increase the production of certain critical commodities. For example, in 1976-75, Indonesia initiated an active intervention program in rice production using subsidization of inputs, price stability, increased research, increased extension inputs, etc.

In 1969-71, the average production of padi was 2346 kg/ha in Indonesia. By 1978, this figure increased to 2921 kg/ha, an increase of 27.5% in 8 years, or about an increase of 3.5% per year. Preliminary figures for 1980 show a dramatic increase and give hope that Indonesia is about to gain self-sufficiency in rice production.

An interesting aspect of Indonesia's successful efforts on a single commodity, rice, is the fact that production improvements in other commodities have not improved, and some have decreased. To me, this is a god sign--meaning that with increased effort, Indonesia could increase production in any agricultural commodity chosen. Their leaders now know this and are putting forth successful efforts to increase production of other selected plants and animals.

REFERENCES

Bronowski, J. 1973. The Ascent of Man. Little, Brown and Co., Boston, Mass.

Burke, J. 1978. Connections. Little, Brown and Co., Boston, Mass.

Foster, P. 1978. Food as foundations of civilization. In Food and Social Policy. G.H. Koerseiman and K.E. Dole (Ed.) Iowa State University Press, Ames, Iowa.

Harlan, J.R. 1975. Crops and Man. Crop Science Society of America, Madison, Wisconsin.

Heiser, C.B. Jr. 1978. Seed to Civilization. W.H. Freeman and Co., San Francisco, CA.

Lee, R.B., and I. DeVore. 1968. Man, the Hunter, Aldine, Chicago, Ill.

Leeds, A., and A.P. Vayda. 1965. Man, Culture and Animals. AAAS, Washington, D.C.

Rostow, W.W. 1978. Food as foundation of civilization. In: Food and Social Policy. G.H. Koerseiman and K.E. Dole (Ed.). Iowa State University Press, Ames, Iowa.

Simon, J.L. 1981. World population growth. The Atlantic Monthly 248 (2):70-76.

Tannahill, Rey. 1973. Food in History. Harper and Row, New York, N.Y.

Thomas, Hugh. 1979. A History of the World. Harper and Row, New York, N.Y.

Willham, R.L. 1980. Historic development of use of animal products in human nutrition. Mimeo. Rpt., Iowa State University, Ames, Iowa.

3

WORLD LIVESTOCK FEED RELATIONSHIPS: THEIR MEANING TO U.S. AGRICULTURE

Richard O. Wheeler,
Kenneth B. Young

INTRODUCTION

Livestock producers both in the U.S. and worldwide have an important stake in the operation of the world grain economy. The continued availability of relatively low-cost grain in the world economy would tend to foster further livestock development in grain-deficit countries and provide competitive advantages in all countries for livestock and livestock production practices more dependent on intensive grain feeding. On the other hand, if world grain supplies become more restricted, grain-deficit countries would be more likely to import livestock products rather than grain, and livestock production practices less dependent on grain feeding would gain a competitive advantage.

Prior to the early 1970s, the world trend was toward increased grain supply and continued buildup of large stocks in the industrialized exporting countries, primarily the U.S. and Canada. These large stock levels helped to maintain relatively low prices and assured a stable supply for importing countries. For example, U.S. wheat export prices deviated very little from $170 per ton from the mid-1950s to the early 1970s. This long period of stable grain supply, sold at very attractive prices, encouraged the widespread use of grain feeding to increase livestock production. In addition, some grain exporters offered other inducements, including liberal credit arrangements and Public Law 480 assistance for countries unable to compete in the international grain market.

During this era, many developing countries adopted intensive grain-feeding practices for poultry and swine production, and cattle feedlots became a prominent feature of the U.S. agricultural system. On a worldwide basis, use of all cereals for animal feed increased from 37% in 1961-65 to 41% in 1975-77 (Harrison, 1981). Eastern Europe and the Soviet Union registered the largest increase in grain feeding of all countries--a change from 48% in 1961-65 to 69% in 1975-77. Use in Latin America increased from 32% to 41%.

At the present time, the world grain market has recovered from the 1972/73 shortfall and world stocks have been restored to the level of the 1960s. Nevertheless, the shortfall did mark a major shift in the supply-demand balance of the world grain market, a situation that had not occurred previously.

The structure of the world grain market has changed dramatically since the 1950s. World trade increased over 200% from 1960 to 1980. Currently, over 100 countries are dependent on grain imports from a few exporters. The U.S. dominates the world grain trade, accounting for about 60% of global coarse-grain exports and 44% of world wheat exports. About 40% of the total U.S. grain production is now exported and the annual rate of increase in exports reached 7% per year during the 1970s. Projections of future U.S. crop exports available from the Economic Research Service of USDA (1981) indicate that the growth in foreign demand will continue through the 1980s, although not as rapidly as in the 1970s (table 1). Annual average export demand for corn and rice is projected to increase about 4½% compared with 2% for wheat and soybeans.

TABLE 1. PROJECTED INDICES FOR U.S. CROP EXPORTS, 1981-1989

Com-modity	1981	1982	1983	1984	1985	1986	1987	1988	1989
					(1981=100)				
Corn	100	106	111	115	123	127	131	135	139
Wheat	100	96	99	101	103	105	107	110	115
Rice	100	109	113	117	120	124	127	131	135
Cotton	100	107	103	103	103	104	106	106	107
Soybean	100	100	101	104	107	111	113	116	119
Peanuts	100	123	140	147	150	153	157	160	163

Source: These projections have been calculated from Problems and Prospects for U.S. Agriculture, ERS-USDA (1981). They are not official USDA projections.

There is some question now about the U.S. ability to keep up the recent pace of expanding exports. Most of the available cultivated land is currently in production and we are losing about a million acres of cropland per year to nonagricultural uses. The rate of soil erosion has increased substantially with more intensive cultivation and use of marginal cropland formerly not used for crop production. The same problem is occurring in other countries and average crop yields are leveling off over much of the world.

PROJECTIONS ON WORLD GRAIN SUPPLY

A Winrock International study was completed in 1981 on world use of grain and other feedstuffs (Winrock International, 1981). The study was designed to evaluate the interaction between the world livestock system and the feed- and food-grain system.

Estimates of current world use indicate that poultry consume 27% of all grain fed; swine--32%; draft animals--4%; sheep and goats--2%; and cattle and buffalo, including dairy animals--35% (table 2). Feed use in table 2 is expressed in terms of megacalories of metabolizable energy

TABLE 2. ESTIMATED ANNUAL WORLD FEED USE FOR DIFFERENT TYPES OF LIVESTOCK, 1977

Livestock category	Livestock Output		Feed use				
	Meat	Other	Grain	Protein meal	By-products	Forage & other	Total feed
	(million metric tons)			(billion mcal ME)			
Poultry	22.8	23.3[1]	387.9	91.1	73.4	51.7	604.1
Sheep & goats	7.3	--	23.6	5.3	35.8	993.9	1,058.6
Cattle & buffalo	46.8	415.0[2]	507.3	42.8	204.2	4,101.0	4,855.5
Swine	41.0	--	460.9	56.0	213.1	157.2	887.2
Draft animals	13.4	--	57.9	5.9	23.5	1,214.9	1,302.2
All livestock	131.3	--	1,437.6	201.1	550.0	6,518.7	8,707.4

[1] Eggs.
[2] Milk.
Source: Winrock International (1981).

rather than metric tons due to variation in the quality and variety of feed used in different countries. For example, grain-feed use in the Soviet Union is reported on a "bunker weight basis" generally containing excess moisture and extraneous matter. The percentage of grain use in poultry rations is estimated to be similar for developed, centrally planned, and developing countries since the technology of modern poultry production has been readily adopted all over the world. Developing countries feed less grain and more forage and by-products to swine. The ruminants in develop- ing countries subsist almost entirely on forages. However, nearly half of the world grain feeding occurs in developing and centrally planned countries dependent on grain imports.

Grain feeding was projected to continue increasing according to recent trends evaluated in the Winrock study. There will be occasional setbacks for countries with severe foreign exchange problems and domestic recession. Most of the centrally planned countries have set target levels of increased livestock production requiring additional grain feeding. The Winrock study projected that total world feed use of wheat and coarse grains would surpass direct human and industrial consumption by 1985. Recent trends also indicate that total world grain use is increasing at a faster rate than world production. World grain demand has been increasing steadily due to continued growth in world population and rising per capita consumption of livestock products dependent on grain feeding while the growth in supply is slowing due to limitations on development of new

cropland and reduced productivity gains on existing cropland. This increased tightening in the world grain market
implies that the grain export price should increase substantially by 1985.

GRAIN SUPPLY OUTLOOK FOR U.S.

Winrock International is currently initiating a study
of both the potential for and implications of additional
crop production in the U.S. The 1977 Natural Resource
Inventory compiled by Soil Conservation Service of USDA
shows a total of roughly 460 million acres of cropland
available in 1977 containing about 70% prime land in the
Class 1 and 2 categories. Heady and Short (1981) of Iowa
State University have projected that the 1977 cropland base
will dwindle to 353 million acres by the year 2000, but that
there are 37.6 million acres of high-potential land and 90.1
million acres of moderate-potential land that could be
converted to cropland. This land area for potential development is located primarily in the South Atlantic, South
Central, Great Plains, and North Central regions of the U.S.
However, other economists in the U.S. have serious doubts
whether it would be feasible to convert this much additional
land to crop production. Some limitations to development of
additional cropland and current use of this land are shown
in table 3; the data indicate that much of this potential
cropland is currently used for pasture and timber production
and that there are definite erosion hazards and probable
high conversion costs to develop this land area for crop
production. Such limitations imply that there will be a
major increase in production cost to bring these new lands
into crop production after we reach full capacity on existing cropland.

TABLE 3.   ESTIMATED POTENTIAL CROPLAND AND LIMITATIONS TO
           DEVELOPMENT IN THE CONTINENTAL UNITED STATES

Type of limitation	Percent of potential cropland	Present use	Percent of potential new cropland
Erosion	59	Pasture-range	79
Drainage	23	Forest	17
Soil	7	Other rural	4
Climate	4		
No limitation	7		
Total	100	Total	100

Source: 1977 Natural Resource Inventory, Soil Conservation
          Service, USDA.

IMPLICATIONS FOR U.S. LIVESTOCK PRODUCTION

World population is projected to increase 50% between 1975 and 2000. There is increasing emphasis on livestock production to improve the quality of human diets, particularly in centrally planned and developing countries, and increasing pressure on cropland worldwide. In the short term, there may be temporary swings between shortages and surpluses in the world grain market. This is expected due to greater year-to-year variation in world crop production as a result of expansion of cultivation on marginal lands with increased drouth stress and other climatic variability. Stability of supply may also be reduced due to mounting pressure on exporting countries to reduce the carryover of grain stocks from year to year. Grain prices are projected to increase with gradual tightening of world grain supplies. Increased export volume will require eventual conversion of at least some pasture and timber land in most of the key exporting countries, with an associated rise in grain-production cost.

Some implications for U.S. livestock producers include increased grain-feeding costs eventually rising above the general inflation rate and the loss of some pasture and rangeland area converted to cropland as indicated in table 3. Higher grain prices will be translated into somewhat higher meat prices, particularly for poultry, swine, and fed cattle because these enterprises are highly dependent on grain feeding. However, it will become more profitable to utilize additional crop residues and by-products in livestock feeding, particularly for cow maintenance, to replace present grain use. The biggest deterrent to using these low-quality feeds is cost--primarily for labor, equipment, and interest. To date, the availability of a stable supply price for grain is analogous to the situation we had 10 years ago for oil and natural gas. It has not been cost effective to utilize many alternative sources of feed energy, although there is an abundant physical supply available in the U.S. The amount of corn crop residue physically available was estimated to be 231 million tons in 1977 (Ensminger and Olentine, 1978). This would support 117 million cows for a 4-month grazing period on a purely physical supply basis. The nutritive value of crop residues can be enhanced with special processing techniques, and some very promising research work has been done on ammonia treatment of straw.

Cattle, sheep, and goat producers could potentially utilize these alternative sources of feed to substitute, at least partially, for grain or to enlarge breeding herds even on a drylot basis if it became more economical to do so. If meat prices increase along with grain prices, some livestock producers may regain a competitive advantage over poultry and swine producers who are more vulnerable to rising grain prices. Under the current regime of depressed grain prices, poultry producers, in particular, have been gaining a significant competitive advantage over beef producers. Poultry meat prices have now declined to 30% of average beef prices compared with 80% a few years ago (National Cattlemen's Assocation, 1982).

A reduction in grain feeding of cattle in the U.S. would mean increased competition for use of existing pasture and range lands. With increased grain exports, there would be an associated reduction in the grazing land area, particularly in the Southeast and Great Plains regions of the U.S. Increased dependence on crop by-products and residues implies that more livestock will be produced in traditional cropland areas to utilize these waste products, as was the practice in the U.S. before the feedlot era began. This is the situation now in most developing countries where the bulk of livestock production is found in mixed crop/livestock systems (Winrock International, 1981).

Increased use of crop residues to reduce the amount of grain feeding would have a significant impact on the management system for livestock, especially cattle. Levels of annual offtake would decline as cattle would have to be nearly a year longer to reach market weight on a less intensive feeding program. The cattle operator would be forced to move cattle to crop-production areas and to lease crop residues from crop farm owners as is now done for wheat pastures. Additional use of feed supplements would be necessary as crop residues are generally lacking in total nutrient requirements. The cattle operator would also have to invest in additional fencing and equipment to utilize crop residues. Thus the overall implications are that major adjustments would be required in the U.S. cattle industry, including shifts in the location of production, the composition of herds, and feeding programs.

CONCLUSIONS

The general outlook for the international grain market points toward continued price variability for feedgrains, a gradually rising price level for grains as the world market continues to tighten, and eventual loss of some grazing land in the U.S. when additional cropland is needed to meet expanding export requirements. It is possible that the problem of price variability may be alleviated by additional government intervention such as paid acreage reduction or other methods of supply control on the market, but this does not appear likely in view of the current emphasis on curbing spending for most agricultural support programs.

Expected consequences of the grain-market outlook for cattle producers include continued fluctuations in feeder cattle prices and returns from cattle feeding during the next few years and a general trend toward higher feeding costs. Although price variability is nothing new to livestock producers, the sharpness and range of price movement will likely be increased as long as we continue to be the shock absorber for the world grain market. The U.S. is one of the few nations that exposes domestic producers to price fluctuations of the international market.

Short-term effects of the expected swings in prices will be of more immediate concern to most livestock producers than the longer term upward trend in feeding cost, particularly for those in a weak financial position. To some extent, producers may be able to reduce the financial risk through greater participation in the futures market or by direct contracting. However, their most urgent need to survive in the livestock business will probably be to secure alternative methods of financing to provide more flexibility on repayment of loans. Other possible options for reducing or spreading the risk of price movement include the development of programs for outside investors to assume partial ownership of livestock and other creative financing schemes to shift at least part of the risk from producers to other outside parties. There may also be an opportunity for further revision of the tax laws to encourage more outside investment in the livestock business.

Long-term implications of changes in the world grain market, as well as expected increases in transportation cost, are that the structure of the U.S. cattle production system will change. Projected world food-system trends suggest increasing prices for all livestock products due to rapidly rising consumption in most countries and upward pressure on grain prices. However, there may not be much increase in livestock-product consumption in the U.S. market because per capita consumption rates have stabilized. A continuing problem for beef will be competition from pork and poultry in the U.S. meat market. To recapture its former market share, beef will require more efficient production and marketing throughout the system.

Increasing grain prices may provide some opportunity for beef producers to improve their production-cost relationship relative to pork and poultry by changing to less intensive grain feeding. Additional research and development is needed on the utilization of crop residues and by-products to reduce cost in cattle production.

REFERENCES

Economic Research Service, USDA. (1981). Problems and
    Prospects for U.S. Agriculture, Washington, D. C.

Harrison, P. 1981. The inequities that curb potential.
    FAO Review on Agriculture and Development. Food and
    Agricultural Organization of the United Nations, Rome,
    Italy.

Heady, E. O. and C. Short. 1981. Interrelationship among
    export markets, resource conservation, and agricultural
    productivity," Agr. J. Agr. Econ. 63:840.

National Cattlemen's Association. 1982. The future for
    beef. Special Advisory Committee Report, Englewood,
    CO.

Soil Conservation Service, USDA. 1977. 1977 Natural
    Resource Inventory. Washington, D. C.

Wheeler, R. O., G. L. Cramer, K. B. Young and E. Ospina.
    1981. The World Livestock Product, Feedstuff, and
    Foodgrain System. Winrock International, Morrilton,
    Ark.

Winrock International. 1981. Report on Livestock Program
    Priorities and Strategy. Winrock International, Mor-
    rilton, Ark.

4

WORLD AGRICULTURE IN
HOSTILE AND BENIGN CLIMATIC SETTINGS

Wayne L. Decker

CHARACTERIZATIONS OF CLIMATE

Health, nutrition, and suffering of the human population are determined, in part, by weather and climate. Regional wealth and the levels of economic development are impacted by the natural resources, including the climate resource. But agriculture and the associated food production industries are more directly affected by weather and climate than any other sector of the economy. Climate determines production potential of both grain and livestock producers, identifies strategies available to the producer for resource allocations and marketing, and determines the feasibility of plans for exports and imports of commodities. An improved understanding by agriculturalists of the nature of the climatic resource is essential if the impacts of climatic risks are to be minimized.

Climate is defined by the space and time distribution of weather events: temperature, precipitation, wind, humidity, and sunshine. In spite of the unpredictability of weather events, climate occurs systematically in both the space and time scale. As a result of these consistencies, climatic zones are easily recognized. For example, in the tropics and subtropics some regions are consistently rain-free in summer, others are smaller areas with an even seasonal distribution and abundant rainfall. In the temperate latitudes, the continental regions also demonstrate regional consistencies in climate. The west coasts of continents are mild with abundant winter rainfall, while the continental interiors tend to exhibit summer maximum of precipitation and marked seasonal temperature variations (Mather, 1974).

In most climatic regions, there are periods during the year with hostile climates for agriculture. These climatic hostilities are associated with temperature stresses (both high and low) and with deficiencies of rainfall. In many of these regions, there are periods of the year during which the weather is consistently dry, thus producing a hostility. This climatic hostility can be removed by irrigation, or avoided by adopting an enterprise with a low water need. For the hostile climates produced by temperature stress, shelters may be constructed to protect animals from the critical temperatures, and crop production can be scheduled to avoid the consistent occurrence of high or low temperatures.

Many regions have climates that are consistently favorable for agricultural production. These climates are usually characterized by dependable water supply (rain or irrigation supply). Benign climates are also characterized by moderate temperatures without a high probability of extreme temperatures during critical times for sensitive plants.

CLIMATIC CHANGE

Climatic change and the impact of climatic change on man have become controversial issues in recent years. Articles on the subjects appear regularly in technical and popular magazines, and both paperback and hardback books have been published dealing with climate change. The written opinion concerning climatic change and its impact are as different as day and night. Even scholars of climatology are confused by the diversity of opinion.

The evidence to support the existence of major climatic changes through geologic time is well documented. Long periods of geologic history are characterized by mild climates, i.e., benign climates. These periods were interrupted by relatively short intervals when glaciers extended into the middle latitudes (the ultimate in hostile climates). It is generally accepted that the current climate of the world is more like that of the glacial period than the warmer "climatic optimum." Climatic researchers do not agree about the mechanisms causing these major climatic changes. Current thinking focuses on long-term oscillations in the slope of the terrestrial axis, but continental uplift and the associate volcanic activity appear to be necessary conditions for the glacial climates.

Variations in the climate of the earth also have been documented from historical records. The rise and decline of civilizations during the past 4,000 years appear to be related to changes in climate. Plagues, famine, and migrations have been linked to shifts in climate. In modern history, the period corresponding to the North American settlement and the establishment of the United States was a period of climatic stress, frequently called "the little ice age." Again, meteorologists do not agree on the physical processes that caused the climatic variations in historical time. Volcanic activity, variability in the solar output, and combinations of both these factors are mechanisms receiving prominent attention.

It was not until the late 1800s that a worldwide network of weather observing stations was established. Although records of weather observations can be traced into the 18th century at selected points, networks of observational stations did not generally exist until the early and mid-nineteenth century. In the United States, for example, it is difficult to find documented weather records prior to the establishment of the Weather Bureau in the Department of Agriculture in 1890. For this reason studies of climatic change based on meteorological observations are confined to the most recent 90 years.

Attempts have been made to establish the trends in climate from the meteorological observations. The best documented estimate of the trend in climate is shown in figure 1 from Waite (1968). During the first 40 years of this century, the average air temperature near the earth's surface increased, but about 1940 this trend was reversed. Figure 1 verifies that these trends in temperature apply to regions of different size and are the most pronounced in the polar and subpolar regions of the northern hemisphere.

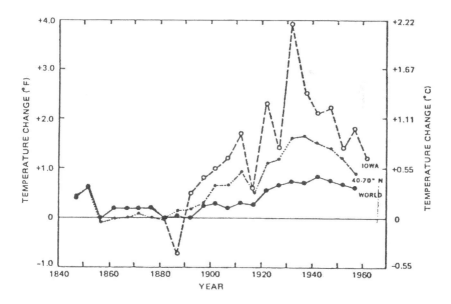

Figure 1.   Worldwide trend in mean annual temperature (Waite, 1968)

For agriculture it is more instructive to look at the climatic variabilities associated with precipitation. In figure 2 and figure 3 the historical records of summer rainfall are summarized for the prairie provinces of Canada and the winter wheat region of Russia. It is difficult to identify trends from these data, although many interesting pulsations in the annual rainfall statistics, lasting for a decade or so, are apparent.

The atmospheric mechanisms producing the climatic trends and fluctuations that extend for a decade or a few decades have probably all been identified. These mechanisms include ocean-atmosphere interactions, volcanic activity, man's interference (CO_2 and particulate matter), and solar activity. The analytical contribution of each individual mechanism and the interaction between the mechanisms have not been defined; a major objective of the meteorological community to mathematically describe these processes based on known physical relationship efforts.

Efforts to research the physical causes for climate change led Dr. B. J. Mason, Director General of the British Meteorological Office, to observe in a recent article in the New Republic (1977):

> The atmosphere is a robust system with a built-in capacity to counteract any perturbation. This is why the global climate, despite frequent fluctuations, is fairly stable over periods of 10,000 years or so. Sensational warnings of imminent catastrophe, unsupported by firm facts or figures, not only are irresponsible but are likely to prove counterproductive. The atmosphere is want to make fools of those who do not show proper respect for its complexity and resilience.

FLUCTUATIONS IN CLIMATE

Climate variability adds an additional stress to the agriculture system and adds a component to climatic hostility. When the weather of one or more years departs markedly from the expected, a farm management strategy that has been successfully used becomes inappropriate for the agricultural enterprise for a particular year or growing season. Several years of drought (such as the 1930s on the U.S. Great Plains) is hostile to the farm enterprise adopted to nondrought enterprises. On the other hand, periods of years with benign climates often lure farmers into strategies not adapted to the hostile and stressed condition that follows. The type of fluctuation most often used by climatologists deals with the variation of climatic events between years. This variability refers to the variation in annual or seasonal temperatures and precipitation totals. McQuigg (1973), for example, demonstrated the low variability in climate between 1955 and 1970 for the major agricultural production regions of the U.S. McQuigg simulated

CANADIAN PRAIRIE PROVINCES

MAY TO JUN BINOMIAL RUNNING MEAN TOTAL PRECIP

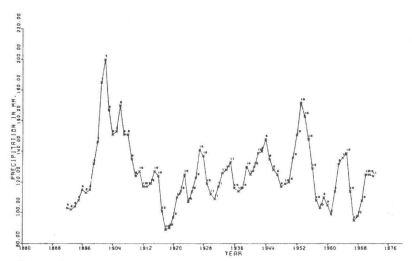

Figure 2. The year to year variability in the total
precipitation for May and June (smoothed by a binomial
technique) in the spring wheat producing area of
Central Canada.

SOVIET REGION WEST OF VOLGA

MAY TO JUN BINOMIAL RUNNING MEAN TOTAL PRECIP

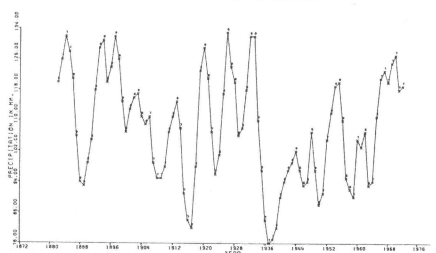

Figure 3. The year to year variability in the total
precipitation for May and June (smoothed by a binomial
technique) in the winter wheat producing area of the
Soviet Union (west of the Volga River).

34

yields of grain throughout this century from climatic data
at a constant technology. He showed (figure 4) that the
yields simulated from climate during the period extending
from the late 50s through the 60s were remarkably constant
and relatively high, i.e., the climate of the U.S. Corn Belt
was benign. The year-to-year variability in climate in the
U.S. has been greater in the 1970s and the early 1980s than
the preceding decade and a half.

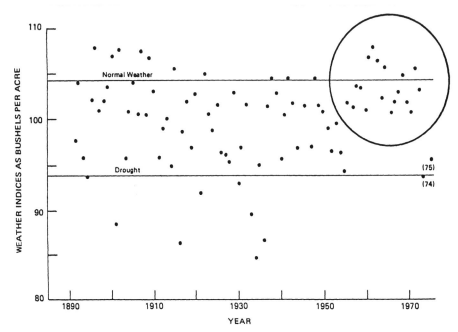

Figure 4. **Year-to-year variability in climate expressed in
simulated corn yields for the U.S. (NAS, 1976)**

        To examine how climatic fluctuation varies through
time, the variances of climatic elements by decades have
been computed for major agricultural production areas in the
world. Figure 5 shows the 10 year variances in the May plus
June precipitation in the Canadian prairie region, while
figure 6 presents the same values for the Soviet winter
wheat region. The May and June rainfall totals are vital to
wheat production in these two regions. In both cases, the
variability in May and June precipitation during the most
recent decades were below average; however, the tendency for
a lower variance does not appear to depart from that expected from the normal variability. The high year-to-year
variability in the May-June rainfall in Canada just after
the turn of the century is quite striking. There was also a
period of high variability for the May and June precipitation in the Soviet Union between 1925 and 1950.

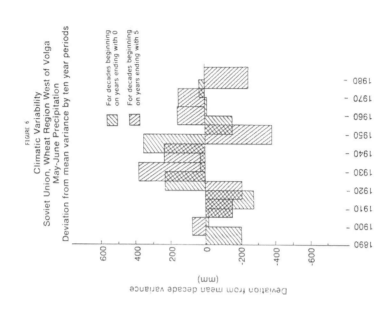

FIGURE 6
Climatic Variability
Soviet Union, Wheat Region West of Volga
May-June Precipitation
Deviation from mean variance by ten year periods

For decades beginning
on years ending with 0

For decades beginning
on years ending with 5

Deviation from mean decade variance
(mm)

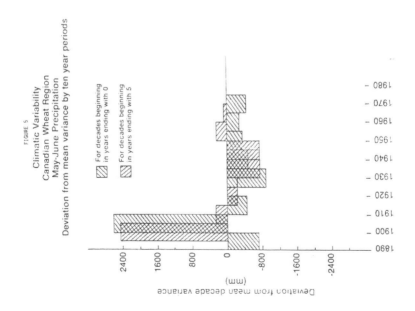

FIGURE 5
Climatic Variability
Canadian Wheat Region
May-June Precipitation
Deviation from mean variance by ten year periods

For decades beginning
in years ending with 0

For decades beginning
in years ending with 5

Deviation from mean decade variance
(mm)

Figures 5 and 6 show that climate did not experience unusually low variability everywhere in the world during the period 1955 to 1970. A second, and more important, lesson concerns the recognition that periods of 1 or more decades in length with high (and low) year-to-year variability in climate do occur.

A report by the National Research Council (1976) identifies sudden and unexpected climatic fluctuations as the greatest climatic hazard to agricultural production. The managers of agricultural systems are forced to use practices that are not adapted to the regional climate. The farm manager has difficulty in finding the "best" strategy during periods with fluctuating climates. A large year-to-year variability forces the manager to choose management options that are "out of step" with the season's weather.

RESPONSES TO CLIMATE CHANGES AND FLUCTUATIONS

The policy makers and agricultural managers could easily respond to weather and climate fluctuations, if events could be anticipated before they occurred. Although several climatologists, with impressive credentials, regularly make seasonal forecasts, most meteorologists agree that the science of meteorology has not advanced sufficiently for these forecasts to be considered valid. The thirty-day outlook, which projects the general trends in mean temperature and precipitation, is correct only about 55% to 60% of the time. Seasonal forecasts (cold winter, dry summer, etc.) have an even lower accuracy. Complex computer models, which simulate the atmospheric circulation, offer the best promise for the development of rational forecasting schemes. As these atmospheric models are improved, weather and climate forecasts will have improved accuracy and be extended for longer periods. The improvement in forecasting skills will be slow. In the next two decades there appears little hope for major and sudden breakthroughs in our understanding and interpretation of atmospheric circulation.

Weather modification provides an additional strategy for removing weather risks. Modification of the surface energy budget through changes in the surface color, drainage of the land, or shaping of the soil surface through tillage, offers many important options for improved technologies for agriculture. Increased rain through cloud seeding is more often considered as a weather modification option. There are several difficulties that reduce the potential for cloud seeding to respond to fluctuations in climate:
- Proof of small increases in amounts of rain are almost impossible to obtain because variations in area and time affect the amount of rain falling over a region.
- Rain-making only augments the natural rainfall, so it is not a "drought stopper."

  - The possibility exists that the manipulation in
    the clouds will reduce the rainfall from some
    clouds within a given weather system.
Cloud seeding appears to be most favorable for use in the
mountainous regions--to increase the winter snow pack.
Increased snow in the mountains improves the water supply
for the agriculture of the adjacent semiarid and arid
regions.

  Disaster insurance spreads the risks of "bad" weather.
An international program for grain and food storage should
stabilize supply between the "lean" and "bountiful" years.
A marketing cooperative, or even the individual farmer, may
establish an "ever-normal granary" by withholding grain from
the market. For the individual farmer, the best opportunity
for spreading the risk is through disaster insurance. A
national food policy must include an insurance program using
both governmental and private agencies as underwriters.
Insurance programs may stabilize farm incomes, but, in the
long run, will not provide increased productivity of food
for the expected increase in world population.

  Technologies developed through private and public
research provide an additional response to provide for a
stable food supply and farm income under a fluctuating climate. Meteorological science cannot be expected to deliver
completely reliable warnings of pending shifts and fluctuations in climate. The meteorologists will not save us from
the adversity of a variable and often hostile climate.
Agricultural strategies and technologies must be developed
to respond to the expected climatic variabilities. These
developments will emerge from integrated, interdisciplinary
agricultural research.

REFERENCES

Mason, B. J. 1973. Bumper crops or droughts. Mimeo.
    NOAA, U.S. Dept. Commerce, Washington.

Mather, J. R. 1974. Climatology, Fundamentals and Appli-
    cations. McGraw Hill, pp 112-131.

National Research Council. 1976. Climate and Food. Report
    on Climate and Weather Fluctuations and Agricultural
    Production. National Academy of Science, Washington.

Waite, P. J. 1968. Our weather is cooling off. Iowa Farm
    Science 23:13.

5

THE IMPACTS OF CLIMATIC VARIABILITIES ON LIVESTOCK PRODUCTION

Wayne L. Decker

CLIMATE AND LIVESTOCK

Regional climates contain factors that need consideration in determining the kind of profitable livestock production for a region. The variability of weather and climate provides a component in determining the profitability of the livestock enterprise. Climate imposes both direct and indirect effects on commercial animal agriculture. The direct effects include weather events producing physical injury (lightning, wind, flood, temperature extremes), occurrences associated with physiological stress (such as heat and humidity), and weather events promoting insect or disease episodes. Indirect climate events are those that impact on availability of forages and the supply of feed grains. These indirect impacts of the weather and climate are generally imposed by chronic deficit in water and occasional droughts.

LOSSES IN ANIMAL PRODUCTION DUE TO CLIMATE EVENTS

Animals, grown commercially on farms and ranches, are normally subjected to ambient environmental conditions. On many occasions, the atmospheric conditions are less than ideal and the animal is subjected to stress. This stress, which is usually related to the heat and energy balance of the animal, reduces the production. The stress causes declines in egg or milk production and reduced weight gains for swine, beef, or broilers.

Over the years there have been repeated attempts to mathematically define the impact of stress imposed by atmospheric conditions (temperature, humidity, wind, etc.) on animal production. The experimental basis of these efforts comes from two sources: (1) barns or chambers with controlled environmental conditions (Brody, 1948) and (2) field experiments measuring animal performance as related to observed ambient environmental conditions. From the observations obtained through the experimentation, functional relationships between the weather event (or events) are derived. The resulting mathematical formulas are usually obtained through standard statistical procedures. Strickly speaking, the expressions are only applicable to the experimental conditions from which the relationship is derived, so each relationship must be tested against independently collected data.

Literature has many examples of relationships between the performance of domestic animals and weather and climate events. Two mathematical expressions for relating cattle performance to environmental conditions are discussed below.

Milk Production

Using data obtained from controlled experiments, Berry et al. (1964) related the decline in milk production to an index involving both atmospheric temperature and humidity. This index, which is called the temperature-humidity-index (THI), is shown in equation (1).

$$THI = T + .36\ T_d + 41.2 \qquad (1)$$

where
T is the temperature in °C,
T_d is the dew point temperature in °C.
The relationship between THI and the decline in milk production (MD) is shown in equation (2).

$$MD = -2.37 - 1.74\ NL + .0247\ (NL)(THI) \qquad (2)$$

where
NL is the normal production of a cow under thermoneutral conditions.
THI is the temperature-humidity-index.

Since negative values for MD do not make sense under the definition in this equation, decline in milk production is assigned the value of zero for all negative values in equation (2). This means that a zero production decline is expected until a critical value of the THI is reached. For higher values of THI, the decline in production decreases linearly. This critical value of THI is between 70 and 74 for normal production levels of between 25 and 30 pounds of milk per day.

Meat Production in Cattle

Bolling and Hahn (1981) and Bolling (1982) report the results of a regression analysis relating climatic variables to rates of gain for beef animals. The functional relationship, which best explains the reduction in weight gains, contained terms related to cold stress, heat stress (THI), precipitation, and wind. The results of the regression analysis are shown in figure 1, as taken from the work of Bolling (1982). The author concluded:

"Analyses consistently suggested that stress resulting from the direct effects of cold, combined with the effects of precipitation, have a greater impact on feedlot cattle in Nebraska than does any other type of atmospheric stress studied. Interestingly, heat stress rarely appeared to be a significant factor affecting cattle performance."

Of course, this conclusion concerning heat stress is counter to the one for milk production. This difference may be due to the difference in climate of the regions where the milk production and rate of gain experiments were conducted.

Hahn et al. (1974) noted that beef cattle were able to overcome heat stress. In the Missouri Climatic Laboratory, beef cattle were stressed by being subjected to 5 weeks of temperatures of 30°C. A marked decrease in rate of gain as compared to a control group resulted; when the animals were returned to optimal conditions, the stressed animals outgained the control group. This result, which Hahn calls "compensatory growth," is demonstrated in figure 2. No such compensation occurred after animals were subjected to a greater stress (35°C). Hahn (1976) indicates that similar compensatory growth occurs with swine and broiler chickens. It is significant that (1) the research indicates that compensatory growth does not occur after exposure to a high degree of heat stress, (2) the laboratory experiments on compensatory growth were done at constant temperature and may not apply to temperatures experiencing a diurnal range, and (3) no experiments have been made to discover whether "compensatory growth" occurs after cold stress.

MORTALITY OF ANIMALS DUE TO CLIMATIC STRESS

Hostile climates do cause mortality to domestic livestock. This hostility occurs as a result of both heat and cold stress. In summer, high temperature and humidities (THI values of 80 or higher) produce stress that can lead to death. This condition is, of course, aggravated by other stress factors associated with handling and/or shipping. In winter, the cold stress can cause tissue to freeze and the animal to die. For U.S. cattlemen of the open ranges in the High Plains and eastern slopes of the Rocky Mountains, the cold stress may be further aggravated by high winds with snow. These blizzards cover the winter food supply, make access to the herds difficult (if not impossible), and bury herds under the drifting snow.

Bolling (1982) presents analyses that document the weather impacts on cattle mortality under feedlot confinement. Strong winds and cold stress were the "best" predictors of mortality under Nebraska conditions. The author was apparently unable to document the mortality due to stress imposed by hot and humid weather. Of course, one would not want to use these results to estimate mortality under range or pasture exposures.

The National Weather Service has established policies for issuing weather advisories to stockmen. These advisories are issued by Weather Service Forecast Offices located

42

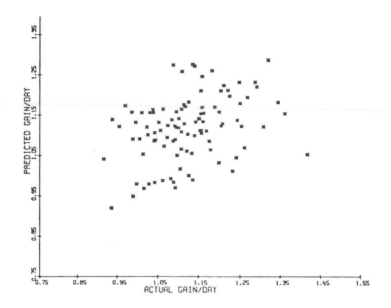

Figure 1.   Relationship between measured gain of beef cattle and that
            predicted from weather data using a statistical relationship
            (Bolling, 1982)

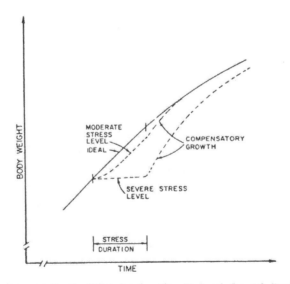

Fig. 2.   Schematic to Illustrate the Principle of Equifinality
          in Animal Growth   (Hahn, 1982)

in each state. These advisories are:
- Heat stress advisories are issued whenever high temperatures combine with high humidities to present danger to livestock. The THI is used as the index for danger to livestock. Two categories are recognized: (1) livestock danger (THI 79 to 83) and (2) livestock emergency (THI higher than 83). When air temperatures are below the body temperature, wind provides cooling. Because of this, wind is mentioned in the stockman's advisories when velocities higher than 20 mph are expected.
- Winter storm watches are issued when there are strong indications of a blizzard, heavy snow, freezing rain, or sleet. A blizzard is defined as a condition with high winds (in excess of 35 mph), with falling or blowing snow (visibilities less than 3 miles). A winter storm warning or blizzard warning is issued when the storm's development is a virtual certainty.
- Chill indices are released routinely. This index combines wind and termperature in an effort to approximate the equivalent temperature under light wind (4 mph) conditions. At Columbia, Missouri, in the winter of 1981-82, there were 398 hours (11% of the hours) with chill factors below 0°F, 181 hours (5%) with chill factors below -10°F and 72 hours (2%) with chill factors below -20°F. The number of days with chill factors below 0°, -10° and -20° sometime during the day was 30, 18, and 7 days, respectively.

44

REFERENCES

Berry, I. L., M. D. Shanklin and H. D. Johnson. 1964. Dairy shelter design based on milk production decline as affected by temperature and humidity. Trans. Amer. Soc. Agr. Engr. 7:329.

Bolling, R. C. 1982. Weight gain and mortality in feedlot cattle as influenced by weather conditions: Refinement and verification of statistical models. Progress Report 82-1. Center for Agricultural Meteorology and Climatology, Univ. of Nebraska, Lincoln.

Bolling, R. C. and G. L. Hahn. 1981. Climate effects on feedlot cattle: Growth and death losses. Proc. 15th Conference on Agr. and Forest Meteorology: 86-89. Amer. Meteorological Soc., Boston.

Brody, S. 1948. Environmental physiology with special reference to domestic animals, I: Physiological background. Res. Bull. 423. Mo. Agr. Exp. Sta., Columbia.

Hahn, G. L. 1982. Compensatory performance in livestock: Influences on environmental criteria. Livestock Environment, Proc. 2nd Internatl. Livestock Environment Symposium. Amer. Soc. Agr. Engr., St. Joseph, Michigan. (In Press.)

Hahn, G. L. 1976. Rational environmental planning for efficient livestock production. Proc. 7th Intern. Biometeorological Cong., College Park, MD. pp 106-114.

Hahn, G. L., N. F. Meador, G. B. Thompson and M. D. Shanklin. 1974. Compensatory growth of beef cattle in hot weather and its role in management decisions. Livestock Environment, Proc. Internatl. Livestock Environment Symposium, pp 288-295. Amer. Soc. Agr. Engr., St. Joseph, Michigan.

Part 2

GENERAL CONCEPTS AFFECTING
AGRICULTURE AND THE INDUSTRY

6
POLITICAL CHALLENGES FOR TODAY'S ANIMAL AGRICULTURE

George Stone

Not all of the challenges to today's animal agriculture are political. Some of the challenges are internal. They relate to the nature of our industry and to the nature of the people in the livestock industry.

We seem to have an allergy to change. And we have a disposition to go our own way as individuals, even if we could improve things for ourselves by working together.

ROLE OF FEDERAL GOVERNMENT

Part of the time, we want the government to leave us alone. Part of the time, we get very impatient when the government is too slow with helping us. This is not the first, nor the last, speech to be made in this nation on the role of the federal government in food and agriculture. We talk about that subject as if we were about to make an original choice. But, as a matter of fact, the choice was made long ago. As early as 1796 and as recently as 1977, and many times in between, the federal government has decided that the family farm should be fostered and encouraged.

Our society has decided that the federal government should take measures to help assure that land remains in the hands of family farmers.

Our society has decided that the federal government should be involved in the conservation and protection of land and water resources.

Our society insists on a federal involvement in environmental protection.

Our society has dictated a federal role in assuring that food supplies are safe and wholesome.

Our society requires federal supervision to see that pesticides and chemicals are safe for farmers and consumers.

Our society provides for federal supervision of the marketing system to try to keep it fair and competitive.

There are more kinds of government intervention in agriculture today than ten, twenty, or thirty years ago. There will probably be more federal involvement in farm and food policy in 1990 or the year 2000 than there is today.

The real question should be in regard to the nature, extent, and purpose of government involvement, and the degree to which farmers have a voice in decisions that will affect them.

Government intervention has often been justified when there was no other effective way to cope with problems. The government's role in agricultural research and education is widely accepted and advocated. Various federal farm credit programs had to be initiated because the private sector could not take the entire risk in financing agriculture. Commodity exchanges and boards of trade had to get federal supervision because they could not be left to police themselves. The federal rural electrification system had to be created because the private sector could not, or would not, do the job.

In many agricultural sectors, government involvement can be good or bad:

- Rules on farmers' handling of pesticides or chemicals can be reasonable or ridiculous, depending on how much farmer input there has been in the process.
- Feedlot pollution abatement rules can be workable and effective, or unrealistic and oppressive.
- Dredge and fill regulations can be a protection or a harassment for farmers.
- OSHA regulations can be a godsend or an aggravation.

Everything depends on how well farmers have involved themselves in the process and how much voice they have had in shaping these laws and regulations. That is not an easy task for us as farmers and livestock producers. The leading codification of laws affecting agriculture now runs to 14 volumes of about 500 pages each. That is 7,000 pages of laws.

MAJOR POLITICAL CHALLENGES

Having said this much in the way of background, let me now turn to what I perceive as some major political challenges for livestock agriculture. One thing that is quite obvious, but not generally appreciated, is that we do not function in a vacuum as livestock people. We depend upon consumer purchasing power and demand for our products, and we cannot expect to be stable and prosperous if there is high unemployment and weak buying power.

Recession and Unemployment

Much of the difficulty that has faced the livestock producer in the past three years has been attributable to recession and unemployment. High interest rates the past three years have diverted about 30 billion dollars a year of consumer purchasing power from food and other necessities. High interest rates have an impact on the cattle producer as well. Some calculations done at Oklahoma State University reveal that the interest cost per head of livestock sold would be $133 per head at 9 percent interest; $237 per head at 16 percent; and $297 per head at 20 percent interest. The difference between the two extremes of 9 percent and 20 percent is $164 a head, more than enough to wipe out any potential profit.

When national economic conditions are difficult and there is a pinch on consumer buying power, the tendency is for a reduction in the higher-priced meat purchases and, to some extent, for the consumer to buy other products. In a time of recession and unemployment, meat purchases are the first thing affected.

Specifically, because of the tendency for the American diet to be hurt by recession and unemployment, the Congress in its wisdom developed and implemented the food stamp program. Of course, one can expect that when times are tough, the food stamp program becomes costly. Each additional one percent of unemployment adds one million people to the food stamp rolls, not because it is a bad program, but because it is doing what it is supposed to do--maintain a healthful diet for the lowest-income people in our society.

About 27 percent of the food stamp benefits are spent by recipients to buy meat and meat products. This means that if the food stamp program makes available 12 billion dollars in food subsidies to low-income families, over 3 billion dollars will be used to buy meat. If the food stamp program is cut by 3.7 billion dollars, as it has been in fiscal budgets for 1982 and 1983, that means a 1-billion-dollar reduction in the demand for meat. That is a political decision and we live with it as livestock producers.

Competition Between Humans and Farm Animals

One of the political challenges that will become more serious as time goes on will be the competition for living space between humans and farm animals. Twenty years ago, a livestock economist at a midwestern land grant university was making the prediction that there would be little room in the world for livestock 100 years in the future. Human population was increasing so rapidly that the land would be needed for living space. There would only be room for enough livestock to provide meat flavoring for synthetic meat substitutes, he predicted. Although there are still 80 years to go to see if the good professor was right, we doubt that he was on the mark. Still, his kind of thinking has surfaced in some other forms.

One noted futurist has looked in his crystal ball and concluded that if we fed the grain to humans instead of to livestock, we could perhaps support a population more than three times greater than at present. Some world hunger activists were suggesting a few years back that if each of

us would give up one hamburger a week, it would help feed
the starving of the world. Still others have suggested that
Americans should quit fertilizing their lawns and golf
courses and send the fertilizer to the developing nations to
help them grow more of their own food supply. These ideas
are simplistic and, even if carried out, would have little
measurable effect on hunger in the world. Basically, outside of famine caused by natural disasters, there is no
actual shortage of food in the world, nor of land or fertilizer. There are more than adequate supplies of land, fertilizer, and food. What is lacking is the purchasing power
to pay for the food. What is lacking is effective consumer
demand. Wherever the cash is available to pay for the food,
the food becomes available.

Politically Prescribed Diets

I rather expect that we will have increasing frustra-
tions ahead with those who wish to try to influence the
diets of the American people by political prescription.
For forty years, the Food and Nutrition Board of the
National Academy of Sciences has been the widely recognized
and respected source of dietary guidance. The Food and
Nutrition Board has been the agency that has issued the
"Recommended Dietary Allowances" or RDA that have been the
basis for most nutritional education aimed at the consuming
public. Just two years ago, the Food and Nutrition Board
issued a report, entitled "Towards Healthful Diets," in
which it advised that the average adult American whose body
weight is under reasonable control should feel free to
"select a nutritionally adequate diet from the foods available, by consuming each day appropriate servings of dairy
products, meats or legumes, vegetables and fruits, and
cereal and breads."
In the same report, the Food and Nutrition Board
recommended that dietary change or therapy should be under-
taken under a physician's guidance. Aware of some of the
political headline hunting being done by self-appointed
guardians of American diet, the Food and Nutrition Board
plainly warned that it is "scientifically unsound to make
single, all-inclusive recommendations to the public regarding intakes of energy, protein, fat, cholesterol carbohy-
drates, fiber, and sodium."
You are aware, of course, of the rash of studies and
reports on diet and heart disease, diet and cancer, and
other topics that have singled out animal fats and meat as
the causes of human health difficulties. There was a
surgeon general's report in 1979 and a study by the Senate
Select Committee on Nutrition and Human Needs. Last summer,
a special panel of the National Academy of Sciences on
"Diet, Nutrition, and Cancer," issued a report which differed in important respects from the position of the Food

and Nutrition Board and that had not, in fact, been submitted to the Food and Nutrition Board for review and evaluation. Like many of the other political diet studies, the cancer study issued sweeping recommendations that included a 25 percent reduction in the consumption of fats, fatty meats, and dairy products. The report recommended avoidance of smoked sausages and fish, ham, bacon, frankfurters, and bologna.

The cancer and diet panel admitted it did not know what percentage of cancer risks are attributable to diet or how much the risk could be reduced by modifying one's diet. Still, while admitting that there was considerable uncertainty about the scientific basis for its findings, the diet and cancer panel issued its recommendations anyway and, as you might expect, the press treated it in a sensational manner. It was entirely in order for the livestock industry to ask that a review be held to reconcile the contradictory advice that was originating at the same time from the National Academy of Sciences. In our belief, the practice of medicine should neither be carried out by politicians or advertising agencies. The practice of medicine should be left to the medical profession.

Livestock Marketing Revitalization

At a time when all other industry seems to be centralizing, livestock marketing is disintegrating--breaking up into bits and pieces--with no central system for determining price.

As a result, live cattle prices are being based on data reported in the Yellow Sheet or the USDA meat news--sources that may represent as little as 2 percent of the market volume. Farmers have been looking at other options, such as electronic auction markets, to restore some competition into the system. But, it will take time, considerable capital, and major organizational efforts to establish an effective producer-controlled system of that sort. We may need some federal help and encouragement to get the job done.

International Trade in Meat and Meat Products

Another political challenge we may face will be in regard to international trade in meat and meat products. We appreciate the desire of some in the livestock industry to expand foreign markets for U.S. meat and related products. Foreign market development ought to be pushed in any constructive way. There may well be some potential for gains, particularly as some of the developing countries increase their purchasing power and seek to upgrade their diets. However, care must be taken that nothing we do in seeking to expand world trade reacts to undermine our own meat-import control laws. We ought to recognize that if we attack the quotas, the nontariff barriers and protectionist devices of other countries, this will certainly expose our own Meat

Import Act of 1964, as amended by the Meat Import Act of 1979, to attack from abroad. You may recall that when the 1979 Act was adopted, it was criticized by some who look to the U.S. market to dump their oversupplies. While we now export almost one billion dollars' worth of meat and meat products, we import 2.2 billion dollars in meat and meat products, plus another 1.5 billion dollars in animals and animal products.

The expansion of U.S. meat exports will be a gradual, long-term proposition. It will take a long time before exports offset imports, and this will be particularly true if we go to a free market in meat trade. At any time that world meat supplies are excessive in relation to effective demand, the U.S. will tend to be the magnet for oversupplies. That situation will tend to prevail most of the time. So, in a free market situation, U.S. meat imports will tend to expand more rapidly than meat exports.

We have a good law in the 1979 Meat Import Act. It is a responsible measure that helps us retain a domestic livestock and meat industry. The countercyclical factor that determines the allowable level of imports is particularly important. When the U.S. cattle industry is in the liquidation phase and beef production is relatively high, the countercyclical factor will tend to reduce the allowable level of imports. When the cattle cycle is in the rebuilding phase and domestic production is low, the allowable imports will be increased. If we can hold foreign imports to the minimum figure of 1,250 million pounds or near to the figure, the situation will remain in control. But, in a free market situation, it is easy to imagine that without any controls, meat imports, now subject to the law, would quickly advance to new all-time record levels.

Animal Welfare Concerns

Another political challenge livestock producers may have to face would be from the animal welfare lobby. Up to a couple of years ago, few farmers were concerned about the animal welfare lobby--many had not heard of it. Thus far, there has been no serious effort in the Congress to delete the several words from the Animal Welfare Act that would end the exemption of farm animals and birds from that statute. However, there have been bills in the 96th and 97th Congresses that would regulate confinement feeding of animals. Some activist groups have emerged and have gotten some coverage from farm magazines and the media but, so far, it seems they have been open to dialogue with farm and livestock groups. We should be realistic enough to expect that concerns of nonfarmers about this area of farm production will increase but, if we can keep some lines of communication open with responsible citizen groups, perhaps the discussions can be kept on a reasonable basis.

National Farmers Union has had some concerns over the harmful effects of excessive concentration of poultry and animals. To be frank about it, our concerns have been more in terms of environmental, health, and economic effects of such concentration, rather than with humane treatment of livestock and poultry. Up to this point, here in the U.S., we have been able to avoid the confrontations between consumers and farmers that have been common in Western European countries over animal rights. Most of the difficulty arises with people who are almost totally lacking in knowledge about livestock production methods.

By attempting to carry on a dialogue, we may be able to avoid misunderstandings that tend to polarize viewpoints on their side or ours. If we can win some understanding, we may be able to keep the issue out of the political arena.

Sensible and Effective Regulations

In the time that we have had on the program today, it has not been possible to touch upon all the political decisions that will affect our business and livelihood. There is a whole array of potentially serious problems, if the hysteria for deregulation is carried to the extreme. Reducing needless paperwork burden is desirable, of course, but to eliminate needed regulatory measures is something else. The Packers and Stockyards Administration covers a whole range of competitive issues. What we need is sensible and effect regulation, not the elimination of regulation altogether.

Most of my adult life was spent in Oklahoma as a farmer and farm leader. We found at times that there were governmental decisions so bad that we had to fight them with all our might. But, we also found that we could reason with people and that if we got involved early enough and got a voice in shaping the decisions, we could avoid situations that otherwise might have become desperate. The key is to get involved.

Rest assured, the government is involved and will continue to be involved in our agriculture and our society. If we don't take part in the process, the results will be worse. If we curse the government, the government will not go away. We will just be destined to live under scarcely bearable regulations devised by someone who doesn't really understand livestock farming. Fortunately, we do have a choice and a voice, if we want it.

7

AN AGRI-WOMAN'S VIEW OF THE POLITICS OF AGRICULTURE: NATIONAL AND INTERNATIONAL PERSPECTIVES

Ruby Ringsdorf

As a city girl turned farm wife some 27 years ago, learning all about the various aspects of agriculture has been very interesting and educational as well as very rewarding for me. Because so many have left either the farm or rural area and relocated in the city, farm community or farm bloc no longer carries the clout it once did.

When we look at the statistics we see that in 1949 25% of the populations were farm folks--today we constitute about 2.7%. From a significant segment of society that was once courted, soothed, and cared for, we have become a minority of the voting public--our "clout" practically nil. Our collective ability to feed and clothe our nation and significantly affect this nation's economic status of favorable contributions to the balance of payments still exists, but now one producer feeds 78 people--not just a simple household. And therein lies the rub. Production capability and output efficiency do not vote. Contributions to the common good carry little political weight. This has created our current predicament. We, as producers, find ourselves in the unenviable position of providing a product with public utility overtones without even the restrictive protection that public utility status would bestow.

From a public relations standpoint we are viewed as either land-hungry, dollar-greedy, unpatriotic, rich farmers (witness the writing coming from some of our church groups--writings such as "Strangers and Guests in the Heartland"), or as not-so-smart farm folk in overalls or print dresses (note the celery farmer and his wife on the Baggies commercial), or as sheltered, dull, farm children, apprehensive of the outside world (M & Ms make friends and the hayseeds are readily won over with a hand full). We are none of these or all of these. Farm folk run the gamut--as do shop keepers, manufacturers, and other American entrepreneurs. We are business people. We love agriculture and perhaps are willing to sacrifice some "return on investment" in monetary measures for the "return on investment" in satisfaction and the personal pleasure that come from involvement in production agriculture. But agriculture is too expensive to keep as a hobby. We are commercial farmers. "Commercial" is not

a dirty word--nor is "profit." The politics of agriculture
has economic overtones and very real economic repercussions
for the producer.

I would like to touch on a few of the political influences with which we have had to deal over the past few
years. I apologize for the negative tone of this presentation, but other than our own attitude and the placement of a
free marketer in the office of Secretary of Agriculture,
there has been little to be positive about for commercial
production agriculture.

Past grain embargoes and trade restrictions have worked
to the detriment of the free-market system and have, as a
by product, encouraged grain and oilseed production in other
countries, in direct competition with the U.S. producer,
thus diverting sales arrangements, agreements, and traditions to these countries and away from the U.S. We have become, in many cases, the residual supplier--the supplier of
last resort. We have a reputation as an unreliable source
of grain, a nation whose past political philosophy has allowed food to be used as a weapon and the farmer as a political pawn in the game of foreign relations. The 1973 embargo sent Japan to South America with the technology and economic necessity to develop a source of soybeans not dependent upon U.S. philosophical and political attitudes. Now,
rather than competing with the protein supplied by the an-
chovies harvest of the coast of Peru, our U.S. soybean farmers must contend with a full-blown, mirror-image harvest in
the southern hemisphere. We welcome healthy competition,
but resent our own government policies encouraging such,
while at the same time inadvertently and directly threaten-
ing the economic livelihood of our American producers.

The most recent embargo reinforced this scenario, left
us a residual supplier to the U.S.S.R., and brought agripo-
litical America to its economic knees. The U.S. farmer cannot continue to produce in abundance without the reasonable
expectation of an adequate return for the labor and capital
involved; prices commensurate with this return are probable
only if the world market is available to the producer. Ours
is a global society. It makes no sense to subject the producer to the unpredictable risks of the world while confin-
ing his free marketplace to the narrow national confines of
the U.S. and placing constraints on his participation in the
world economy.

The political intrigue surrounding the long-term trade
agreement with the U.S.S.R. and the threat to employ trade
sanctions against our enemies have kept the farmer in the
dark. Any long range capital plans, any cash flow projections are subject not only to the economic ebb and flow of
supply and demand but to the implementation of foreign policy philosophies either to further our political ideologies
in another country or to achieve some internal goal. our
soybean farmers held their breath for possible repercussions
when our government talked of imposing import taxes on Jap-
anese automakers.

In 1979 we felt the political influences nipping at our heels as the Secretary of Agriculture Bob Bergland called for hearings on the "Structure of Agriculture." (Our official response as American Agri-Women is included as an Appendix.) We had women testifying at hearings held all over the country. Our basic premise was that agriculture does have some problems but we're basically sound. We were profoundly disturbed as we heard witness after witness talk of agriculture's problems and propose solutions bordering on political and economic systems inconsistent with our free-enterprise democratic society.

Other players in the political game are the food lobbyists. We, as producers, are part of the picture. With representatives of both general farm organizations and commodity groups in our state capitals and in Washington, we pay individuals to serve as our eyes and ears and also to pass along our concerns to the powers in political and economic decision-making.

Often working at cross-purposes with us, and just as often uninformed as to the probable repercussions of actions they propose, are various church groups, hunger organizations, social activists, and do-gooders. A recent example is the animal-rights lobby and their campaign against proven and protective commercial agricultural practices. In recent years we've dealt with land groups who favor land redistribution. There is also a strong school of thought that land belongs to everyone and should not be privately held nor used to make a profit.

We've had farm price-support programs since 1933. Although the concept may have initially been a necessary emergency measure, the fantastic productive capability of the agriculture sector today has been accomplished in spite of government interference via the farm programs, not because of them.

We would rather see the emphasis of government involvement redirected toward programs that will encourage a healthy, contributory agriculture. We see government commodity programs as an alternate source of credit rather than as a guaranteed price. The trend should be toward market development and a broader use of trade barriers and guarantees (supported by action) that the U.S. farmer is a reliable supplier.

Food is the primary need of all mankind. Efficiency in agricultural production has allowed portions of the labor force to enter industry, the arts, government, religion, and other human endeavor while depending on others to provide their daily requirements of food. The U.S. has always held individual and religious liberty to be a right of all its citizens and has chosen to protect this liberty by establishing a democratic republic and by fostering the private ownership of land and other resources. The ownership of this land is openly accessible to all who will work for it. American agriculture is the most efficient in the world, and its productivity has reduced the cost of food to the point

that only about 17% of our average disposable income is now
spent for this most basic human need. This monumental productivity has relieved 97% of our population of any need to
produce food, leaving less than 3% of our people directly
employed in farming. This separation of most people from
everyday agricultural production has raised a great deal of
public questioning of agricultural practices, farmland ownership, labor, and similar issues. These are political
issues and affect every individual in this country. They
deeply affect farmers and their methods of production and
could ultimately affect both the structure of American society and the total supply of food available to the people of
the world.

APPENDIX A

AMERICAN AGRI-WOMEN
POSITION ON STRUCTURE OF AGRICULTURE HEARINGS

Whereas, R. Bergland, Secretary of Agriculture has requested
input on the Structure of Agriculture in twelve areas, we,
American Agri-Women, submit positions on the following subjects:
1.   Land ownership, tenancy and control.
     -   Land ownership, control and tenancy belong
         in the private sector.
     -   The majority of ownership will be limited
         to U.S. citizens.
     -   Since unlimited leasing is not restricted
         in any other business, ownership and pro-
         ductive use of agricultural land need not
         be with the same party.
     -   Agricultural landowners need the flexibil-
         ity to manage their land as their exper-
         tise guides them. A residency requirement
         for agriculture is archaic - just as it
         would be for any other business.
2.   Barriers to entering and leaving farming.
     -   Barriers to entering are basically deter-
         mined by the lending institution.
     -   Leasing can be a vital avenue for entry
         into farming and therefore must be unre-
         stricted.
     -   We object to government support of "life-
         style farmers" (non-commercial) because
         productive incentive will be lacking.
     -   Capital gains taxes and environmental zon-
         ing restrictions are barriers to farming.
3.   Production efficiency, size of farms, role of
     technology.
     -   Each farmer should be able to determine
         his farm size. Technology should be uti-

            lized to the fullest. Agriculture must
not be singled out in use of size technology, and methods of production.

4. Government programs.
   - Much of the success of American agricul-
     ture is due to agricultural research.
     This represents only a small portion of
     the USDA budget. To feed a hungry world,
     we vitally need accelerated research pro-
     grams.
   - Food assistance programs represent 56% of
     the total USDA budget. A cut in agricul-
     tural research, extension, soil and water
     conservation programs would not be in the
     national interest.
   - We oppose USDA funding of persons to tes-
     tify at regulatory hearings.

5. Tax and credit policies.
   - Inheritance tax laws need changing, par-
     ticularly the carry-over provision.
   - More tax credit incentives for conserva-
     tion measures are necessary to justify the
     required capital outlay.
   - No taxes on transfers between spouses.

6. Farm input supply system.
   - Leave in the private sector. The current
     system works well. Leave it alone.

7. Farm product marketing system.
   - Leave in the private sector. The current
     American marketing system is the best in
     the world.

8. Present and future energy supplies.
   Develop multi-faceted energy supplies and aggressively work to decrease our foreign dependence.
   - Agriculture uses only 2.9% of U.S. energy
     supplies and we needn't apologize to any-
     one.

9. Environmental concerns, including conservation and the use of soil and water.
   - Farmers were the first environmentalists
     and conservationists as their income and
     livelihood depend on it. Let us continue
     with less regulatory interferences.

10. Returns to farmers.
    - Farmers need to develop collective bar-
      gaining. But farmers, like any other in-
      dependent businessmen, accept the respon-
      sibilities and assume the risks.

11. Cost to consumers.
    - The American farmer produces the least ex-
      pensive, most bountiful, and highest qual-
      ity food and fiber for the consumer any-
      where in the world.

- Farmers must be allowed to pass on their costs as in any other business.
12. Quality of rural life.
  - There is a difference between rural America and farming America. The economic quality of life of agripolitician America is dependent upon agriculture - just as the economic quality of life of metropolitan America is dependent upon the central business district.
  - Quality of life in rural America is as good as the people who live there and make up the rural community. As people from rural areas, we can attest to the fact that it is a good life if you work at it, and are allowed to work at it without government interference.

Whereas, we American Agri-Women, representing 24,000 people, present these proposals with a diversified history of farming success and failures in each of our 50 states; that the aforementioned proposals are based in fact and are the positions American Agri-women takes on the structure of agriculture.

Adopted November 3, 1979 by the governing body of A.A.W. at annual convention, San Diego, California.

8
OREGON WOMEN FOR AGRICULTURE TALK ABOUT THE STRUCTURE OF AGRICULTURE

Ruby Ringsdorf

Let it be known that Oregon Women for Agriculture do believe that it is the inherent right of every child born today to have adequate nourishment; and that the American farmer will continue to feed the hungry if not strangled with bureaucratic rules and regulations.

We furthermore feel that it is neither our duty, nor even our right, to enter into the internal policies of a foreign country whose political system, or local corruption, are preventing food from reaching their hungry masses. Neither are we prepared to let those countries' Marxist-oriented political ideologies creep into and destroy our free enterprise system. The free enterprise system is the propelling factor that has made American agriculture the envy of the world!

LAND OWNERSHIP, CONTROL, AND TENANCY

Fifty-five percent of the land in the state of Oregon is already publicly owned (L.C.D.C.). The number of all commercial farms (farms with sales of $2,500.00 or more) in Oregon increased rather than decreased from 1969 to 1974.

The number of commercial farms in Oregon with sales greater than $40,000.00 increased 67% from 1969 to 1974 and comprised 30% of all of Oregon's commercial farms. At the rate of inflation over the past 10 years it is surprising that this percentage is not greater. It doesn't take much of a farm to produce $40,000.00 in sales today, but the net probably isn't enough to keep the family dog in dog food for the year.

According to an Oregon State study (EM N:23), family farms, nonincorporated, comprise 96.4% of all commercial farms with 3.6% being corporate farms. Of the corporate farms, 87.3% are family farms (94% have 10 or fewer share-holders, 44% are controlled by one stockholder).

Many family farmers own some land and lease more from retired farm relatives or neighbors in order to make their units more economical.

BARRIERS TO ENTERING AND LEAVING FARMING

Inflation, high interest rates, inheritance tax laws, FHA regulations, EPA regulations, and our national cheap food policy are all barriers to entering and leaving farming. Inflation is not only driving up the cost of land to an impossible price for a beginning farmer (the interest alone for each acre of ground is much more than what he would have to pay to lease the ground), but also the cost of the equipment needed to start farming. Inheritance tax laws, especially the carry-over provision, make it almost impossible for children to inherit a farm without selling a portion of it or splitting it up to meet the tax obligation so that they no longer have an economical unit.

EPA regulations are becoming increasingly more difficult to cope with and discourage many young people from even thinking of farming. FHA regulations restrict the amount of money available to a young farmer for a small acreage because it is not an efficient, economical unit that could produce enough net income to service debt and provide a decent living for his family without off-the-farm income. And yet, if the unit is large enough to do both, the cost is far more than FHA is allowed to cover for one farm.

It is difficult for a farmer to retire and leave farming. The land cannot be sold because of strict zoning laws, because of capital gains tax on the appreciated land value (often the only net savings realized from the farmer's investment in time and labor), and because his acreage is no longer large enough to be an economic unit for a family farmer. He stays on the land, rents to others, and does the best he can, too often becoming another rural-poor statistic.

PRODUCTION EFFICIENCY AND SIZE OF FARMS

Production efficiency and size of farm are tied together. Our Oregon State study shows that the average size of all commercial farms in Oregon increased very little from 1969 to 1974. It would seem the trend to larger farms has already peaked because of production efficiency.

Size and number of farms in Oregon vary greatly from one geographical area to another. In the Lake Labish area near Salem, where land sells for $10,000.00 per acre and up, a 20 A farm is considered large. In the southern end of the Willamette Valley, in the grass seed capitol of the world, a thousand acre farm is not considered large, and in the cattle grazing lands of eastern Oregon a 5,000 A ranch is not large. A dairy farmer can have only 100 A but milk 500 cows and be considered a large farmer, but another dairy farmer can have 50 cows on 500 A and be considered a small farmer.

Because Oregon produces over 170 different marketable commodities, it is unfair to use a gross dollar amount

figure for sales to define large and small. Different crops
show different net results. It is impossible for a 20 A
berry farm to net more than a 500 A wheat ranch. Most often
a farmer cannot convert his acreage to higher value crops
because of soil types, marketing limitations, increased
risks, and increased operating and capital requirements.
The efficiency of the American farmer is the envy of
the world. After American farmers feed the U.S. they export
60% of their wheat and rice, 50% of the soybeans, one-fourth
of their grain sorghum and one-fifth of their corn. The
U.S. provides half of the world's wheat (Oregon Grange
Bulletin 9-4-78).
Agricultural products are the second largest category
of U.S. exports. Agricultural exports returned $23 billion
to our country in 1976. In 1975 agricultural exports provided the foreign exchange to cover 83% of our petroleum
imports.
American farmers provide all this despite the fact that
the number of U.S. farms and farm workers has decreased by
two-thirds since 1940.
One American farmer can now produce enough to feed 60
people. From 1950 to 1978, farm productivity increased at
an annual rate of 5.3%--more than twice as much productivity
as compared to any other nonagricultural business (Oregon
Grange Bulletin 9-4-78).

GOVERNMENT PROGRAMS

Much of the success of American agriculture is due to
agricultural research. This represents only a small portion
of the USDA budget. Food assistance programs, including
Food Stamps and Child Nutrition (programs benefiting from
past research programs), represent 56% of the total USDA
budget. A cut in agricultural research, extension, and soil
and water conservation programs would not be in the national
interest.
As to having a national or world grain reserve, why not
establish a worldwide monetary food fund (a required UN
fund, a contribution fund with participation by churches and
other interested groups, internationally funded and administered) to be used for international food crises. Reserves
have a history of depressing prices to producers and stabilizing prices at the lower levels. Reserves have also acted
as a disincentive to production so that farmers change (if
possible) to producing crops that will hopefully yield more
net return. Any grains in reserve should be isolated from
world markets and used as aid rather than trade. The cost
of this grain reserve should be shared by all people internationally.
Another program that is under consideration for USDA
funding is the plan to pay the expenses of low-income and
nonprofit groups that testify at regulatory hearings. We

are very much opposed to such a plan; even though Women for
Agriculture would qualify for funding.

TAX AND CREDIT POLICIES

Inheritance tax laws need changes, particularly the
carry-over provision. Prior to enactment of the carry-over
provision, beneficiaries inheriting appreciated property
received a stepped-up tax basis on property at the time of
inheritance, and each generation of a farm family was subject to capital gains tax only on the appreciation that
occurred while they owned the property. This procedure was
radically changed by the carry-over provision that bases
capital gains on inherited property on the descedent's
acquisition price and not the market value at the time of
transfer.

There should be more tax credit incentives given for
conservation practices, since these can be very costly for
one individual.

FARM INPUT SUPPLY SYSTEM

Government regulations and inflation have had a strong
influence on the farm-input-supply system. Labor costs have
spiraled, inventory taxes have prevented smaller manufacturers and suppliers from keeping a full inventory; land
costs have spiraled and it is becoming more difficult to
obtain necessary capital.

FARM PRODUCT MARKETING SYSTEM

There is a lot of talk about direct marketing from
farmer to consumer. This works only in agricultural areas.
Most crops sold in this manner are perishable, thus limiting
choice, variety, and quality. It takes much more time and
energy to drive all over to pick up vegetables here, eggs
there, milk at that place, and fruit at still another stop.
We already have the most energy-efficient distribution
system.

Farmers are being told, "You can get a better price for
your produce than the processor gives and the consumer can
get it for less. Let's cut out the middle-man." Just who
is the middle man? If we cut out all the middle men in the
food-processing chain, our unemployment rate would probably
be closer to 30% or 40% than the 8% or 9% it is now. Also,
at the same time we are hearing rumblings about vertical
integration (selling your own produce as a finished product
and cutting out all middle men). This is what multinational
corporations are accused of doing. Direct marketing is the
same.

Probably close to 100% of Oregon's fruits (not including tree fruits) and vegetables are sold by forward contracting or contractual arrangements with a processor. This method is preferred over freedom of decision-making at harvest time. There are not many farmers who would care to wait until harvest time to search for a market for their perishable products. They would be at the mercy of the processor who would then know the farmers have no choice but to take whatever price is offered. If an equitable price is not offered at planting time, the farmer has the freedom of decision to plant or not to plant.

Contractual arrangements and forward contracting on seed and grain crops isn't bad either. It can certainly provide some freedom from fear. The farmer who stays in business all his life usually is the one who contracts ahead whenever he feels the price is such that he can make a fair return. It takes some of the gamble and risk out of farming.

PRESENT AND FUTURE ENERGY SUPPLIES

In the early 1920s we had 25 million horses to pull the plow, the wagon, and the carriage. We fed about one-fifth of our grain and roughage to those horses. (Today it would take one-third of our crop land plus 20 years to breed enough horses and mules for today's needs.)

It is time we look into using biomass or agricultural products as a future energy source. The liquid energy that we import is priced at $1.50 to $2.50 per gallon in most major industrial nations. We are nearing these world prices now, which will make the production of biomass fuel profitable. There are tons and tons of grass straw in Oregon alone that can be used for fuel pellets or biomass conversion.

We feel it is unfair for American farmers to be told to conserve fuel and energy when farming uses only 3% of the total energy consumed in this country. It takes more energy in the home for food preparation than it does for agricultural production (including fertilizer and other energy-intensive inputs).

Productivity per man-hour in agriculture has been increasing about twice as fast as the rate of productivity per man-hour in manufacturing.

ENVIRONMENTAL CONCERNS, INCLUDING CONSERVATION AND THE USE OF SOIL AND WATER

Oregon farmers are also environmentally concerned. We are also concerned because we who are engaged in agriculture are such a minority. Even though Department of Environmental Quality tests gave proof to the fact that smoke from field burning, a practice used in the Willamette Valley to sanitize our seed crop grass fields, really had little or no effect on the quality of the air in the Valley, the EPA would not allow easing of regulations that are now strangling the Grass Seed Industry in Oregon. Why? <u>Because all</u> <u>other sources of pollution are increasing yearly</u>!

Bureaucratic rules and regulations often conflict with each other. In the Silverton Hills area, for example, 25 to 30 years ago farmers were losing tons of top soil every winter as a result of water run-off on cultivated fields. They discovered that their soil and climate was suitable for perennial grass seed production, which held the soil on the hillsides. Now EPA has restricted field burning because of air quality. Without burning, the grass fields become diseased and seed production is no longer economically feasible. We are again faced with soil erosion and resulting probability of water pollution.

Oregon has been working on Water Quality Management Programs, or non-point-source pollution. There are six major agricultural pollutants: sediments, nutrients, salts, organics, pesticides (including herbicides) and disease-producing organisms. So far only sediments have been found in our streams and these have been coming primarily from nonagricultural sources.

America's agricultural engineers are coming up with new and better seed drills that utilize low-till and no-till methods. Oregon farmers have been using grassland drills for many years, but we cannot use a grassland drill in a grass field that <u>has not been burned</u>. Here again, air quality versus water quality.

RETURNS TO FARMERS

Because our government has long endorsed the cheap-food policy, returns to farmers for the last 35 years have not kept up with the rest of the economy.

The American farmer was forced to become more and more efficient and only those who were efficient, excellent managers stayed in business. (Often even the most efficient were wiped out because of extraordinary conditions, such as weather. The margin in good years was so slender that a poor year wiped them out.)

Naturally, when the returns per acre became less and less, we had to expand our acres if we were going to live off the land. Now, suddenly, we are all called "commercial farmers" and that seems to be the wrong kind of farmer to be. "Commercial farmer" seems to be a dirty word in many circles. At the Rural America Conference in Washington, D.C., in June of this year, commercial farmers were being blamed even for dope addiction because farm mechanization caused these people to be out of work.

The people in Rural America meetings defined the family farm as a unit that produces food only for the people on the farm unit and who do not sell any farm product for profit.

In 1945 our fathers sold rye grass seed for 12 cents a lb. (They also sold wheat for $100 a ton or $3.33 per bu.) A tractor at that time cost around $3,000 and the first self-propelled combines came out for around $5,000. Today a tractor costs between $30,000 and $60,000; a combine from $65,000 to $70,000. Guess what the price of rye grass seed is today? Twelve and one-half cents per lb. Last year it was only 10 cents per lb so there are big headlines in the newspapers that agricultural wholesale prices increased almost 25% over a year ago! It is usually buried in the back pages when our farm product prices decrease or simply stay at the level of 30 years ago.

COSTS TO CONSUMERS

We as farmers are always hearing from our city friends about farm subsidies and how we are being paid to keep land out of production, etc. The truth is, the American farmer has been subsidizing the consumer for the last 40 years.

The trend toward greater efficiency in farming has benefited the consumer most. People have never had a greater variety of safe and nutritious food at so low a cost. The consumer can purchase more food for his hour of labor than at any time in history or in any other country. When the housewife complains about high food prices, she is often paying for a maid-in-a-box by buying prepared and semiprepared foods that cut down on food preparation time at home. It all depends on your priorities--time versus cost.

We also feel that if it is a policy to serve the public interest, by drastic and disruptive actions (such as export controls and import floods), then the general public, not just the producer, should pay or help pay for this policy.

QUALITY OF LIFE IN RURAL AREAS

This is interesting! Rural areas, not agricultural areas! Rural America: Educational Needs of Rural Women-- these programs are not referring to farmers or farm women. Rural America means the urban population that has moved to the less-populated areas of the U.S. from all socioeconomic levels and who most likely have never farmed and never will. The problems of these people should come under the jurisdiction of H.E.W., not USDA.

We farmers have noticed that this migration to the rural areas is causing problems. They tell us, "We are moving out to the country because we want peace and quiet and we want to be one with nature." And then they complain about the noise and dust from farm machinery and activities; so now we have noise pollution and dust-control laws. Now rural residents are suing because of smells coming from swine production and cattle feeders, forcing farmers to put in costly equipment to take care of the smell.

The environmentalists insist we use too much commercial fertilizer, but they complain about the smell when the local dairy farmer puts the liquid manure on his fields with an irrigation gun. There are over 100 known toxic substances occurring "naturally" in the environment, yet they complain any time they see our spray rigs come out of the yard.

Rural America says there are 131 rural towns in the U.S. without a doctor, that many people have to drive from one to two hours to a medical center. This is not always bad. We would rather drive for two hours to a good medical center with full facilities than fifteen minutes to a small facility where sometimes a local doctor tries to do only what a very specialized doctor should do.

Rural America also says one-half of the maternal deaths occur in rural areas. We believe that. Much is due to the back-to-nature trend, which is currently popular along with do-it-yourself childbirth. This is fine if everything proceeds normally. In the home we do not have the back-up facilities to aid a difficult delivery; it is too late then to rush to the hospital. Many of these same people have no prenatal care of any kind and their diet is often very inadequate because of their chosen life style and eating habits.

If we talk about the quality of life of real farm and rural people, we are talking about something entirely different. Genuine farmers seem to have fewer divorces per thousand population and they generally have a very strong family unit because everyone learns to work together.

A farmer has a lot of respect for his Creator; he is too closely involved with growing and living things to think that we are here purely by chance.

Generally speaking, the quality of life depends on the individual involved. The socialite who grew up in the large city might be quite disenchanted and bored living in a small rural community. She might complain bitterly about the lack of culture! On the other hand, the people who grew up in that same small area are quite content and feel that they have the good life. If they were to move to the large city they might then cry bitterly about the hustle and bustle, the unfriendliness, the foul air, the crime rate, etc., etc.

The kind of life we want in America can be found by any one who intends to earn it. If we expect it to be given to us, we will never find it.

Oregon Women for Agriculture

9
THE CURRENT STATUS OF THE
FAMILY FARM IN AMERICAN AGRICULTURE

George Stone

Almost everyone knows what a family farm is, but hardly anyone is able to define it on paper to the satisfaction of other people. A particular farm may meet most of the criteria that might be suggested, but there will always be some differences of opinion on such things as size, ownership, and control. There is not time for a debate on the fine points of a family-farm definition. But if I am to talk to you today on family-farm agriculture, you are entitled at least to know what I think I am talking about.

FAMILY FARM DEFINED

I like the National Farmers Union's definition of a family farm. It says:
"A 'family farm' is, ideally, one which is owned and operated by a farm family, with the family providing most of the labor needed for the farming operation, assuming the economic risk, making most of the management decisions, and depending primarily on farming for a living."
That is probably as well as it can be explained in less than fifty words.

PUBLIC POLICY RELATED TO FAMILY FARMS

Our national public policies have endorsed and advocated a family-farm structure of agriculture for almost 200 years, dating back to the Ordinance of 1785, the Land Act of 1796, the Pre-emption Act of 1841, the Homestead Act of 1862, the Reclamation Act of 1902, and a half dozen major statutes in this century as recently as the Food and Agriculture Acts of 1977 and 1981.
The 1977 Act includes a declaration by the Congress that it "firmly believes that the maintenance of the family-farm system of agriculture is essential to the social well-being of the nation"...and that "any significant expansion of nonfamily owned large-scale corporate farm enter-

prises will be detrimental to the national welfare."
The 1977 Act also mandated that the Department of Agriculture should issue an annual report on the "Status of the Family Farm." This has been done, supplying some continuing data on the structural trends and changes in agriculture.

In March 1979, speaking at the national convention of the Farmers Union, Secretary of Agriculture Bergland called for a national dialogue on the structure of agriculture, declaring:

"We are at a point in our history where a broad-based public discussion of the issues that shape national policies is needed to promote the kind of agriculture and rural living this nation wants for the future."

In that Kansas City speech, Secretary Bergland observed that "we really don't now have a workable policy on the structure of agriculture," and warned:

"We can act now to insure the kind of American agriculture we want in the years ahead. Or we can let matters take their course, with the probable result that we will wake up some morning to find that we have forfeited our last chance to save those characteristics of the farm sector we believe are worth preserving. I, for one, do not want to see an America where a handful of giant operators own, manage, and control the entire food production system. Yet that is where we are headed, if we don't act now."

In late 1979 and early 1980, Secretary Bergland conducted this national dialogue at a series of regional hearings. Numerous economic papers and a comprehensive report were eventually published. It was a worthwhile exercise in stimulating Americans to think about what they want in an agricultural system and what they want their federal government to do to assure such a system. But, while there may be better public understanding of our agricultural system, there is little in the way of agricultural legislation or administrative decision-making that can be attributed to the Bergland study.

ASSESSING CURRENT FARMING CRISIS

We find ourselves here, early in 1983, still trying to assess the current situation of family farms. Through our 80-year history, National Farmers Union has been totally dedicated to the family-farm system. We believe that the family farm represents the best choice for the American people on every score:
- Assured abundance
- Efficient production
- Best care and use of land and water resources
- Rural employment
- Quality of life in rural communities

- Highest export earnings
- Most favorable balance of trade

During the 80 years of Farmers Union, the family farm has proven its durability and staying power. Family farms have survived wars, natural disasters, and a total of 14 recessions, panics, and depressions, including the most recent. During most of this century, it was usually assumed that the family farm would survive as the dominant form of agricultural structure. Now, although it is readily acknowledged that the family farm is the most efficient agricultural production unit, it is no longer that certain that it will survive much longer.

In a spirit of candor, one must admit that there have been other times when the survival of the family farm appeared to be in doubt. Calamity seemed at hand. Yet, while some farm operators were lost in the crises of the past, and their loss was regrettable, the system as a whole survived and continued to produce for the nation.

Having expressed the caution that at times in the past things have appeared worse than they turned out to be, there are signs that the current challenge to the survival of family farms is the most dangerous, at least since the years of the Great Depression.

Not since the early 1930s has the nation had three such bad years in agriculture in succession. Many signs indicate the magnitude of the farm crisis. Net farm income dropped from $32 billion in 1979 to $19.8 billion in 1980, $18.9 billion in 1981, and while we do not yet have final figures, appears destined to be still lower in 1982. In terms of purchasing power, the farm parity ratio in much of 1981 and into 1982 has been the most unfavorable suffered by farmers since 1933. In 1933, U.S. farmers had $3 billion in net income, but only $9.1 billion debt. That was a ratio of $3.10 in debt for each dollar of net income. Today, we have something over $11.00 in debt for each dollar of net income.

In 1981, for the first time in recorded history, U.S. farmers paid out a total of $19 billion in interest outlays on their debt, a sum that exceeded their net income for the year. At the worst of the Depression of the 1930s, the interest rates paid by farmers averaged 6.4%, while recently the rates paid by farmers on loans to commercial banks averaged over 18%, as reported in the Federal Reserve Bulletin.

In these last three years, farmers have been substituting credit for income at an alarming rate. Years ago, farmers were able to generate much of their capital needs internally. In 1970, for example, farmers depended on borrowed capital for only 5% of their cash operating funds. By 1980, the proportion was up to 21% and, in 1981, it was almost 23%.

Another important measurement is the liquidity ratio of farmers. In 1950, as an example, U.S. farmers had $13.8 billion in cash assets such as deposits, currency, and savings bonds. Against this, they had $12.4 billion in debts. That was a liquidity ratio of 111%. In 1960, there were cash assets of $13.9 billion and $24.8 billion in debts, a liquidity ratio of 56%. By 1970, there were $15.6 billion in cash assets and $53 billion in debts, a liquidity ratio of 29%. In 1981, the cash assets totalled $19.9 billion against a total debt of $194.5 billion, a liquidity ratio of about 10%. Behind all those statistics are human families trying to earn a living in a productive and useful endeavor.

FACT-FINDING HEARINGS

To document the human side of the farm crisis, the National Farmers Union held a series of nine regional fact-finding hearings in March and April of last year. We heard testimony from 230 witnesses, including farmers, farm wives, main street businessmen, cooperative officials, teachers, bankers, and community leaders. The summary report, which we published on these hearings, is entitled "Depression in Rural America." It did not deal just in generalities or endless statistics, but told the personal story of families beset with hardship and despair because of conditions over which they had no control. The report showed how the desperate economic conditions were affecting the lives and survival of working farmers, their business communities, and the fabric of life in rural America.

The purpose of the hearings and the report was, of course, to mobilize opinion and develop a sense of urgency about farm legislation that would help family-scale farmers survive.

Of course, there are some who say that the federal government should not intervene on behalf of family farmers --that we should just let nature take its course. The theory is, if we just let the decline in family farms continue, then, after a while, just the efficient farmers will remain and they will be able to prosper.

But those who have been involved in agriculture for a lifetime have yet to see such a scenario work. In 1960, for example, there were about four million farming units and they were earning farm income and purchasing power equal to 80% of parity. In the decade of the 1960s, the nation lost one million farmers, but farm income did not go up. We have been as low as 57% and 58% of parity. At this rate, it might be asked, how long will it take to get to 100% of parity?

The truth is you won't reach some sort of ideal economic situation for farmers by that route. The truth is that the economic hardship is not weeding out small, marginal, or innefficient farmers. The farmers who have been hurt most in these past three years of low farm prices and high interest rates have been the good, efficient operators in the middle of the scale in farm size.

Further, the projections are that this kind of attrition of our best farmers will continue. In 1980, USDA economists did a projection of what will happen to farm size and structure by the year 2000. The report projected that the number of farming units would drop by 30% to 1.8 million by the year 2000, with most of the decline in middle-sized farms. Small farms, with less than $20,000 annual gross, will still make up 50% of the total farming units, with large farms, with $100,000 or more in gross sales, edging out the middle-sized operators.

Along with these structural changes, the USDA officials foresee an increase in concentration in both farmland ownership and production. The USDA specialists project that it will take $2 million in capital assets to run a farm capable of grossing $100,000, and that these large capital requirements will tend to concentrate farm wealth in the hands of a relatively few.

Young beginning farmers will have increasing difficulty entering the industry. USDA projects that there will be fewer than 300,000 farmers under the age of 35 years in the year 2000, a drop of 200,000 from the current level. The number of individual ownerships and partnerships in farming will decline by the year 2000, while the number of corporate farms and multiownership units will increase, the report indicates.

In another report associated with the farm structure dialogue, entitled "Another Revolution in U.S. Farming," USDA economists predict that there will be further declines in the number of farms, but not at as sharp a rate as in the 1950s and 1960s. However, there will be increasing concentration of production among the largest producers, along with strong pressures for the separation of ownership and use of farming resources. Because of taxes and other factors, off-farm investors can get higher overall returns by investing in farmland than they can by investing in common stocks of business and industry.

In the Farmers Union, we view the separation of land ownership and farming operations with a great deal of concern. It seems to us that such a trend will have the tendency to create a new generation of sharecroppers--people who have little control over their own destiny. That is why Farmers Union in the past several years has taken the leadership in seeking to limit absentee ownership of agricultural land. The threat has come from three different sources:

- Investments in U.S. farmland by American business corporations, conglomerates, and off-farm investors. This has included efforts by individual business firms and such spectacular schemes as the Ag-Land Trust Company.
- Investments by foreign corporations and investors.
- Proposals to invest pension fund assets in U.S. farm cropland.

In regard to the threat of domestic corporations and investors to take over farmland, several midwest states in the past several years have enacted limitations on corporate ownership of farmland and corporate farming. Most of the states in the Mississippi Valley now have restrictions of some sort on corporate farm ownership. Largely because of the vigorous campaign by the Farmers Union and the opposition raised at a Congressional hearing in Washington, D.C., the Ag-Land Trust proposal was dropped.

Because American farmland had become a magnet for foreign corporations and investors, the Farmers Union successfully won adoption of the Agricultural Foreign Investment Disclosure Act of 1978, under which foreign persons or foreign-controlled firms acquiring U.S. farmland must report such holdings to the USDA. Although there may be some evasion of the disclosure law, we now have some hard data on the extent of foreign holdings. It is now clear that there has been more foreign investment in American farmland in the past five years than in the previous fifty years. The latest annual disclosure report by USDA shows that almost 5 million acres of land were acquired by foreign persons during 1981. The foreign acquisitions were equal to about 25% of the total of 18.1 million acres of farmland sold during the year.

The third proposal, that of the American Agricultural Investment Management Corporation of Chicago, proposed to facilitate the investment of nonprofit pension funds in U.S. farmland. We became concerned because pension funds represent a huge pool of capital earning very modest returns on the order of 3% to 4% a year. We thought the opportunity to invest and take advantage of the rapid appreciation of farmland values would be irresistible, even if the profit from farming were modest. As a matter of fact, pension fund assets now total about $700 billion and are expected to rise to $1.5 trillion by the year 1990. Obviously, there would be enough capital to buy all of the farmland in the nation.

We don't expect that to happen. But even if only 3% or 4% of the pension funds were invested in farmland, that would total $18 to 24 billion a year--about the total of farmland sales values in recent years.

The ability of beginning farmers--or existing farmers seeking to expand their operations--to bid for land would certainly suffer by the presence of institutional investors who would not have to pay for the land from their agricultural earnings. Young farm couples who hope to acquire a viable farming unit would be virtually fenced out of the competition for farmland by vast amounts of absentee capital. Of course, the promoters of the pension fund scheme claim they would be doing farmers a huge service by relieving them of the necessity to own farmland. Such a separation of land ownership and farming operations would enable farmers to use all of their limited capital in production.

This is a phony argument. The operating farmer pays a land cost whether he owns or rents. He pays land costs whether he is a cash renter or a share renter. We take this attitude in the Farmers Union because we believe that public policy, whether federal or state, ought to be helping families become owners of the land they farm, not separating them from that possibility.

In conclusion, we regard our efforts to keep farmland in the hands of operating farm families as very important. But, it should be pointed out, the challenge of these outside forces is most damaging because of the weak economic position of our farmers. If farm prices and income were maintained at a more satisfactory and stable level, farmers would be able to withstand more easily the competition of outside investors. Low farm prices and income, accompanied by high interest rates, compound the problems of farmers in sustaining themselves in land ownership and farming. Because this is true, we cannot simply go on as we are and let nature take its course. We must act positively on farm income and other measures to assure that we continue to have a predominantly family-owned, family-operated farming system in our nation.

10

LIVESTOCK PRODUCTION
ON NEW ENGLAND FAMILY FARMS

Donald M. Kinsman

INTRODUCTION

The Northeast region of the United States encompasses the 12 states of Delaware, Maryland, New Jersey, New York, Pennsylvania, West Virginia, Connecticut, Maine, Massachusetts, New Hampshire, Rhode Island, and Vermont; with the last 6 named constituting the New England states. Table 1 cites the basic facts and figures for the total Northeast region, which contains 128 million acres of land or 5.6% of the land area of the U.S. The 6 New England states represent 40 million acres (about one third of the Northeast) or 1.7% of the U.S. land area, yet the Northeast contains 21.6% of the nation's population or about 49 million people. New England has 12.3 million people or 5.4% of the U.S. total. The average annual precipitation is 40 to 46 in., and the mean temperature variation is from 20°F to 40°F in January to 70°F to 80°F in July. Temperature and snowfall vary considerably with elevation, which is dominated by the Appalachian Highlands. The Atlantic Ocean serves as a moderating influence along the coast. The frost-free period ranges from 90 to 150 days.

Sixty-four percent (64%) of the area is forested, compared with a U.S. average of 32%. The Northeast contains 11.4% of the nation's forest lands; New England represents 4.5%. The land suitable for agricultural production is primarily gray and brown podzolic soils, and agriculture on these lands is intensive.

Being the most highly urbanized region in the U.S., over 12% of the land is city, urban, and industrial compared to 9% U.S. average. Grassland pasture represents 3.2% of this area versus 26% for the U.S., and 15% is crop land as compared to 21% for the nation.

LIVESTOCK PRODUCTION

Against such a background, one might wonder about the livestock potential for this Northeast region of which New England is a microcosm. Approximately 6.1% of the U.S.

TABLE 1. COMPARATIVE AGRICULTURAL DATA* FOR NEW ENGLAND, THE NORTHEAST, AND THE U.S.A.

	New England	% of U.S.	Northeast	% of U.S.	U.S.A.
States	6	-	12	-	50
Land area (million acres)	40	1.7	128	5.6	2,264
Population (million)	12.3	5.4	49	21.6	227
Number of farms (thousand)	26.4	1.1	179	7.7	2,333
Average farm size (acres)	171	-	183	-	450
Cropland (million acres)	2.2	0.5	19	4.2	456
Forestland (million acres)	32	4.5	82	11.4	718
Livestock numbers (thousand):					
Cattle (beef & dairy)	753	0.6	5,750	5.0	115,013
Sheep	42	0.3	450	3.6	12,492
Swine	106	0.2	1,687	2.6	64,520
Agricultural cash receipts ($ billion):					
All commodities	$1.4	1.1	$8.0	6.1	$131
Livestock products	$0.98	1.4	$5.4	7.8	$69
Dairy products	$0.55	3.7	$3.1	20.7	$15
Cattle & calves	$0.08	0.2	$0.7	2.0	$35

*Data selected from USDA reports.

agricultural commodity value is produced in the Northeast. The major portion of the livestock receipts is from dairy products--3.1 billion dollars. Beef cattle and calf receipts account for $700 million annually or approximately 2.0% of the U.S. total. All cattle in the Northeast represent 5.0% of the U.S. total, sheep 3.6%, and hogs 2.6%; for New England alone, those percentages are an infinitesimal 0.6%, 0.3%, and 0.2%, respectively. Therefore, it goes without saying, the Northeast is a meat deficit area, and New England in particular produces but 4% of its meat consumption, thus it imports 96% of its meat supply, chiefly from other sections of the nation. Livestock production, exclusive of dairy, in New England must be considered as consisting primarily of small-livestock farm operations. Using the USDA definition of a small farm ($20,000 or less gross sales), 42% of all farms in New England are considered small. Most producers are part-time farmers. These part-timers are on family farms and do contribute to their owners' well-being in addition to producing meat for a ready market.

CAN NEW ENGLAND LIVESTOCK FARMS CONTINUE?

Although dairy farms in New England do continue to decrease in numbers and dairy cow numbers diminish accordingly, the milk production per cow and per herd or farm continues to climb. This, in turn, maintains the milk supply, perhaps at too high a level, but also makes available land, facilities, and expertise for other agricultural pursuits. To protect the better agricultural land, several northeastern states, namely Connecticut, Massachusetts, and New Jersey, have instituted Farm Land Preservation Acts that set aside for perpetuity the best agricultural lands to remain forever available for food production. This needs to be done nationally before any more of our precious, highly productive agricultural land comes under control of the developers and their paved jungles. We in New England have felt the pressure first and gladly share our experience with all to maintain a viable agriculture. Our aim is to achieve greater self-sufficiency in producing more of our food requirements as we recognize the danger of possible isolation in the paths of energy crises, weather, transportation strikes or failures, and our climate restrictions. With the courage, fortitude, industry, and imagination of our early forefathers, we are moving toward narrowing the gap between dependency and self-sufficiency.

Multiple land use is important to livestock production in New England. In the one instance, where our forestlands account for twice the proportion of total acres compared with the U.S. average, we have a large potential to utilize forages within these forestlands for grazing. Secondly, with the decrease in dairy numbers, there is a tendency to replace them with beef cattle, or sheep, or even hog operations--thus utilizing existing land, facilities, and labor for a combined or replacement operation. Additionally, the availability of inexpensive by-product feeds and the use of unconventional feedstuffs encourages these livestock operations. Because the ruminant especially can be maintained on lower quality or by-product feeds, New England as a natural, cool-weather grass country can produce forage-fed beef and lamb with a minimum of purchased feed through wise grazing and forage harvesting management.

Furthermore, the markets are prevalent in New England, with its population density (albeit concentrated in large cities along the sea-coast), thus allowing rural production within easy access to the consumer. This proximity also permits direct marketing from producer to packer or to consumer. This proximity also permits direct marketing from producer to packer or to consumer, development of a freezer trade, and utilization of farmers' markets as well as the existing auction markets. New England does lack a major terminal market within its confines. Livestock marketing pools are becoming more prevalent. Specialty marketing, catering to the natural or organic food interests, also is practiced to some degree. Some cater to specific ethnic demands, often of a seasonal nature, such as Easter lambs.

THE TIME IS NOW

In general, New England does have much in its favor for the production and marketing of livestock for meat purposes. New England farmers historically have been excellent livestock men. The first U.S. meat packer was Captain John Pynchon (established in 1645) in Springfield, Massachusetts. The Brighton (Massachusetts) Stock Yards were developed to feed George Washington's Continental Army in 1775--the nation's first and oldest terminal market. "Uncle Sam" Wilson followed suit supplying the U.S. troops with meat during the War of 1812. The first agricultural and livestock show or fair in the U.S. was held in Pittsfield, Massachusetts, in 1810. Many of the early imports of livestock from Europe funneled into the New England states, and in time this breeding stock of all species was disseminated west. The first major U.S. importation of Merino sheep was to Weathersfield Bow, Vermont, in the early 1800s by Hon. William Jarvis, then U.S. Consul to Lisbon, Portugal. By 1865, there were 1.5 million sheep in Vermont alone. These sheep were the foundation of the great Merino flocks of Ohio and now Texas. In 1875, Herefords from the Bodwell and Burleigh herd of Vassalboro, Maine, sold to the Hon. William F. Cody of Scout's Rest Ranch, North Platte, Nebraska, and to other prominent breeders of that day.

Although there are some large-acreage livestock farms in New England, most are small, family farms. The average farm size in New England is 171 acres compared to a U.S. average of 450 acres. Some of these are registered, purebred breeders supplying breeding stock to the area and throughout the U.S. A limited few even sell breeding stock or semen internationally. New England has long been a seedstock producing area, and its livestock compete very successfully in the show and sales rings of the nation's major expositions and sales.

The New England farmer has often been faced with the quandary of how to make a living under ofttimes less-than-desirable conditions. Frequently he has survived by living on "not what he earned, but what he did not spend." Through the vagaries of climate, weather, topography, land capabilities, and pressures, the New England livestock producer has developed a unique capability in growing, managing, harvesting, and preserving forage in the form of grass or legume hay, haylage or silage, and corn silage where possible. The producer has realized maximum TDN per acre through wide pasture management and has obtained maximum livestock production on his precious land. New England has been forced to take the lead in forage production of meat and dairy animals as its ability to raise grain and protein supplement has been very limited to practically nonexistent. This has perhaps been the salvation of the New England livestock producer and especially the family farm where homegrown labor and homegrown feed have been the major resources for survival.

Some New England farmers combine livestock operations with other agricultural pursuits such as:

Major Enterprise	Supplementary Enterprise
1. Cucumbers (pickles)	Hogs
2. Dairy	Sheep, feed cattle
3. Forestry, firewood	Beef cattle, sheep
4. Hogs	Sheep
5. Landscape and bedding plants	Beef cattle
6. Maple products (syrup, sugar)	Sheep, beef cattle
7. Orchard	Sheep, beef cattle
8. Poultry	Beef cattle, sheep, hogs, veal
9. Tobacco	Beef cattle
10. Vegetable gardening	Hogs, sheep, cattle
11. Vineyard	Beef cattle

Generally, these are family farm operations that have diversified to utilize surplus feed, labor, facilities, or alternatives that best fit the existing situation and provide additional homegrown products for family use as well.

NEW ENGLAND, WHAT'S AHEAD?

With the advent or resurgency to greater self-sufficiency, we have already noted a greater number of "backyard" meat animals being produced for the home meat supply. More family farms and part-time farmers are turning to this program--not only for their own meat requirements but also for producing "a few extra to sell." Dairy farms are replacing some of their cull cows with dairy steers or other livestock to utilize homegrown roughage that can be marketed through these animals. Some operators are expanding their programs to satisfy the continuing and expanding demands for fresh and processed meat of all species, as well as from the fast-food chains. Most of these increasing needs require leaner meat that favors a forage-fed program. The challenge is to develop animal-forage management systems that will maximize the utilization of forages through grazing. Additionally, with the recognized growth efficiencies and greater muscle production of intact males, and with the great availability of dairy bull calves in this region, New England has the opportunity to utilize these surplus (to the dairy herd) bulls and feed them out for a specialized market. These bull calves provide an alternative veal production system that presently serves as a viable program for the small family farm that may feed out 100 to 500 vealers in confinement systems. Over one million dairy bull calves are produced annually in the Northeast with approximately 161,000 of these being New England-reared.

The future belongs to those who prepare for it. New England livestock producers, though small and often diversified, are facing the future with courage, adaptability, innovation, and confidence that they will continue to do a respectable job in maintaining their families and farms and contributing to the nation's meat supply.

(Statistical data presented herein has been derived from "Beef Research Program for the Northeast," [in progress, 1982], of which the author is a member of the Steering Committee.)

11
REGULATION OF AGRICULTURAL CHEMICALS, GROWTH PROMOTANTS, AND FEED ADDITIVES

O. D. Butler

Agricultural chemicals, from fertilizers to pheromones, help make U.S. agriculture the most productive in the world. Discovery, testing for efficacy and safety, manufacturing, marketing, and proper use all represent the ultimate in biological sciences, in ingenuity, and in exercise of the free enterprise system.

Some say that in this case the enterprise system is not very "free." Thalidomide, DDT, aldrin, dieldrin, arsenic, and many others did not pass safety tests. The thalidomide tragedy may have aroused the most fear in public minds, but the diethylstilbestrol use in the 1950s for sustaining pregnancy in women, which apparently resulted in increased incidence of cancer in their daughters 20 or more years later, would have to be rated a close second in the world, and first in the U.S.

Public demand expressed through members of Congress the last couple of decades caused ever-more-strict federal regulations on development and use of agricultural chemicals. During the past year, however, the Food and Drug Administration, the USDA, and the Environmental Protection Agency (the major responsible agencies) have shown good evidence of more reasonable postures concerning laws, regulations, and interactions with manufacturers and users of agricultural chemicals.

President Reagan's appointment of a cabinet-level committee chaired by Vice-President George Bush with a mission for reducing burdensome regulations, gave an unmistakable signal to the agencies. Now we see Congress considering revision of the Federal Insecticide, Fungicide, Rodenticide Act (FIFRA), and the Food Safety Laws, especially the extremely strict 20-year-old Delaney anti-cancer clause. This clause was made obsolete by almost unbelievable advances in assay procedures that now detect parts per trillion of materials in foods that were considered to have zero residue with the parts-per-million capability of assays in the 1960s. Assays are now as much as a million times more sensitive.

Strict laws that were formerly written to ban toxic substances on the basis of risk alone are being reconsidered. A couple of reasons derive from the issue of essential elements--such as selenium required by the body at a low level, but toxic at higher levels, and nitrite used for centuries in meat curing to give the characteristic color. Derivatives--for example, nitrosamines that may be developed during cooking of bacon--have been shown to cause an increased incidence of cancer in susceptible laboratory animals. More recently, the finger of suspicion has been pointed at nitrite itself, in a highly disputed experiment with laboratory animals. Nitrite produces color, but more importantly, it protects against the deadly botulism bacteria, so use of nitrites has not been banned, but has been strictly limited. Critics of the regulations point out that many natural foods contain nitrites and that human saliva does also. Avoiding cured meats would reduce nitrite consumption by a very small and negligible amount, critics say. But the "scare" stories certainly reduce demand for ham, bacon, and hot dogs.

What are producers' primary concerns about agricultural chemicals? I believe that you should have a general idea of how they are discovered, tested for efficacy and safety, and used in a safe and effective way. You should also know the direct cost of materials, as well as the indirect cost, if consumer concerns affect demand for products marketed.

Good basic biological research done primarily by public institutions, such as the Land Grant Universities, usually provides the foundation for development of an effective product. The need for products to control pests or diseases usually is expressed by producers reinforced by producer organizations, by extension specialists and research workers who interact with producers, and by supplier representatives.

Because of the similarity of all living cells, there must be a good understanding of the biology of both species affected to be able to kill a parasitic living organism without consequent toxic effect on the host. Then, for food producing plants and animals, there must be great concern about residues that might have an effect on consumers.

Animal producers are served well by a group of competing companies seeking profit by manufacturing and marketing drugs, biologicals, pesticides, and related materials. Most of the companies belong to an industry trade association, the Animal Health Institute (AHI), headquartered in the Washington area. It serves the industry the same as the many other trade associations there, trying in every way to protect the opportunity for the industry to produce products that customers will buy and use because of benefits and thereby earn a profit for investors.

Almost inevitably it seems, any position taken or change advocated by the AHI is opposed by one or more organizations that classify themselves as consumer protectionists. Lawmakers and regulators usually have to make decisions between opposing viewpoints without the benefit of absolutely conclusive evidence. In the last decade such controversy has been a major stimulant to the formation of the American Council on Science and Health (ACSH) and Council for Agricultural Science and Technology (CAST), both of which I support.

"The American Council on Science and Health (ACSH) is a national consumer education association directed and advised by a panel of scientists from a variety of disciplines. ACSH is committed to providing consumers with scientifically balanced evaluations of food, chemicals, the environment, and human health." This is quoted from their March 1982 publication, "The U.S. Food Safety Laws: Time for a Change?"

The Council for Agricultural Science and Technology (CAST) is an organization sponsored and managed by twenty-five scientific agricultural societies. Its major purpose is to assemble and report the scientific information on important issues of national scope for the benefit of law-makers, regulators, and the general public. It is not an advocacy organization. Most of its task force reports, now numbering about a hundred, were prepared at the request of members of Congress, some by government agencies, and some because the 47 officers and directors, all representing the scientific societies, decided that there was a need to assemble and print the scientific evidence on an important issue. CAST celebrated its tenth birthday anniversary in July 1982 at a directors' meeting at its headquarters. I have the privilege of serving as president of CAST in 1981, as did Frank Baker, the Director of this International Stockmen's School, in 1979. (I want to especially recommend CAST task force reports mentioned in the references.)

Some of the scientific societies work directly with regulatory agencies. I served as chairman of the Regulatory Agencies Committee of the American Society of Animal Science for about 10 years until 1981. The Institute of Food Technologists, like the American Society of Animal Science, has been very active in identifying and nominating qualified scientists to serve on CAST task forces and has also produced independent papers on various aspects of food safety.

Drug manufacturers have been very critical of the Food and Drug Administration (FDA) for taking so long to consider new animal drug applications (NADAs) before approval. A recent report entitled "The Livestock Animal Drug Lag" by the AHI describes the problem and suggests solutions. U.S. manufacturers have been able to obtain approval to market their products in the United Kingdom and European countries in a fraction of the time required for U.S. approval. An example is albendazole, a broad spectrum anthelmintic effective against gastrointestinal roundworms, lungworms, tapeworms, and liver flukes in cattle. Approval was obtained in 5 months in England in 1978. The same application filed in the U.S. in 1977 is still pending, though strong producer pressure resulted in limited approval in 1979 under a special investigative New Animal Drug authorization in a limited number of states. After the Food and Drug Admini-

stration banned hexachloroethane for liver fluke control,
cattlemen had no approved drug. Texas and Florida producers, with pastures along streams and low-lying areas that
have snails (the intermediate fluke host), just had to have
an effective drug. Cattle producer organizations rallied to
the cause and helped obtain the limited approval.

The AHI sponsored a Forum on Regulatory Relief in
Alexandria, Virginia, in June 1982. Dr. Arthur Hull Hayes,
Commissioner of the FDA, announced there that "I've decided
that all activities in the Review of Animal Drug Applications, including issues of Human Food Safety, will be con-
solidated within the Bureau of Veterinary Medicine." That
is certain to allow faster decisions. The Bureau of Food
review has been blamed for much of the delay in the recent
past.

Dr. Hays gave a definition of safe as "a reasonable
certainty of no significant risks based on adequate scientific data, under the intended conditions of use of a substance." More and more we are realizing that there is no
such thing as absolute safety, or zero risk. His speech
gave some reassurance concerning "sensitivity of method"
regulations that have been under consideration for several
years by FDA. The bureau now seems willing to accept
foreign data in support of New Animal Drug Applications
under certain restrictions and also to consider cross-
species approvals. It is not a good investment for drug
companies to spend several million dollars to obtain
approval of an anthelmintic for goats, for instance, that is
very important in Texas (which has about 95% of U.S. goats)
because the market is so limited. Other minor species, even
sheep, fall in that same category. I believe it will be
necessary for publicly supported institutions like the Texas
Agricultural Experiment Station to assist in developing
drugs and obtaining approval for use in such minor species.

The FDA is also considering some liberalization of
restrictions on feed manufacturers. Dr. Lester Crawford,
recently reappointed to the position of Director of the
Bureau of Veterinary medicine (BVM) of FDA spoke to the
American Feed manufacturers' 74th Annual Convention at
Dallas in May 1982. He reported that "The Subcabinet
Working Group, chaired by USDA Assistant Secretary Bill
McMillan, has proposed the total elimination of FD 1800s,
the notorious application required to authorize manufacturing and sale of medicated feeds. Instead, the BVM
would have authority to deny registration of feed manufacturers that lacked adequate facilities and controls to
assure safety."

Even the Environmental Protection Agency (EPA) is
trying to "simplify the regulatory burden on industry and
reduce unnecessary costs." So said John A. Todhunter,
Assistant Administrator for Pesticides and Toxic Substances,
at the 1982 Beltwide Cotton Production Mechanization Conference, January 1982, at Las Vegas. He described a
reassuring response to Vice-President Bush's task force,

especially that FDA has instituted a plan to improve the quality of scientific assessment, including a peer review system for major scientific studies and reports. There is, therefore, hope for maintaining availability of the herbicide 2, 4, 5T and even reapproval of compound 1080 for predator control. The states also have regulatory authority and enforcement responsibility. We are all aware of Governor Brown's reluctance in California to institute effective control measures for the Mediterranean fruit fly because of the political pressure of environmentalists.

Food Chemical News, a weekly publication, keeps you up-to-date on what is happening in Washington.

For those of you mixing your own feed, and for feed distributors, I recommend the annual Feed Additive Compendium, a guide to use of drugs in medicated animal feeds with monthly, up-to-date supplements.

In conclusion, I want to make a plea to agricultural producers for closer adherence to label requirements and restrictions on use of agricultural chemicals. The Agricultural Extension Service in every state has a responsibility for assisting producers in the proper use of chemicals. More attention is being devoted to that. Very few people deliberately break the laws, but many are not aware of the precautions necessary to prevent cross contamination of products and elimination of residues in feeds and foodstuffs. The USDA state producers' effort to eliminate sulfa drug residues in pork is an example of the kind of cooperation required to maintain availability of chemicals so important to modern food production. Let us resolve to intensify the effort for safe use of agricultural chemicals in order to gain greater public confidence in the safety of our abundant food supply.

88

REFERENCES

American Council on Science and Health (ACSH). 1982. U.S. food safety laws: Time for a change? 1995 Broadway, New York, N.Y.

Council for Agricultural Science and Technology (CAST). 250 Memorial Union, Ames, Iowa 50011.

CAST. 1977. Hormonally active substances in foods: a safety evaluation. Report No. 66.

CAST. 1981. Antibiotics in animal feeds. Report No. 88.

CAST. 1981. Regulation of potential carcinogens in the food supply: the Delaney clause. Report No. 89.

CAST. 1982. CAST-related excerpts from U.S. House of Representatives hearing on the Federal Insecticide, Fungicide, and Rodenticide Act (FIFRA). Special Pub. No. 9.

CAST. 1982. CAST-related testimony on the food safety amendments of 1981. Special Pub. No. 11.

Feed Additive Compendium. Miller Publishing Co., 2501 Wayzata Boulevard, P.O. Box 67, Minneapolis, Minnesota 55440.

Food Chemical News. 1101 Pennsylvania Ave., S.E., Washington, D.C. 20003.

Part 3

GENETICS AND SELECTION

12
THE VALUE OF DHIA TO PRODUCERS AND TO THE INDUSTRY

Richard S. Sechrist

Our society has moved into the "Information Age." Any business, including dairy farming, needs all pertinent information to make wise management decisions. As farms have gotten bigger, records of all kinds have become essential to profit and success. Once, with simple operations, a dairy farmer could keep details in his head and "on the back of an envelope," like the Gettysburg Address, but no longer. A "cow-accounting" system is required. Some farmers "make it" without DHIA, but they could "make it" easier by using the first-class management system that Dairy Herd Improvement Association (DHIA) provides.

SCOPE OF DHIA

The rapid growth of the DHIA System in recent years, the diversity of computer management information systems currently available, and the expansion of DHIA-related services, such as mastitis screening, protein testing, pregnancy diagnosis, forage-soil testing, and animal identification attest to its importance to dairy farmers.

The DHIA System in the U.S. employs 3,000 persons who service 65,000 dairy farms with 4,750,000 cows--about 43% of all the dairy cows in the country. The System has been gaining 2,000 herds and 200,000 cows per year in recent years. Herd size has been increasing less than 2% annually but cows on test are going up by over 5% annually (table 1).

DHIA computer-management information systems available to dairy farmers through nine regional DHIA computer centers have improved remarkably. Computer analysis came to DHIA 30 years ago. As computer technology has improved, so has the scope and practicality of information available to make on-farm management decisions. Biggest changes have occurred in optional reports and "action" sheets. Table 2 is an example. Reports are available to DHIA dairymen in routine hard copy delivered by mail or by remote on-farm computer terminal on a daily basis. About 300 farms use terminals to supplement mailed reports. DHIA monthly management reports received by mail generally arrive about a week after "test-

promptly. As computerization on the farm takes place, management decisions will be implemented even more swiftly.

TABLE 1. DHIA SYSTEM GAIN PER YEAR, 1976-82

Year	Herds (thousands)	Cows (millions)	Herd size	% On test	% gain
1976	53	3.5	66	31	--
1977	56	3.7	66	33	5.9
1978	58	3.9	67	35	4.7
1979	59	4.0	68	37	3.4
1980	61	4.2	69	38	5.0
1981	63	4.5	71	40	6.9
1982	65	4.7	73	43	5.8

TABLE 2.

B R E E D	COW SIRE OR DAM NAME OR NO			CONTROL OR CHAIN NUMBER	CURRENT STATUS		C O D E
		STRING			MO.	DAY	
		P	T				
COWS TO BREED							
H*	40H2115			4106	8	21	1
H*	VENTURE			4020	9	07	1
H*	PETE			4380	9	15	1
H*	7H58			4320	10	08	1
H*	11H105			4661	11	03	2
H*	1H130			4677	11	12	2
COWS TO CALVE							
H*	40H2025			3946*	7	17	6
H*	29H1771			2842*	7	27	6
H*	ROCKET			3629	11	18	6

Expansion of DHIA-related services has increased the overall usefulness of DHIA to participating dairy farmers. The number of DHIA programs has proliferated to 12 to meet growing demand for DHIA System record-keeping under many farm conditions and sizes. Although Standard DHIA continues to be the predominant program chosen by farmers, AM-PM is close to 1 million cows (about 20% of all cows on test) and

is growing faster than any other. The DHIA System also includes 1,400 herds of dairy goats with 13,400 does.
Four million cows are tested for butterfat electronically, virtually eliminating the sulphuric acid-based Bab-
cock Test in vogue since DHIA's beginnings in the early part
of the century. Infra-red butterfat-protein analyzers now
in common use in DHIA labs are fast and accurate. With
efficient instruments, protein testing has zoomed to 1.4
million cows per month (table 3).

TABLE 3. DHIA PROTEIN TESTS

	1982	1980	1976
Number	1,400,000	700,000	10,000
% of DHIA	29	17	--

A few years ago the only practical system for detecting
mastitis was CMT (California Mastitis Test), a satisfactory
method, but electronic instrumentation in DHIA labs has made
routine monitoring of cows more practical, with 2 million
cows now checked monthly (about 42% of all cows on test).
Assimilation of 60 somatic-cell counters into DHIA operations accounts for most mastitis screening, with herd and
individual cow somatic-cell-count (SCC) reports reaching
DHIA dairymen every month (table 4).

TABLE 4. DHIA MASTITIS SCREENING (MILLIONS OF COWS)

	1982	1980	1976
Total	2.0	1.4	0.4
SCC	1.8	1.0	0.04
CMT	0.2	0.4	0.3
Other	---	---	0.02
% of DHIA	42.0%	34.0%	11.0%

Another DHIA optional service, National DHIA's Verified
Identification Program (VIP), fills a need not provided
otherwise for DHIA dairymen as the first all-breed grade
identification system. As an opportunity to enhance cow
management and to improve cattle-marketing potential, VIP
has been accepted by dairymen to the extent the program now
issues more identification certificates annually (19,000)
than three of the dairy breed registration societies (table
5).

TABLE 5. VIP HERDS AND CERTIFICATES

	1982	1980	1976
Certificates	19,000	13,000	1,300
Herds	1,200	800	70

Some DHIAs, in an attempt to provide needed services for DHIA dairymen, have added other optional programs (table 6).

TABLE 6. DHIA OPTIONAL PROGRAMS

Feed testing	40,000 samples/year
Soil testing	23,000 farms/year
60,000 soil samples/year	
180,000 soil recommendations/year	
Farm enterprise analysis	1,000 farms/year
Progesterone (pregnancy) testing	5,000 analyses/year

DHIA has become what dairymen expect it to be: an information management system including special services not available elsewhere in agriculture--a system geared to ensure that dairy farming is a profitable venture.

WHAT DHIA FOR THE DAIRY INDUSTRY

The DHIA System costs dairymen about $70 million, but provides annual returns on investment to participating dairymen exceeding $1.5 billion! These returns are the result of direct benefits from improved management at the farm level, with consequent increase in production per cow (table 7).

TABLE 7. DHIA AVERAGES BY YEAR

Year	Cows (millions)	Milk (pounds)	Fat (pounds)
1941	0.6	8,133	331
1951	1.2	9,195	370
1961	1.9	10,796	418
1971	2.2	13,000	489
1981	3.2	14,960	553

There are many additional indirect benefits to individual producers, both DHIA and non-DHIA dairymen. DHIA is the backbone for the best educational system in agriculture --the dairy extension program. It is the backbone for a genetic evaluation enterprise that is the envy of the rest of the world. It is the foundation upon which the artificial insemination (AI) industry has been built; without it AI would be nothing more than a "cow-freshening" operation. And breed association survival could be threatened without improvement programs like DHIA.

The foundation for artificial insemination is based on a record-keeping system that is totally dependent on the DHIA System. Without performance records upon which to determine genetic selections, artificial insemination of 7.5 million dairy cows annually would be an "exercise in futility," eliminating the need for a bull on the farm to fertilize cows, but with small hope for true genetic improvement in succeeding generations.

When AI seriously began to take shape in the 1940s, selection of bulls depended on hand-calculated lactation information supplied by DHIA supervisors from only one-fifth as many cows as today. This spotty information forced a heavy emphasis on pedigree rather than performance. The daughter-dam comparison was a step forward, but comparisons often were misleading due to environmental opportunities afforded daughters of some bulls--and numbers are small!

As DHIA computerized in the 1950s, more and better information became available from the DHIA System, so a truly scientific analysis of genetic potential become possible; gradually DHIA dairy records processing centers, the U. S. Department of Agriculture (summarizing genetic information), and the AI industry developed new and progressively more sophisticated schemes to locate and develop the animals necessary for true genetic improvement generation after generation. Genetic progress generated in recent years is the pride of the dairy industry but AI success and genetic improvement ultimately are a result of progress and success in the DHIA System.

Breed associations accept DHIA records for registered cattle as Dairy Herd Improvement Registry programs (DHIR). Breeds add special requirements for animals participating in DHIR over and above standard DHIA requirements, but breed associations are users of DHIA records. DHIR cows are a relatively minor segment of the DHIA System: 350,000 cows (9% of the total).

However, DHIR records as part of the DHIA System are essential to sound breed-improvement programs and promotion -marketing of dairy cattle locally, nationally, and internationally. Performance testing provides reasonably scientific evaluation of individual animals not humanly possible when basing individual merit on "eye-ball" decisions. Over and over, experience says outstanding-type cattle attract more attention, but seldom do "fancy" animals get very serious attention unless there is documented evidence of producing ability. "Fancy" cattle command high prices if DHIA records demonstrate they can milk, too! With DHIA, "typey" cows are worth thousands of dollars more. From a genetic improvement standpoint, no breed association can

justify its existence without tying to an impartial performance recording system. DHIA provides the credibility
necessary for breed promotion and marketing efforts.

Extension education programs in dairy, sponsored by the
land-grant universities of America, have contributed heavily
to the on-going process of improving the technology for production of milk. DHIA and extension have had close
relationships over a long period of time, each depending on
the other for the ultimate benefit of producers. In the
view of extension, DHIA demonstrates clearly the effects on
farm profit of good (and bad) management practices. Part of
the educational process is exactly that! But, DHIA also
provides the "monitor" to evaluate new techniques so that
the educational process can move ahead. In addition, extension investigation into individual dairy-production problems
is futile without a "cow-accounting" system to evaluate.
Clearly, the reciprocal efforts of DHIA and extension ultimately result in profit for dairy farmers (both DHIA and
non-DHIA).

Ultimately dairy farmers can prosper in a fiercely competitive food business. After all, North America is competing with low-cost production in Australia and New Zealand
and huge subsidies in the European common market. American
dairy farmers must be the most efficient in the world or the
competition (whether fair or unfair) will beat us at our own
game. We have the kind of first-class milk recording system
needed for efficient, successful milk production.

DESCRIPTION OF PROGRAMS AVAILABLE TO DHIA DAIRYMEN

The DHIA testing program best suited to an individual
farm depends on the size of the herd, the labor available
on-farm at milking time, cost, objectives for herd performance testing, and traditions of the area or region. For
example, low-cost Owner-Sampler (OS) is particularly suitable to a young dairy family milking 30 to 50 grade cows,
particularly if the wife can be available to record milk
weights and sample milk on test day. Owner-Sampler is not
suitable for a highly automated milking operation, especially if the herd is registered and the owner has aspirations to market surplus cows and heifers. In other words,
type of testing program should be selected rationally to fit
the needs of the farm. However, the really important decision is to test!

Whatever program is selected, some guidelines to keep
in mind include:
- DHIA programs (all types) provide an estimate of
  the production of the herd and individual animals; accuracy of the estimate varies slightly
  according to type of test, but any program provides information sufficiently accurate to make
  management decisions.

- "Official" programs provide indirect benefits to all dairy farmers because records are included in national genetic evaluations and become part of research and extension-education efforts.
- There are slight regional differences in computer management information reports, but all programs (DHI, OS, etc.) provide the same computer reports for management decisions.
- Optional computer reports and lists are available and should be considered to complement management needs.
- Optional analyses, such as protein testing and somatic cell testing, are available almost everywhere and can be part of any DHIA program.
- Relative cost is less important in selection of the right DHIA program for a farm than is making certain the program is practical, convenient, and useful.

Official DHI (also called Standard DHIA). The DHIA supervisor weighs and samples milk two consecutive milkings per month and records feeding information, fresh dates, dry dates, breeding dates, etc. Milk samples are analyzed subsequently by a DHIA lab; barn information goes to DHIA computing center for analyses and print-out; management information report is mailed directly to the farm.

Official DHI AM-PM. Official DHI AM-PM is the same as Official DHIA except that the DHIA supervisor weighs and samples only one milking per month (alternating the "tested" milking from month-to-month) requires an electronic interval recorder to determine milking times so appropriate factors can be used to estimate the day's total milk weights.

Official DHIR. Conducted like any of the three programs described above, it requires acceptance by the appropriate breed association for the registered cattle in the herd; Holstein and Milking Shorthorn breeds accept only two milkings and two milk samples (DHI) for DHIR.

Nonofficial Owner-Sampler. Milk weights and milk samples are collected by the farmer to reduce cost; DHIA supervisor picks up samples and information to complete the same lab and computer steps as described in plans described above.

Nonofficial AM-PM. Generally the same as DHI AM-PM, except that there is no requirement for an electronic interval recorder. Milking times are recorded.

Nonofficial milk-only-records. Same as Official DHIA except that no milk samples are taken and no milk analyses are made; however, some DHIA computing centers use the breed average in making calculations.

The above are the major plans, but several variations also may be available. Cost of DHI (Standard DHIA) generally is $1 to $2 per cow per month; AM-PM plans usually cost about 20% less; Owner-Sampler costs 40% less. Major cost in testing is the labor and travel cost of the DHIA supervisor at the farm during one or two milkings; milk analyses and computer reports are relatively inexpensive.

In addition, an on-farm computer terminal can complement any DHIA program by providing up-dated lists of cows to calve, cows to watch for heat, cows to dry off, etc., on a daily basis. The on-farm computer also can supplement DHIA by providing inventory information, payroll, ration balancing, etc., not directly available in DHIA.

The following table shows the breakdown of various DHIA programs on January 1, 1981 (table 8):

TABLE 8. DHIA SYSTEM JANUARY 1, 1981

Program	Herds	Cows per herd	Cows (thousands)
Official			
DHI	32,207	78	2,516
DHIR	4,284	80	342
DHI-AM-PM	3,033	112	339
Subtotal	39,524	81	3,197
Nonofficial			
OS	16,596	45	745
AM-PM	5,587	56	313
MOR	748	142	106
COM	262	370	97
Other	258	86	22
Subtotal	23,451	55	1,283
TOTAL	62,975	71	4,480

PROFITS FROM DHIA COMPUTER REPORTS

The primary reason a dairyman should pay for DHIA service is to receive computerized management information reports and summaries that depict what is happening in the herd and what can be expected to happen in the future. Naturally, these reports are of little value unless a member of the farm management team studies them and acts accordingly.

Basically, DHIA reports provide production information:
- Test-day milk weight and milk analyses (% butterfat and protein) including a persistency comparison to the previous month, daily income estimate, and grain information (table 9).

TABLE 9

DAYS IN MILK	MILK DAILY	% BUTTER FAT	FAT DAILY	PERSIS-TENCY	C M T	+ − $	% SNF OR P.	GRAIN FED	GRAIN REQ'D
				TEST DAY DATA					
28	87.0	3.53	.05	128	N	8.61	8.5		31
46	76.5	2.51	.91		1	6.53	8.5		14
28	44.6	3.41	.52	92	T	4.34	8.4		8
		3.71					8.66		
FRESHENED BUT NO MILK REPORTED									
28	86.5	3.22	.77	112	N	8.31	8.5		27
28	82.3	3.32	.72	95	N	7.66	8.2		26
28	69.1	3.72	.56	82	N	7.08	8.7		23
28	37.5	4.71	.76	98	N	4.35	9.2		11
25	48.3	3.01	.45	89		4.50	8.5		
28	78.8	3.93	.07	111	N	8.05	8.4		31
28	73.5	3.42	.50	98	N	7.14	8.4		23
		3.16					8.62		
28	18.3	2.9	.53	65	1	1.60	7.8		2
28	102.5	2.22	.26	98	1	8.51	8.3		19
28	92.3	3.12	.86	101	2	8.61	8.5		28
28	95.5	3.02	.87	102	N	8.73	8.4		28
28	91.0	2.92	.64	90	N	8.40	8.4		25

- Lactation to date including age at calving, total days in milk, and income estimate (table 10).
- An estimate of the expected lactation for each cow standardized for age and length of lactation so that a comparison can be made from cow to cow (Relative value %: 100 = average for herd) in table 11.

The same information also is printed on an "Individual Cow Record." As an example, table 12 illustrates some of the features.

As shown, each cow is provided a complete history to date, which can be up-dated every month.

The DHIA Monthly Herd Summary organizes herd information in such a way that herd patterns can be determined that can lead to needed management changes, such as shown in table 13.

- Lactation number: Are the first-calf heifers genetically equal, worse, or better than the older animals?
- Cows calving: Is the "spread" of calvings throughout the past year and the projected calvings by month during the next year what you need to maintain cash flow (or milk base, etc.)?
- Breeding records: Are cows being bred soon enough after calving? Are there a significant number of problem cows? Is the average calving interval satisfactory?

- Herd production: Is production improving or deteriorating?
- CMT (or Somatic Cell Count): Does the herd mastitis-monitor information suggest mastitis control is adequate? Is there an indication that more stringent control measures are needed?

Management Lists are an important feature of DHIA Records. They are provided in a variety of ways, such as special lists and pocket-size "Action Sheets." Lists also adapt well to on-farm computer use on a daily basis. Illustrations of these features are in table 14.

Each dairy farmer uses DHIA records in ways specifically geared to his own operation. It is not likely that he will use all of the information provided, but the more carefully he analyzes the implications of the reports and summaries the more likely DHIA records will return his investment many times over. The first big step is to start a DHIA program; the second is to spend adequate time with DHIA records to ensure that rational management decisions are made. By taking these two steps, success in dairy farming isn't guaranteed, but history says the chances are a lot better.

TABLE 10

CURRENT LACTATION						
AGE AT CALVING ·RS | VC	-ACT NC	DAYS IN MILK	MILK	FAT	SNF OR P.	+ $ −
303	2	85	5860	255	499	628
⚹ 803	5	523	36290	1437		
1005	7	46	3240	86	281	300
⚹ 900	6	391	25900	9562240		2608
		365	24740	9172143		2499
⚹ 800	5	486	30270	11412623		3099
809	7	158	15970	5071373		1538
704	5	370	19740	8761726		2144
800	6	56	4200	157	345	420
703	5	201	18470	7091613		1912
⚹ 607	4	294	17510	7811592		1969
⚹ 602	4	255	19320	7151618		1917
602	4	231	22250	8881842		2260
506	3	323	29870	9472573		2862
		305	28550	9022462		2733
506	4	252	12590	4311049		1212
510	4	69	6360	164	535	568
505	4	74	5880	219	506	602
503	4	79	7000	223	605	679
408	3	252	24330	8022064		2337

TABLE 11

EXTRAPOLATED 305 DAY M.E.		% REL VALUE
MILK	FAT	
18120	764	99
24520	989	127
17120	436	71
23000	863	117
23020	859	121
24570	801	116
17160	774	100
17640	633	88
23050	910	121
17860	781	100
19710	715	98
25250	1027	135
29410	911	135
14110	483	68
21220	538	88
18530	675	93
20970	648	96
28050	925	133

TABLE 12

JULY 80 TEST HERD
INDIVIDUAL COW RECORD
OFFICIAL DHIR

TABLE 13

DHIA MONTHLY HERD SUMMARY

TABLE 14

SPECIAL HERD MANAGEMENT REPORT

SIRE #1	SIRE #2	TRAITS ABCDEFGHIJ		COWS TO BREED		
11H2006	11H1012	0+000-+0	H	11H1215	34908	4172
23H206	11H464	0+000000	H	11H151	54454	2081
40H2294	3H602	+C0C0000	H	CARL	64725	5041
11H1056	24H702	10+00000	H	11H161	54780	6211
7H821	24H702	-+000000	H	24H224	64354	6241
			H	15H123	64790	6301
			H	11H2028	54968	7162
40H2518	7H821	1C00-000	H	11H84	14742	2141
3H602	7H155	+000---0	H	7H196	64891	5142
24H702	Y S	-000+000	H	LENNIE	5 501	6131
11H1056	24H702	0++++000	H	E REX	64393	6191
11H1056	294	00000-00	H	11H1073	44307	6291
			H	11H2028	55034	7092
			H	23H178	55025	7102
15H123	11H1124	++0+0000	H	40H2217	43980	7121
11H1056	Y S	-++000-0	H	PAYOFF	64523	7191
11H166	11H142	++--0000	H	1H92	43428	7221
			H	29H2502	54954	7262
			H	11H2028	54973	7302
		# END #				

		COWS TO DRY		
8748962H	11H145	24214	10191
5724070H	9H107	24865	11222
5724067H	11H151	14845	12022
5498723H	11H161	54829	11122
5472941H	23H198	14649	8081
8619598H	KING	44149	1191
# END #			

		POTENTIAL CULLS		
7116363H	SGB	3003	7296
8748934H	11H145	14259	12291
5472940H	11H84	14742	2141
# END #			

		CHECK FOR PREGNANCY		
9472976H	11H151	34673	11101
9035753H	15H120	34487	2071
7778443H	11H151	13498	8121
9812004H	9H143	34948	3242
5812007H	11H2028	34960	3182
8444684H	40H2181	14048	11171
5222007H	11H151	14577	5042
9472918H	15H140	54697	3151
# END #			

13
FRIESIAN CATTLE STRAINS COMPARISON: DAIRY PERFORMANCE

Henryk A. Jasiorowski

Friesian cattle are the leading dairy strain in the world. The strain originated in Friesia, a very small province of northern Europe, which now is divided into 2 parts belonging to Netherlands and the Federal Republic of Germany. Because of proven high quality, Friesian cattle are popular worldwide; they have acclimatized very easily in new countries and have been bred both to obtain pure bred cattle and to create new breeds.

Because of different methods of selection and different environmental conditions, considerable differences are found among Friesian cattle from the different countries, although the strain originally stemmed from the same population.

Since ancient times, breeders have made use of their neighbors' animals to improve their herd. In recent times, however, this process has accellerated.

The constant trend towards increasing cattle productivity, the development of artificial insemination,' and especially the possibility of longer storage and easy transportation of semen, has rapidly expanded the world trade of the frozen semen of Friesian sires. As a matter of fact, it is now almost impossible to find a country that does not import semen and that does not cross cattle with Friesians to improve the genetic values of its own breeds or strains, especially for milk production.

Cattle crossbreeding with Black-and-White cattle now occurs in Africa, Asia, and Latin America, and some of the traits of the most famous Black-and-White breeds in the world continue to be improved by crossing with other Friesian strains.

The good milk performance of American Holstein-Friesian cattle has boosted their popularity among all stock farmers in the developed world and has intensified the importation of strain semen for crossing with other Black-and-White strains.

Such a situation calls for a thorough evaluation of many traits of different strains of Friesian cattle. The production records available proved to be of limited value for comparisons due to differences in environmental and feeding conditons. Thus, an important problem of genetic value evaluation and comparison of different strains has emerged with regard to important productive features. Since the early 1970s, many scientific works have been published dealing with the results of these crossing programs and genetic value comparisons of different strains of Friesian cattle. The largest number of these works were reported in Europe. The studies comparing Holstein-Friesian cattle with local Black-and-White strain or its crosses were published in Holland (Oldenbroek, 1980), in West Germany (Gravert, 1975), in Ireland (O'Ferral and Cunningham, 1978), in USSR (Farzaliev, Nramcev, Nedava, 1980), in Yugoslavia (Konezar, 1979), in Poland (Nahlik and Szelag, 1979). All of these works noted the superiority of Holstein-Friesian cattle as compared with other tested Black-and-White strains—in terms of milk yield, body gain, and weight. However, their performance was not as good as the European strains in terms of dry matter, fat content in milk, and musculature. These works usually dealt with two or three Friesian strains, and were done on the base of relatively small populations. Besides various environmental and feeding conditions made it impossible to compare the results from different countries. Therefore, specialists have long discussed the need to compare different Friesian strains on a large scale.

In 1973, the Food and Agriculture Organization UN took up this problem (at that time author served at the FAO as Director of the Animal Production and Health Division in Rome). It was obvious that the comparison should include all significant Friesian cattle strains. It was decided that the experiment should be done in Poland where the Friesian cattle dominates and where there are large state-owned farms. The Polish scientists charged with this task (H. Jasiorowski, Z. Reklewski, M. Stolzman) worked with a Technical Advisory Committee representing all the countries involved in the experiment to analyze the yearly results.

This paper is based on the results endorsed by the Technical Advisory Committee that have been published to date.

ORGANIZATION OF THE PROJECT

The experimental part of the project began in 1974 and was planned for ten years. Ten countries cooperated in the project providing semen from a random sample of young bulls that was used to produce firstcrosses and backcross generations in the northern part of Poland (figure 1).

Nine countries (Canada, Denmark, The German Federal Republic, Great Britain, Israel, The Netherlands, New Zealand, Sweden, and the U.S. cooperated in the test. Polish cattle also were involved as a control group.

The comparison of Friesian strains is made between the first crosses (F_1) on the Polish Friesian cows inseminated with semen provided by the participating countries. This is followed by backcross to the paternal strains and Polish bulls (figure 2).

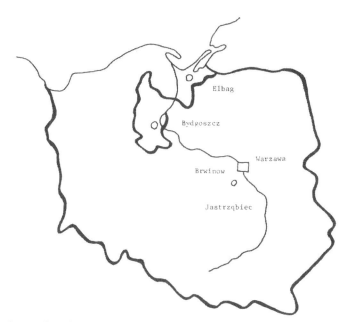

Elbag
Bydgoszcz
Warzawa
Brwinow
Jastrzqbiec

Figure 1. Poland: The provinces of the project

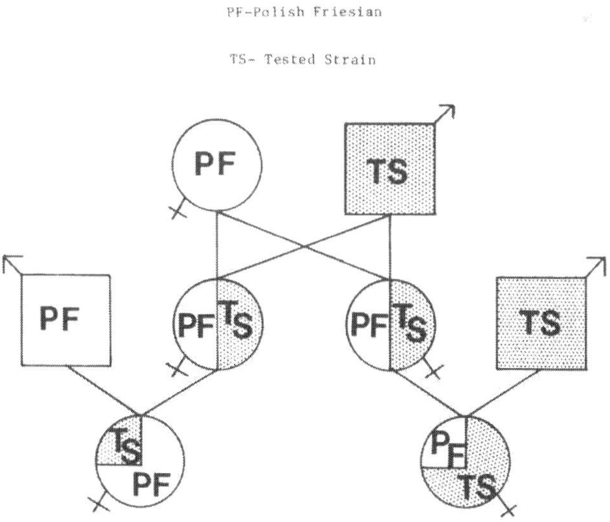

PF-Polish Friesian
TS- Tested Strain
PF
TS
PF
PF TS
PF TS
TS
TS PF
PF TS

Figure 2. Breeding plan in the Polish Friesian strain comparison

For this purpose, 4,000 doses of semen coming from young untested bulls were sent to Poland during the years 1974 and 1975 by each of the nine counries mentioned above. The same amount of semen was taken from young Polish bulls that served as a control group. Therefore, about 80,000 doses of semen from 400 bulls from 10 countries were used in the experiment. The semen was used with more than 30,000 cows on 70 state farms in northern Poland. At the farms, cows were inseminated at random, with semen from bulls of each strain used on each farm.

The progeny of both sexes was tested under the field conditions on the 70 state farms where they were born.

All progeny were raised without selection. Bulls were fattened to 450 kg live weight. Heifers stayed in the test until the end of the first lactation. The experimental procedure included extensive observations of progeny, starting with birth and ending with the lactation performance of the females and slaughter value of the males.

Apart from the field studies, it was decided to make more detailed observations connected with dairy production and fattening under conditions of intensive and controlled feeding. Thus, representative groups of cattle were chosen at random from the obtained population and put on two experimental stations with equal feeding conditions.

The total experimental design is illustrated in figure 3.

Figure 3. Experimental design

		female milk performance
	field conditions	
		male beef performance
Progeny		
	intensive and controlled feeding conditions	female milk performance
		male beef performance

MILK PERFORMANCE UNDER FIELD CONDITIONS

The data discussed here are based on 1,581 standard first-lactation records. The least-square means for milk and fat yield, as well as fat content, are presented in table 1 with the effects of age, year and season of calving, and herd controlled.

TABLE 1. LEAST SQUARE MEANS FOR MILK PERFORMANCE (FIELD TRIALS

Country	n	Milk production kg	SE	Fat yield kg	SE	Butterfat %	SE
USA	120	4,183	83	163	3.4	3.89	0.042
Poland	186	3,397	70	138	2.9	4.05	0.036
Canada	163	3,984	76	156	3.1	3.92	0.039
Denmark	154	3,730	75	150	3.1	4.01	0.038
Great Britain	127	3,717	80	148	3.3	3.98	0.041
Sweden	195	3,865	71	154	2.9	3.98	0.037
FRG	144	3,629	76	144	3.2	3.97	0.038
Netherlands	134	3,588	77	145	3.2	4.04	0.039
Israel	147	4,102	75	162	3.1	3.94	0.038
New Zealand	211	4,023	67	162	2.8	4.03	0.035
Average	1,581	3,820	40	152	1.7	3.98	0.021

Source: M. Stolzman et al. (1981).

The overall mean for 305 days lactation yield was 3,820 kg of milk, 152 kg of fat, and 3.98% fat test. In spite of this rather low production level, large differences in milk yield were observed. Milk production of the F_1 cows from the U.S. sires was almost 800 kg more than that from cows with Polish sires. The Israeli, Canadian, and New Zealand F_1 also were in the leading group. Among European strains, the progeny of Swedish and British sires produced average yields, whereas progeny with Dutch and German sires were below average. The Polish group had the lowest production.

The data concerning milk yield of the cows examined under the field conditions came from the farms with a feeding system that did not use intensive methods of feeding with the large amounts of concentrates. These data represent the F_1 generation, i.e., from 50% share of the genotype of the tested strain.

The fat content in the milk of the groups examined amounted on the average to 3.98%, which is a high ratio for Friesian cattle.

The greatest difference (.14%) in fat content was between daughters from the New Zealand bulls (highest) as compared to those sired by U.S. bulls. The Dutch, Danish, and Polish cows were above the average in fat content.

In fat yield, F_1 progeny from New Zealand, Israeli, and U.S. sires ranked highest, with progeny from Canadian and Swedish bulls ranking above average.

MILK PERFORMANCE UNDER INTENSIVE FEEDING CONDITIONS

The F_1 groups' milk production under intensive feeding conditions was compared with a control group of 400 randomly chosen heifers, i.e., 40 per each tested strain. It was assumed that 40 heifers from one strain should come from as many sires as possible.

For the test, the cows were fed with: hay, beets, pulp, corn silage, and concentrates. The feeding was intensive (0.5 kg of concentrates per 1 kg of milk) and was adjusted every week according to the production. Milk production was recorded every week and estimates made of the content of fat protein and dry matter. Additionally, records were kept on milking ability, feed utilization, weight of cows, and health.

Table 2 shows the least-square means characterizing the calving age, weight, and body measurements for 389 tested cows. The mean age of first calving for the whole population was about 28.5 months.

TABLE 2. LEAST SQUARE MEANS CHARACTERIZING THE TESTED COW POPULATION (INTENSIVE FEEDING TRIAL)

Strain	n	Age of calving in months		Life weight in 10 days after calving		Height in withers	
		LSM	SE	LSM	SE	LSM	SE
USA	39	28.57	0.38	486.64	6.20	129.48^X	0.52
Poland	38	28.37	0.39	468.46	6.39	125.45^X	0.53
Canada	39	28.86	0.38	476.90	6.27	129.80^X	0.52
Denmark	40	28.74	0.37	470.41	6.19	126.95	0.51
Great Britain	40	28.53	0.37	468.15	6.21	127.11	0.52
Sweden	39	28.36	0.38	480.41	6.31	127.81	0.52
FRG	38	28.92	0.38	473.58	6.35	127.75	0.53
Netherlands	39	29.28	0.38	460.68^X	6.29	125.79^X	0.53
Israel	39	27.74	0.38	489.55^X	6.31	129.26^X	0.53
New Zealand	38	28.79	0.38	469.90	6.36	127.56	0.53
Average	389	28.62	0.12	474.47	2.02	127.70	0.17

Source: M. Jasiorowski, Z. Reklewski and M. Stolzman (1981).

The average weight of cows was 474 kg on the 10th day after calving. The cows from Dutch bulls had the lowest weight, although they were the oldest; their weight was significantly lower than Israeli, American, and Swedish cows. Israeli strain, which had the highest weights, were significantly heavier than were the animals from Polish, Danish, English, and New Zealand bulls.

The mean weight of cows was 565 kg after 305 days of lactation. During lactation period, the animals' weight increased about 90 kg on average for all groups. In this period, the weights of cows from all different strains

become almost equal--the difference between two extreme
strains did not exceed 20 kg and was not significant.
The mean height at the withers of all tested cows just
after calving was 127.7 cm. The shortest cows were daughters of Polish bulls; they were 4 cm shorter than the tall-
est cows, which were from Canadian strain. Polish cows were
significantly shorter than all strains except the Dutch.
Dutch cows were significantly shorter than were the off-
springs of Canadian, Israeli, New Zealand, Swedish, and German bulls. Canadian and American strains were the highest
at the withers (more than 129 cm).

Table 3 provides the least-square means for the milk
yield. The overall average yield was 4,970 kg for 389 cows
and 305 days of lactation. The highest milk yield (5,402
kg) was achieved by Holstein-Friesians (U.S.). The groups
from Canadian and Israeli sires had milk production levels
only slightly lower than those of American group. Milk
yields of Polish cows were 1,000 kg lower than those of the
U.S. strain. U.S. cows produced significantly more milk
than did cows from Polish, Danish, Dutch, Swedish, German,
and New Zealand groups. Polish cows produced significantly
less milk than did all other genotypes.

Daughters of New Zealand sires had the best fat yield
(206 kg) and the Polish strain produced the least amount of
fat (177 kg). This difference was significant in comparison
to all strains except the Danish.

The mean protein yield was 174.9 kg for the total tes-
ted population. The best mean protein yields were from the
U.S. sires (182.7 kg) and Israelite sires (181.8 kg).
Polish cows produced 25 kg of protein less than did the
American group, and ranked last. This production was signi-
ficantly lower than that from all other groups, except the
Danish and German. However, Danish and German cows produced
significantly less protein than did daughters of U.S. and
Israeli sires.

Total solids production in the milk of the first lactation was the highest in groups from U.S. (659 kg) and Canadian sires (646 kg); yields of Polish cows were 95 kg lower
and significantly less than all other groups except the
Danish. However, daughters of Danish sires produced significantly less dry matter in milk than did Israeli, Canadian,
and U.S. groups.

Table 4 shows the milk composition for tested groups.
The cows from New Zealand and Dutch fathers produced milk
with high fat content (4.14%). Daughters of U.S. and
Israeli sires produced milk with less fat content, and the
difference between the U.S. and other groups was significant. Significant differences in fat content were found
between milk produced by Israeli cows that were produced by
New Zealand, Dutch, Polish, Danish, English, and Swedish
strains.

TABLE 3. LEAST SQUARE MEANS FOR MILK PRODUCTION (INTENSIVE FEEDING TRIAL)

Strain	Milk, kg		Butterfat, kg		Protein, kg		Dry matter, kg		SNF, kg	
	LSM	SE	LSM	SE	LSM	SE	LSM	SE	LSM	SE
USA	5,402x	144	197.2	5.67	182.7	4.78	649.57x	17.17	451.65	11.88
Poland	4,397x	145	177.1x	5.71	157.2x	4.81	555.04x	17.30	377.29x	11.96
Canada	5,225	145	204.8	5.71	178.5	4.81	645.90	17.29	441.83	11.95
Denmark	4,706x	141	187.9x	5.56	168.8x	4.68	592.25x	16.82	404.77x	11.63
Great Britain	5,065	143	200.6	5.66	177.3	4.76	635.01	17.14	482.94	11.85
Sweden	4,909	142	194.3	5.59	175.2	4.71	619.20	16.95	424.05	11.72
FRG	4,933	146	192.5	5.77	172.9	4.85	616.76	17.45	423.37	12.07
Netherlands	4,839	144	199.2	5.75	174.6	4.79	617.57	17.22	418.78	11.91
Israel	5,222	144	198.0	5.65	181.8	4.80	645.83	17.25	446.73	11.93
New Zealand	4,996	145	206.0	5.72	180.1	4.81	633.60	17.31	428.12	11.97
	4,970	-	195.8	2.16	174.9	1.82	621.08	6.56	429.95	4.54

Source: H. Jasiorowski, Z. Reklewski and M. Stolzman (1981).

TABLE 4. LEAST SQUARE MEANS FOR MILK COMPOSITION (INTENSIVE
FEEDING TRIAL)

Strain	Butterfat %		Protein %		Dry matter %		SNF %	
	LSM	SE	LSM	SE	LSM	SE	LSM	SE
USA	3.69[x]	0.05	3.40[x]	0.03	12.11[x]	8.13	8.42[x]	4.21
Poland	4.06	0.05	3.59[x]	0.03	12.65	8.20	8.58	4.24
Canada	3.94	0.05	3.47	0.03	12.42[x]	8.19	8.48[x]	4.24
Denmark	3.99	0.05	3.50	0.03	12.59	7.97	8.60	4.18
Great Britain	3.98	0.05	3.53[x]	0.03	12.54	8.12	8.55	4.20
Sweden	3.97	0.05	3.57	0.03	12.63	8.03	8.64	4.16
FRG	3.93	0.05	3.55	0.03	12.52	8.27	8.58	4.28
Netherlands	4.14[x]	0.05	3.62	0.03	12.78[x]	8.16	8.68[x]	4.23
Israel	3.83[x]	0.05	3.49[x]	0.03	12.39[x]	8.17	8.56	4.23
New Zealand	4.14[x]	0.05	3.59	0.03	12.71	8.20	8.57	4.25
Average	3.97	-	3.54	-	12.53	-	8.56	1.61

Source: H. Jasiorowski, Z. Reklewski and M. Stolzman (1981).

Similar differences were noted in protein content of
the milk; genotypes from U.S., Canada, and Israel produced
milk with the lowest protein content. The greatest amount
of protein was contained in milk produced by daughters of
sires from Holland (3.62%), New Zealand (3.59%), and Poland
(3.59%).

American cows produced milk with significantly less
protein content than that produced by other genotypes, except Canadian. However, Canadian and Israeli cows produced
milk with significantly less protein content than that produced by Dutch, New Zealand, Danish, and Polish groups.

The average dry matter content in milk was the lowest
in the U.S. breed (12.11%). Dutch-sired daughters produced
milk with 12.78% dry matter, and similar production was
achieved by New Zealand and Polish breed.

The mean milk flow during milking was 1.81 kg/min for
the examined population and maximum milk flow was 2.66 kg/
min (table 5). The highest speed of milking was recorded by
the English cows (max. 3.01 kg/min); the Polish cows had the
lowest speed (max. 2.35 kg/min). In terms of milking speed,
cows from English sires were significantly better than were
cows from Swedish, Danish, and Polish sires. Danish daughters had the best udder index.

RANKING

Final results presented in this paper are from the intensive feeding trials only. Results of the field test,
which involves many more animals, are only preliminary. The

evaluation of the heterosis effect has not been completed.
Therefore, any ranking of the compared strains must be
treated with maximum caution.

TABLE 5. LEAST SQUARE MEANS OF THE MILKING ABILITY TRAITS
(INTENSIVE FEEDING TRIAL)

Strain	Mean milk-yield/min LSM	SE	Max. milk-yield/min LSM	SE	Udder index LSM	SE
USA	1.81	0.09	2.69	0.14	45.15	1.07
Poland	1.57	0.10	2.35	0.14	43.18	1.08
Canada	1.80	0.10	2.65	0.14	44.91	1.10
Denmark	1.67	0.09	2.43	0.14	46.58	1.06
Great Britain	2.02^X	0.10	3.01^X	0.14	44.55	1.08
Sweden	1.71	0.10	2.45	0.14	43.13	1.09
FRG	1.80	0.10	2.66	0.14	45.99	1.09
Netherlands	1.82^X	0.10	2.68	0.14	43.74	1.09
Israel	2.02^X	0.09	2.91^X	0.14	45.62	1.07
New Zealand	1.85^X	0.10	2.76^X	0.14	43.02	1.11
	1.81	0.04	2.66	0.06	44.59	0.43

Source: H. Jasiorowski, Z. Reklewski and M. Stolzman (1981).

Nevertheless, some conclusions can be stated.
Milk yield is the most important feature. Table 6 presents the least-square constants for milk producion under
both field and intensive feeding conditions. Generally, the
rankings for milk yield and fat content are very similar
under field and intensive feeding conditions, which seems to
indicate that there is no interaction between genotype and
environment.

TABLE 6. LEAST SQUARE CONSTANTS FOR MILK YIELD

Country	Milk yield Field	Intensive feeding	Fat content Field	Intensive feeding
USA	+ 298	+ 432	- 0.08	- 0.28
Israel	+ 210	+ 252	- 0.06	- 0.14
Canada	+ 171	+ 255	- 0.04	- 0.03
New Zealand	+ 144	+ 26	+ 0.11	+ 0.17
Sweden	+ 24	+ 61	- 0.01	0
Great Britain	+ 13	+ 95	0	+ 0.01
Denmark	- 148	- 264	+ 0.02	+ 0.02
FRG	- 152	- 37	- 0.02	- 0.04
Netherlands	- 154	- 131	+ 0.06	+ 0.17
Poland	- 406	- 573	+ 0.02	+ 0.09

The strains of Holstein-Friesian type (U.S., Israeli, and Canadian) showed the highest milk yields and lowest fat contents both under field and intensive feeding conditions. Under both conditions, the New Zealand and Dutch strains were the best producers in terms of fat content. The cows of the Polish strain produced the lowest milk yield in both feeding treatments. The ranking of the tested strains according to the live weight can be estimated on the basis of the least square constants presented in Table 7. Here again, ranking in the field and intensive conditions is similar, with the exception of the Canadian and Dutch strains. The heaviest cows were the U.S. and Israeli types. This experiment seems to provide much new information that can be used by the breeders of Friesian cattle worldwide, especially those who wish to import semen for further improvement of the dairy characteristic of their cattle.

TABLE 7. LEAST SQUARE CONSTANTS FOR BODY WEIGHT

Country	Field	Intensive feeding
USA	+ 9.2	+ 12.1
Israel	+ 6.1	+ 15.1
Canada	+ 6.1	+ 2.4
Sweden	+ 1.7	+ 5.9
FRG	+ 0.8	- 0.9
Netherlands	- 1.1	- 13.8
New Zealand	- 5.5	- 4.6
Great Britain	- 6.1	- 6.3
Poland	- 6.8	- 6.0

REFERENCES

Gravert, H. O. 1975. Possibilities for crossbreeding in dual-purpose cattle--Holstein x Black Pied Zuchtungskunde 47:404.

Jasiorowski, H., Z. Reklewski and N. Stolzman. 1981. International testing of different strains of Friesian cattle in Poland. Milk performance of F_1 paternal Friesian strains crosses under intensive feeding conditions. EAAP Annual Meeting, Zagreb.

Koncar, L., M. Simic, A. Antov, J. Vucinic and S. Jovicevic. 1980. The effect of crossbreeding Yugoslav Pied with Holstin-Friesian cattle on milk production and reproduction. Animal Breeding 48:11 (Abstr.).

Nahlik, K. and B. Szelag. 1979. The effect of crossbreeding Polish Black-and-White Lowland cows with Holstein-Friesian bulls on growth and development of male F_1 crosbreeds. Roezniki Naukowe Zootechniki 6:89.

Nedava, V. E. 1980. Holstein-Friesian cattle in the Ukraine. Vestnik Sel'skohozyaistvennoi Nauki 1:56.

O'Ferrall More, G. J. and E. P. Cunningham. 1978. Holsteins better than Friesian for milk, almost as good for beef. Farm Food Res. 9:52.

Oldenbroek, J. K. 1980. Breed and crossbreeding effects in a crossing experiment between Dutch Friesian and Holstein-Friesian cattle. Livestock Production Science 7:235.

S. Stolzman, M., H. Jasiorowski, Z. Reklewski, A. Zarnecki and G. Kalinowska. 1981. Friesian cattle in Poland--preliminary results of testing different strains. World Animal Review 38:8.

Stolzman, M., H. Jasiorowski, Z. Reklewski and A. Zarnecki. 1981. Friesian strains crosses in the field. EAAP Annual Meeting, Zagreb.

NEW ZEALAND DAIRY CATTLE BREEDING RELATED TO HOLSTEIN-FRIESIAN STRAIN COMPARISON RESULTS AND USE OF EMBRYO TRANSFERS FOR GENETIC IMPROVEMENT

R. L. Baker

INTRODUCTION

Man has for many centuries practiced selective breeding of his dairy cattle. Today we have an animal capable of sustaining a high production over 85% of the days in each year, of producing a calf annually, and of staying in the herd for five or more lactations under New Zealand pastoral farming conditions. This has all resulted from sustained selection for a more desirable dairy animal.

Today we have available a much better understanding of the mechanism by which the ability to produce milk is passed from cows and bulls to their progeny. Two important pieces of knowledge are: (1) each animal receives a random half of the genes carried by its sire and dam, and (2) an average of about 25% of the variation between cows in milk fat yield is genetic in origin. These two pieces of knowledge, in combination with the development of artificial insemination, have had a profound effect on the way we select and breed dairy cattle. This paper discusses the factors that influence the rate at which a population can be genetically improved, and identifies bull selection and progeny testing as the major source of progress. Data are presented to demonstrate that the New Zealand dairy cattle can match the genetic merit of other populations around the world. The potential use of induced multiple ovulations and embryo transfer to increase the rate of genetic improvement in dairy cattle is also discussed.

OBJECTIVES

In designing an effective breeding program, it is important to specify a clear objective at the outset. Under New Zealand economic and climatic conditions the objective could be--"to maximize the yield of milk solids from a cow that is easily milked and managed in large numbers, has a long productive life, and is adapted to pasture feeding." This objective clearly places major emphasis on production while at the same time keeping other characteristics of the cow at a satisfactory level. This objective also reflects the high economic importance of milkfat yield relative to milk yield, under New Zealand conditions. In North America, however, primary selection emphasis is usually put on milk yield.

THE NEW ZEALAND DAIRY INDUSTRY

The Jersey and the Friesian (Holstein) contribute almost equally to New Zealand's 2.1 million dairy cows (with small numbers of Ayrshires and Milking Shorthorns). Table 1 documents some pertinent statistics for the dairy industry in New Zealand; average farm size, herd size, and milk fat production are divided into those farms using artificial breeding (AB) and those that do not. The New Zealand dairy industry is characterized by large herd sizes but, by world standards, relatively low per-cow production (table 2). On the other hand, New Zealand produces its dairy products with low labor costs (30 hours per cow vs about 65 hours in the U.S.) and low feed costs due to the use of the pasture system (table 2); therefore total milk production costs are estimated to be about 40% of those in the U.S. and Great Britain (table 3). The net result is that dairy products from New Zealand can be landed in the U.S. or Great Britain at about two-thirds the cost of locally manufactured products.

The New Zealand Dairy Board (NZDB), a farmer cooperative, acts not only to provide dairy farmers with essential services but also acts as a marketing cooperative for all dairy products. Herd improvement services are made available to dairy farmers throughout New Zealand by six Livestock Improvement Associations. The Associations are licensed by the NZDB to carry out herd testing and work with it in providing an AB service. In 1980-81, some 909,000 cows, approximately 45% of the national total, were herd tested. Milk fat production per tested cow was 160 kg. A little over 1 million dairy cows were inseminated in 1980, representing 53% of all cows in milk and 64% of the total dairy herds. Ninety-six percent of Premier Sire Service inseminations (liquid semen from proven bulls in a bull-of-the-day service) used semen from dairy-breed bulls (58% from Friesians, 36% from Jerseys and 2% from Ayrshires). The remaining 4% of inseminations were from Polled Hereford bulls--the only beef bulls offered under the Premier Sire Service, although other breeds and dairy bulls are offered under Nominated Service (frozen semen of a particular bull).

The NZDB also offers an extension service, carried out by a consulting officer team whose services are available free of charge to all dairy farmers. They advise on all aspects of dairy farm management and the integration of Herd

TABLE 1.   NEW ZEALAND DAIRY INDUSTRY STATISTICS*

Statistic	A.B. user	Non-A.B. user
No. of farms	8638	1918
Average area - hectares	72	65
-acres	178	161
Average herd size	138	112
Average milk fat per cow (kg)	148	137

*Abstracted from the 57th Farm Production Report of the New Zealand Dairy Board (1980-81).

TABLE 2.   PRODUCTION PER COW AND CONSUMPTION*

	Production per cow (4% FCM) (lb)	Feed consumed per cow per year	
		Roughages (lb. dry	Concentrates matter)
California	15,700	8,250	8,140
Israel	15,330	4,950	9,130
Wisconsin	11,670	8,800	4,356
England/Wales	10,280	7,920[1]	3,190
New Zealand	7,645	8,350[2]	

*From a paper by J. D. J. Scott to the XIV International Grasslands Conferences.

[1] Includes an estimated 3,000 pounds per cow from pasture.
[2] All pasture and pasture products (hay and silage).

Improvement Services with other aspects of herd and farm management.   In 1981-82, there were 22 consulting officers in the field.

Traditionally, the dairy industry has been sharply separated from beef production (apart from cows culled for slaughter); most male and surplus female dairy calves are slaughtered for export as 3- to 4-day-old veal (bobby calves).   However, the number of dairy calves reared for beef is on the increase and in 1980-81 was estimated to have been 366,000.

An essential part of the NZDB's artificial breeding services is the Sire-Proving Scheme.

SIRE-PROVING PROCEDURES

The method of sire evaluation used in New Zealand has been developing over a long period of time.   In the early 1920s bulls were ranked according to the number of their daughters that produced milk above a certain minimum level.

TABLE 3. COSTS AND RETURNS OF DAIRY PRODUCTION IN DIFFERENT COUNTRIES*

	California 1978)	Wisconsin (1978)	Great Britain (1978-79)	New Zealand (1978-79)
	(dollars per hundredweight milk)			
COSTS:				
Feed	4.95	4.14	5.18	1.14
Labor			1.41	1.41
Other direct costs			1.82	0.55
Labor and other direct costs	2.00	2.91	____	____
TOTAL	6.95	7.05	8.41	3.10
RETURNS:				
Milk	10.27	9.82	9.68	3.95

* From a paper by J. D. J. Scott to the XIV International Grasslands Conference. At the time this table was prepared, the U.S. and N.Z. dollars were approximately equal and $2 = 1 British pound.

This was followed by the use of daughter average yields and, prior to 1949, the daughter-dam comparison. Developments between 1949 and 1964, as described by Searle (1962, 1964), included the use of contemporary comparisons. At that time the use of artificial breeding was widespread, and the assumptions needed for contemporary unbiased comparisons were no longer valid. The genetic merit of herdmates' sires varied systematically (1) between herds, because of differential usage of artificial breeding, and (2) between years, because of the genetic gain resulting from artificial breeding. In an attempt to overcome these problems, the procedures described by Evans (1969) were introduced. This procedure, in its original form, was the same as that described more recently by Thompson (1976) as having best linear unbiased prediction properties. Since 1970, the procedure has been further modified as faster and larger computers became available.

The basic structure of the artificial breeding organization and the sire-proving scheme has been described in some detail by Searle (1962) and Evans (1969) and can be described diagrammatically as in figure 1. Since the New Zealand dairy industry follows a distinct seasonal production pattern with calving and mating in the spring, advances in semen dilution (Shannon, 1968) have led to a much wider annual use of proven bulls over a short mating period (i.e., about 2 months). In 1960, the average number of inseminations per bull was 4,000; in 1967 more than 20,000; and by 1980 about 40,000, with a top-rated Friesian bull averaging about 94,000 inseminations in a single season.

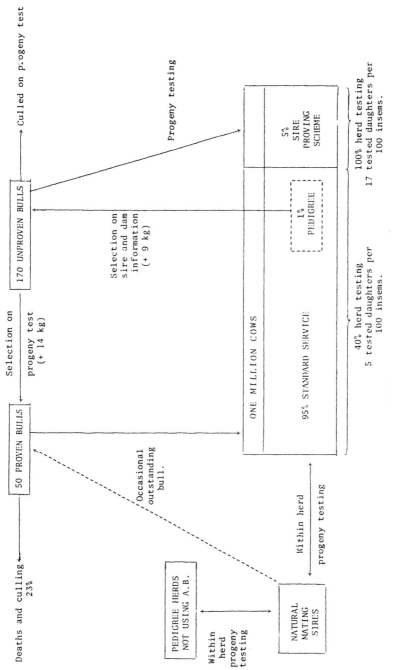

FIG. 1. THE SIRE PROVING ORGANIZATION

The genetic implications of this increased coverage have been discussed by Stichbury (1968). In spite of an increase from approximately 500,000 cows inseminated in 1960 to about 1 million today, the number of proven bulls from the two major breeds has declined from about 100 to about 50. As a result the proportion of inseminations made to unproven bulls has been reduced; at the same time intensive selection from progeny-tested bulls through AB has doubled production. The smaller number of bulls has, however, increased the possibility of inbreeding and possibly has reduced the gene pool available for future selection. The implications of this have continued to be monitored by the NZDB but to date have not presented serious problems.

The 'engine' of the system shown in figure 1 consists of the 50 proven bulls that are used to upgrade the genetic merit of the cow population by mating them to the 1 million or so cows in standard service AB herds throughout the country. The 'fuel' necessary to keep the engine performing efficiently is the 170-odd young bulls that are progeny tested each year in sire-proving herds. These young bulls are selected on ancestry, usually have an AB proven bull as their sire, and until recently, were required to have a Highly-Commended or a Very-Highly-Commended pedigreed dam with 2 to 5 lactation records.

The Sire-Proving Scheme started in its present form in 1961. Members agree to use the semen from unproven bulls and herd test the resulting daughters as 2-year-olds. All inseminations in the Sire-Proving Scheme are provided at no cost, and the Sire-Proving Scheme heifers qualify for a rebate when in milk as 2-year-olds. Five percent of the semen used in the Sire-Proving Scheme is from proven (marker) sires that provide a link between bulls proven in different years.

The purpose of the Sire-Proving Scheme is to provide a continuous supply of proven bulls available for selection for widespread use in the Premier Sire Service. The standard of the bulls entering the Sire-Proving Scheme directly influences the standard of bulls available for selection when proven. Evans (1969) investigated the efficiency of young sire selection on pedigree. He found that, although the sire information fairly accurately predicted young-sire performance; the dam ratings gave little indication of their expected usefulness in predicting young-sire ratings. Since, at that time, dams of young sires had to be pedigreed cows with high breed classifications, it was concluded that rearing, feeding, and other management influences existed that could not be allowed for in the usual contemporary comparison method. Evans, therefore, suggested that young sire selection should be based on sampling bulls from the youngest available AB dams. He also concluded that additional short-term genetic gains would result from sampling young bulls from herds using AB heavily for several years, regardless of pedigree status.

123

Comparison of Unregistered and Breed-Society-Nominated Bulls

In 1971, the NZDB initiated a trial with the Friesian and Jersey Breed Societies, each agreeing to nominate five bulls per year for inclusion in the Sire-Proving Scheme. Since 1972, five unregistered bulls, each with at least 4 generations of proven AB sires in their ancestry, also have been included in the proving scheme. The objective of the experiment was to examine more critically alternative sources of young bulls for progeny testing. Full progeny test results for the Jersey bulls were reported by Wickham (1979a).

The main characteristics of the sires and dams of the two groups of Jersey bulls are shown in table 4. The bulls nominated by the Jersey breed society had higher producing dams and were classified as 'excellent' or 'very highly commended' on type. Likewise, the average daughter yield of the sires of breed-society-nominated bulls had a 20 kg advantage. In contrast, the unregistered bulls were out of dams with higher breeding indexes (BI) and by sires with higher breeding indexes. The BI is an estimate of an animal's genetic merit for milk or milk fat production. For a young bull, this is calculated as one-half of the BI of its sire and dam; for a cow BI is based on her own production relative to that of her herdmates and the BI's of her sire and dam. A sire, on the other hand, has his BI calculated from the production of his daughteres relative to their herdmates and on the BI's of his ancestors.

These two groups of bulls thus provided a unique opportunity to compare the results of two selection criteria--one that emphasized type and absolute production and one that emphasized genetic merit. The experiment did not set out deliberately to examine this question, but the methods of selection in the two groups of bulls permitted the comparison to be made.

TABLE 4. ANCESTRY INFORMATION OF JERSEY BULLS COMPARED BY THE NEW ZEALAND DAIRY BOARD

	Number of bulls	Dam Fat (kg)	BI* (%)	Sire Dgts Fat (kg)	BI (%)	Ancestor BI (%)
Unregistered (U)	14	204	118	167	127	122
Breed society (B)	12	213	110	187	112	111
Difference (U-B)		- 9	8	-20	15	11

* BI = Breeding index.

The Sire-Proving Scheme tested these bulls by comparing the productive and nonproductive traits of their daughters in the same herds. Because the daughters were compared in the same herds, in this case absolute yields were considered without excessive concern with the influence of management on the differences between bulls. The production figures are shown in table 5. The daughters of the unregistered bulls produced 13 kg more fat than the daughters of the breed-society-nominated bulls. In production index (PI) units, there is a 4% advantage, which agrees very closely with the predicted difference of 5.5% (11/2) based on the ancestry information for the two groups of bulls. There was a significant difference between the daughters of the two groups of bulls compared for ease of putting on cups, ease of keeping on cups, speed of milk let-down, speed of milking, shed temperament, or occurrence of mastitis. But the daughters of the breed-society-chosen bulls had superior udder and overall conformation scores. Although the appearance differed, the results under 'workability' traits indicated that both groups were equally desirable in commercial farming situations.

TABLE 5. PROGENY TEST RESULTS OF JERSEY BULLS COMPARED BY THE NEW ZEALAND DAIRY BOARD

	Ancestry BI (%)	Daughter Number	Production Fat (kg)	PI* (%)	BI (after progeny test) (%)
Unregistered (U)	122	644	159	106	121
Breed society (B)	111	480	146	102	111
Difference (U-B)	11		13	4	10

* PI = Production index.

This study clearly illustrates the potential danger of placing too much emphasis on absolute production levels. It must be concluded that well-bred grade bulls with high breeding indexes are capable of breeding high-producing cows that are acceptable to commercial dairy farmers.

Current Developments in Sire Proving

Partly as a result of the study just reported, contract mating started in 1973 as a means for breeding bulls for entry into the SPS but was restricted in the first year to the Friesian breed using imported semen. Contract mating was extended to the Ayrshire and Jersey breeds in 1974; since 1976 the majority of the bulls purchased for the SPS have resulted from contract matings.

In 1978, a modified procedure for recording the identification of grade animals was introduced with the objective
of expanding the population of cows that could be considered
as dams to produce young bulls. By 1981 the number of
Friesian cows with ancestry records on the NZDB files was
60,287, of which 57,413 were pedigreed animals (average BI
of 110) and 2874 were grade (average BI of 119). Even with
these relatively small numbers of grade cows identified to
date, in relation to the pedigreed population, removing the
registration selection has had substantial impact on
bull/dam selection. In 1980, 70% of the Jersey cows and 44%
for Friesian with BI's of 126 or greater were classified
grades. In both breeds, grade yearlings accounted for a
substantial percentage (21% of Friesian and 39% of Jersey)
of the high breeding index females. The net effect to date
of the removal of the breed society registration requirements for bull/dams has been to approximately double the
'active' (bull/dam) population size (3,000 to 7,000 for
Jerseys and 4,500 to 6,000 in Friesians). It is expected
that the active population size will increase further in the
future.

From the inception of the AB service until the start of
the present SPS, reports were solicited from farmers who
were milking daughters (particularly from unproven bulls).
With the introduction of the SPS, daughters of unproven
bulls were limited to fewer herds so that a more formal
method of reporting could be done on traits other than production. The system of reporting and the traits to be
reported on have been modified over the years. Wickham
(1979b) estimated genetic parameters and economic values of
traits other than production in the NZDB files. While it
appears that many of these traits are of low heritability,
there is no justification for including them in a selection
objective if they are of no economic significance. The
results of the study indicated that the nonproductive
traits, such as ease of putting cups on, mastitis, speed of
let-down, shed temperament, and bloat susceptibility are
economically important and should be considered for inclusion in an overall breeding objective. Work is continuing
in this area and in the meantime progeny test results of
some of the above nonproductive traits, as well as udder and
overall conformation scores, are being reported in sire
catalogs.

SIRE EVALUATION-INTERNATIONAL COMPARISONS

The breeding scheme operated by the NZDB is paralleled
by a number of similar schemes in other countries. With the
possibility of importing and exporting semen for New Zealand

it is of considerable importance to be able to evaluate the relative genetic merits of sires proven in different countries.

The very large differences in feeding and management methods between countries make it impossible to use production averages as a guide to genetic merit. On the basis of production averages alone New Zealand animals would rank below almost every other major milk producing country in the world.

By milking animals from different strains in the same environment and comparing their production it is possible to obtain estimates of the difference in genetic merit between the animals representing each strain. By then relating these results to the genetic merit estimates of the same animals (or their sires) in their home country it is possible to derive formulae for converting sire proofs between countries.

New Zealand is involved in a number of experiments designed to enable breeding indexes to be converted to the index units used by other countries. The most significant of these experiments is being conducted in Poland under the auspices of FAO and involves 10 strains of black and white cattle.

This project was started in 1974 and involves ten countries, namely Canada, Denmark, Federal Republic of Germany, Israel, Netherlands, New Zealand, Poland, Sweden, United Kingdom, and the United States with the objective of comparing the overall productivity (milk and meat production) of their respective black and white (Holstein-Friesian) cattle. Semen from about 40 young bulls from each of the participating countries was used on a total of more than 30,000 black and white cows in Poland. The young bulls were selected at random among young bulls entering AB service in their respective home countries. First-cross and back-crosses to the paternal sire strain as well as the Polish strain have been produced. Poland was chosen for the trial because of its large population of Friesian cattle (8 million), its centrally planned economy, and its large state farms--all of which facilitated the organization of a trial on a large scale.

The first results of this trial have been reported by Stolzman et al. (1981). The first lactation production figures (table 6) show that milk production was highest for the U.S., Israeli, and New Zealand heifers. The Polish heifers, which were the lowest strain, produced approximately 700 kg less than the highest producers. Fat yield was highest for the U.S., Israeli, and New Zealand heifers, which was 24 kg above the Polish heifers. Milk fat percentage was highest for the Polish, Dutch, and New Zealand heifers, which was 0.15% above the U.S. with the lowest milk fat percentage of the 10 countries compared.

TABLE 6. AVERGE YIELDS OF DAMS[a] IN HOME COUNTRY AND GRANDAUGHTERS IN THE POLISH TRIAL

Country (strain)	No. of lactations	Dams of bulls			Daughters of bulls Average 1st lactation yield[b]		
		Milk (kg)	Average yield Fat (kg)	Fat %	Milk (kg)	Fat (kg)	Fat %
U.S.A.	4.1	10,883	400	3.67	4,183	163	3.89
Poland	5.0	5,408	227	4.20	3,397	138	4.05
Canada	5.4	8,836	356	4.02	3,984	156	3.92
Denmark	5.9	6,478	279	4.31	3,730	150	4.01
Great Britain	5.1	6,220	252	4.06	3,717	148	3.98
Sweden	5.9	7,048	297	4.21	3,865	154	3.98
Germany	5.7	7,063	301	4.27	3,629	144	3.97
Holand	3.9	7,287	317	4.35	3,588	145	4.04
Israel	4.3	9,297	311	3.34	4,102	162	3.94
New Zealand	6.5	5,016	209	4.17	4,023	162	4.03

[a]Of first 20 bulls supplied for each strain.

[b]Based on 1,581 P_1 progeny. From Stolzman et al. (1981).

Another interesting result of the trial is a very good demonstration of the way in which genotype and environment act together to determine milk yields. This is illustrated in table 6, which shows the average yield of the dams of bulls included in the trial (provided by Dr. B. W. Wickham) alongside production figures of the bull's daughters under Polish conditions. These are unofficial averages that are intended to illustrate the size of the differences, rather than be accurate estimates. The dams, when producing in their home countries, show almost a two-fold variation in milk and milk fat yields.

The extremes for fat yield are the U.S. dams at 400 kg and the New Zealand dams at 209 kg. Both of these averages are well above national means. In Poland, daughters of these bulls are producing at approximately the same level (163 and 162 kg, respectively). An even better demonstration is provided by comparing the figures for the Dutch and New Zealand strains. The dams of the Dutch bulls outproduced the dams of the New Zealand bulls by 108 kg milk fat in their home environment. The data on the bull's daughters producing in the same environment in Poland show that the New Zealand animals out-produced the Dutch animals by 17 kg.

Another interesting result as far as New Zealand is concerned is that the 12-month live weights for New Zealand-sired bulls are above average (although the total range among the means for the ten strains was only 9 kg), but two-year-old cow live weights are below average. It appears that New Zealand may be breeding a small, highly productive cow, while still maintaining good growth potential in males.

MULTIPLE OVULATION AND EMBRYO TRANSFER IN DAIRY CATTLE

About 17,000 bovine pregnancies were produced by multiple ovulation and embryo transfer (MOET) in North America in 1979 (Seidel, 1981). The major use to date of these techniques has been to increase the productive rate of valuable (usually pedigreed) cows. Other applications include circumventing infertility, exporting embryos and testing potential carriers for Mendelian recessive alleles.

Several studies have concluded that embryo transfer would contribute little to the rate of genetic improvement in dairy cattle in traditional progeny testing schemes. Increases in the selection differential of cows to breed young bulls would be small in an efficiently operated AI improvement scheme where most of the genetic progress comes from selection of sires (e.g., McDaniel and Cassell, 1981). However, it has been estimated that, if embryo transfer could be applied in all commercial herds, the rate of genetic improvement in a progeny testing program could be increased by about 15% (Van Vleck, 1982).

Nicholas (1979) suggested an alternative method of using MOET to increase the rate of genetic response. It is based on reducing the generation interval and tolerating less accuracy in selection and requires much smaller numbers of recorded cattle. One scheme, his pedigree scheme, involves selection among transferred sons and daughters when they are 12 to 13 months of age and on the basis of their dam's (the donor) first lactation record. In his other scheme (the sib scheme) selection of tranferred males is delayed until their female sibs have completed their first lactation so that males can be selected on an index using sib and dam performance.

Nicholas and Smith (1982) have expanded on these proposals by adding information on other relatives in selection of both males and females. The time schedule for each scheme is shown in table 7 where the term "juvenile scheme" is used for selection at 1 year of age and "adult scheme" for selection at 3 years of age. The generation intervals for the juvenile scheme is about 1.8 years and for the adult scheme about 3.7 years. These compare with the average generation interval in a conventional progeny testing system of about 6.3 years. No details of the genetic parameters or the methodology used in investigating these schemes are presented here, but the conclusions are discussed below.

Rates of genetic improvement in conventional progeny testing schemes of 2% of the mean per year are usually quoted as the maximum possible rate (Robertson and Rendel, 1950; Skjervold and Langholtz, 1964; Lindhe, 1968). It is theoretically possible to obtain higher rates by more intense selection and by more accurate selection of females (including information on sires and grandsires); Van Vleck (1982) suggests rates as high as 3% per year. With large pedigreed recorded populations, the rate of increase of inbreeding with such schemes is low--from 0.1 to 0.2% per year.

In practice the rates of genetic change in milk yield are much lower than the possible rates, and range from 0.5 to 1.0% of the mean in different countries (Van Vleck, 1977). This is partly due to inefficiencies in the system (e.g., in identifying genetically superior bull/dams as discussed earlier) and also to selection for other traits, especially type and functional traits.

The results derived by Nicholas and Smith for the MOET schemes they investigated show that the adult scheme is preferred where the number of embryo transfers possible is limited and inbreeding is important. A scheme with 1024 embryo transfers per year and 512 females milk-recorded per year will sustain a rate of genetic improvement some 30% above that possible by a conventional national progeny test scheme. Because of the relatively small number of animals involved, it is suggested that the greater control over recording, breeding, and selection should be possible and should lead to a larger proportion of the genetic gains

TABLE 7. TWO POSSIBLE MULTIPLE-OVULATION AND EMBRYO-
TRANSFER SCHEMES (from Nicholas and Smith, 1982).

Month	Juvenile scheme	Adult scheme
1	Born	Born
2		
3		
4		
5		
6		
7		
8		
9		
10		
11		
12		
13	Select on pedigree, MOET	
14		
15	Mate	Mate
16		
17		
18		
19		
20		
21		
22	MOET progeny born	
23		
24	Calve	Calve
25		
26		
27		
28		
29		
30		
31		
32		
33		
34	Complete lactation,	Complete lactation,
35	select MOET progeny	select and MOET
36	for MOET	
37		Mate for further
38		lactations
39		
40		
41		
42		
43		
44	MOET progeny born	MOET progeny born
Generation interval	22 months = 1.83 years	44 months = 3.67 years

Nicholas, F. W. and C. Smith. 1982.

being realized in practice. Other advantages of this kind of scheme include the possibility of measuring feed intakes and thus estimating the lifetime efficiency of milk production. In countries where dairy cattle contribute to meat production, males could be performance tested for beef traits, and selected within sibships, with no loss in selection intensity for dairy traits.

Nicholas and Smith (1982) suggest the role of a MOET scheme may not be to replace progeny testing but to provide a nucleus herd of the type envisioned by Hinks (1978) for breeding young bulls for progeny testing. This is because producers may prefer well-proven bulls rather than young bulls of high-average merit but with much variation. Establishing such a nucleus, using genetically superior national and available international breeding stock, will in itself provide a useful short-term genetic lift.

ACKNOWLEDGEMENTS

The Farm Production Division of the New Zealand Dairy Board, and specifically Dr. B. W. Wickham, provided me with up-to-date information on their dairy cattle breeding program. Dr. Charles Smith kindly provided the manuscript of his paper with Dr. Nicholas that describes their MOET schemes.

REFERENCES

Evans, D. A. 1969. Research on current sire proving procedures. Proc. New Zealand Society of Animal Production 29:77.

Hinks, C. J. M. 1978. The use of centralized breeding schemes in dairy cattle improvement. Animal Breeding Abstracts 46:291.

Lindhe, B. 1968. Model simulation of AI-breeding within a dual-purpose breed of cattle. Acta Agric. Scand. '18:33.

McDaniel, B. T. and B. G. Cassell. 1981. Alternative uses of embryo transfer and their effect on the rate of genetic change. J. Dairy Science 64:2484.

Nicholas, F. W. 1979. The genetic implications of multiple ovulation and embryo transfer in small dairy herds. 30th Annual Meeting of the European Association of Animal Production, Harrogate, England, CG 1.11.

Nicholas, F. W. and C. Smith. 1982. Increased rates of genetic change in dairy cattle by embryo transfer and splitting. Animal Production (submitted).

Robertson, A. and J. M. Rendel. 1950. The use of progeny testing with artificial insemination in dairy cattle. Journal of Genetics 50:21.

Searle, S. R. 1962. Bull sampling programs in the artificial breeding of dairy cattle. Proc. New Zealand Society of Animal Production 22:54.

Searle, S. R. 1964. Review of sire-proving methods in New Zealand, Great Britain, and New York State. J. Dairy Science 47:402.

Seidel, G. E. 1981. Superovulation and embryo transfer in cattle. Science 211:351.

Shannon, P. 1968. Advances in semen dilution. Proc. New Zealand Society of Animal Production 28:23.

Skjervold, H. and H. J. Langholtz. 1964. Factors affecting the optimum structure of AI-breeding in dairy cattle. Zeitschrift fur Tierzuchtung und Zuchtungsbiologie 80:26.

Stichbury, J. W. 1968. Genetic implications of advances in semen dilution. Proc. New Zealand Society of Animal Production 28:32.

Stolzman, M., H. Jasiorowski, Z. Reklewski, A. Zarnecki, and G. Kalinowska. 1981. Friesian cattle in Poland. Preliminary results of testiing different strains. World Animal Review 38:9.

Thompson, R. 1976. Relationship between the cumulative difference and best linear unbiased prediction methods of evaluating bulls. Animal Production 23:15.

Van Vleck, L. D. 1977. Theoretical and actual genetic progress in dairy cattle. Proc. International Conference of Quantitative Genetics. Edited by E. Pollak, O. Kempthorne and T. B. Bailey. Iowa State University Press, Ames, Iowa.

Van Vleck, L. D. 1982. Potential genetic impact of artificial insemination, sex selection, embryo transfer, cloning, and selfing in dairy cattle. In: New Technologies in Animal Breeding. Edited by G. Seidel, Academic Press.

Wickham, B. W. 1979a. Using unregistered bulls. Proc. Ruakura Farmers' Conference. p 147.

Wickham, B. W. 1979b. Genetic parameters and economic values of traits other than production for dairy cattle. Proc. New Zealand Society of Animal Production 39:180.

15
DAIRY CATTLE MANAGEMENT
BY **DHI** OBJECTIVES

Jack D. Stout

Dairy Herd Improvement (DHI) records are to the dairy herd manager as a road map is to a traveler. A road map allows the traveler to determine the best way to get to a desired destination and also provides the details about where he has been. DHI records, through their various summaries, provide a complete production, reproduction, and genetic history of each cow, of selected groups of cows, and of the complete herd. Dairymen should show steady improvement if they use these data to set objectives for their herd and then to analyze direction and manage to meet those objectives.

All DHI record processing centers provide the basic information needed for analyzing herd situations and for setting management objectives. Each center has its own forms and the method of summarization that meets the desires of the dairymen they serve. The examples and data used in this paper are from the Mid-States Dairy Records Processing Center, Iowa State University, Ames, Iowa, and much of the criteria was developed by their very capable Extension Dairy Staff. This center serves a 9-state area in the mid-U.S. and processes records on approximately 500,000 cows per month.

Objectives must be established in all phases of the dairy operation including production, reproduction, genetic merit, and feed utilization. Emphasis must be placed on each area simultaneously to accomplish maximum herd potential. It has been my observation that reproduction usually is the most neglected of the management areas. When individual cow production is high or dairy herd production is at a sufficient level, dairymen tend to not be as aware of missed estrous intervals. Field work at planting or harvest time also is reflected in reproduction efficiency; in many cases, heat detection is left to other members of the family or crew.

To avoid laxness of management, objectives must be set, both short-term and long-range, with constant evaluations made to chart the course to those objectives. A suggested format for setting objectives could be:

MEASURABLE CRITERIA	CURRENT AREA AVERAGE	MY HERD CURRENTLY	12 MO. OBJECTIVE	5 YEAR OBJECTIVE

Any item that warranted the setting of objectives could be plugged into a similar format. The area average allows for a comparison to the status of a herd's current conditions before setting objectives. The format must also contain the analysis of records to determine the herd status in each area, the problems or limitations to correct, and a schedule for action needed.

Herd objectives should be realistic and obtainable. They should deal with only those criteria that are measureable and of economic importance. Annual objectives as well as long-range objectives should be established and the annual objectives should provide for step-wise progress toward the long-range objectives.

Table 1 contains the spring 1982 Mid-States average production data. Using the area average as an example, we'll work through examples of setting objectives designed to raise the level of milk production by 20% in five years. To increase production by 2800 lb may seem a big increase but a 4% increase of only 560 lb for the annual objective does not seem so far out of reach. Also, when put on a daily basis, an increase of only 1.6 lb per day makes the objective appear much more practical. Remember that all items should not increase. Butterfat percentage, for instance, will probably decrease, or at best stay the same, if milk increases 20%. The feed items will vary as costs change, types and ratios vary, etc. The important thing here is income above feed costs not total costs.

TABLE 1. CURRENT PRODUCTION STATUS AND OBJECTIVES

	Mid-States average	From most recent herd report (202)	12 mo Objectives	5 yr Objectives
Milk production/cow/yr	14010	_____	+ 560	16800
Fat production/cow/yr	515	_____	+ 20	638
Milk production/cow/day	38	_____	+ 1.6	46
% B.F.	3.86	_____	_____	not<3.7
% days in milk	85	_____	_____	87
Value of production/cow	1821	_____	_____	_____
Total feed cost/cow	700	_____	_____	_____
Income over feed cost	1121	_____	_____	_____
Lb milk/lb grain	2.5	_____	_____	_____
Feed cost/100 lb milk	5.34	_____	_____	_____
Return/$ feed cost	2.83	_____	_____	_____

Following the setting of overall objectives, the component parts of the management program must be analyzed and objectives set that will make each component part move the herd toward the main goal.

REPRODUCTION

To determine the best approach to meeting these objectives, DHI records must be analyzed to determine the number of cows in various reproduction problem areas.

Some common problem areas in reproduction include:
- Heat period not observed.
- Waiting too long to breed cows after calving.
- Poor AI breeding techniques.
- Poor semen quality.
- Poor reproductive records.
- Health problems--metritis, retained placenta, etc.
- Unsanitary calving conditions.
- Heifers not grown well enough to breed by 15 months of age.

By analysis of the DHI Herd Summary (202) our first area of concern would be Average Days Open (preg. animals) -- 123 days, and % Preg Cows Open > 120 days -- 40%. If these two items were corrected to meet the objectives, most other criteria of reproduction would fall into reasonable range of the objectives set. These problem areas probably are caused by estrous periods not being observed or waiting too long to breed after calving.
As an example:
123 days open (pregnant cows)
-50 days for normal recovery period calving to 1st
     service
73 days time for Breedable Estrus (BE) to occur
73 days divided by 21 day estrous cycle = 3.5 BE
1.8 services/conception divided by 3.5 BE = 51% BE
     used
If the available BE cycles were used with the same conception rate, the calving interval would have been reduced by 35 days. The percentage of Pregnant Cows Open > 120 Days would have been reduced to fewer than 10%.

PEAK MILK FLOW AND PERSISTENCY

Total milk yield for a lactation is a combinatin of how high a cow peaks and how persistent she is throughout the lactation. High lactation records require both a high peak and persistent flow. Both categories must be analyzed when studying production problems and setting DHI management objectives. They will give an indication as to the genetic potential of the herd, effect of the long-term and current feeding program, and the day-to-day management of the herd.

DAIRY HERD SUMMARY

Measurable criteria	Mid-states average	My herd currently	12 mo. objectives	5 year objectives
Projected calving interval (days)	400			385
Avg days open (preg animals)	123			100
Avg services/ conception (cows)	1.8			1.5
Avg services/ conception (heifers)	1.4			1.3
Avg % in milk	85			88
Avg days dry	63			50
% preg cows open > 120 days	40			< 10
Avg age fresh, lact 1 cows	2-04			2-02
Avg age fresh, producing cows	4-03			4-00

Cows that fit in the following categories should be listed:

1. Cows in the bottom 50% of the herd and still open days after calving.
2. Cows that will have dry periods of over 75 days. ____
3. Older cows with poor breeding histories. (Reproductive efficiency under 80%.) ____
4. Cows that have injured part of their reproductive tract during their last calving. ____
5. Cows with chronic uterine infection, cystic ovaries or other reproductive diseases. ____
6. Heifers that are small for their age and open heifers that have been bred several times and are still open at 20 months of age. ____

Research has indicated that each additional lb of milk produced at peak of lactation means 215 lb of milk more for the
year. Research also indicates that first lactation cows
normally peak at levels about 16 lb less than to older
cows. Table 2 shows the average peak production levels for
various expected herd averages.

Some common limitations to peak performance and persis-
tency are:
- Genetic ability
- Improper body condition at time of calving.
- Mastitis and other diseases.
- Metabolic disorders (ketosis, milk fever).
- Improper handling of cow at time of calving and in
  early lactation.
- Incomplete milking.
- Inadequate energy in early lactation.
- Inadequate length of dry period.
- Individual cow care at time of calving.
- Pregnancy.
- Improper milking practices.

TABLE 2. EXPECTED HERD AVERAGES AT VARIOUS PEAK PRODUCTION
LEVELS

| Expected herd average (lb) | Peak production | |
	1st lact cows (lb)	Other cows (lb)
8000	21.7	37.7
10000	30.6	46.9
12000	40.0	56.2
14000	49.4	65.4
16000	58.7	74.7
18000	68.1	84.0
20000	77.5	93.2

FEEDING

There are as many different forage programs as there
are areas of dairy concentration. In general, the highest
quality forage program possible for a given situation is the
most practical. High milk production requires good quality
forage. Time of harvest, methods of harvest, and storage
methods all must receive strict attention to ensure
quality. A concentrate or grain ration must be supplied in
sufficient quality and quantity to properly supplement the
forage available.

It is difficult to suggest feeding program objectives that would fit many varied forage and concentrate programs. A more appropriate approach would be to suggest forage dry matter (DM) intake per 100 lb body weight (DM/cwt BW) and grain to milk ratios that have been successful for dairymen in the Mid-States area.

Forage Intake:	Mid-States lb DM/cwt BW		Your Herd	
	Milking cows	Dry cows	Milking cows	Dry cows
Recommended minimum	2.0	1.8	_____	____
Absolute minimum	1.5	1.5	_____	____
Practical maximum	2.5	2.2	_____	____

Grain Intake:	Grain:Milk Ratio	
	Recommended	Your herd
Average production	1:2.5-3.5	_____
Peak production (% of BW)	2.5%	_____
Early dry period (% of BW)	0.5 or less	_____
2 wks prior to freshening (% of BW)	0.5-1.0%	_____

Some common problems in the area of feeding include:
- Too little grain fed in early lactation.
- Too little grain fed in late lactation.
- Too little grain or corn silage fed during dry period.
- Not enough protein fed.
- Too little fiber fed in the ration.
- Not enough mineral, or mineral imbalance.

BREEDING PROGRAM

Identification (ID) is the base for any breeding program. Without proper ID of each animal in a herd, there can be no basis for selection. All animals should have permanent IDs such as tattoos, brands, or color markings by sketch or photo. These permanent IDs should be cross-referenced with some type of easily visible ID such as ear-tag, neck chain, leg band, etc.

	% Identified				12 month objective		5 year objective	
	Mid-States		Currently					
Age group	Sire	Dam	Sire	Dam	Sire	Dam	Sire	Dam
0-6 months	57%	100%	____	____	100%	100%	100%	100%
7-12 months	54%	100%	____	____	100%	100%	100%	100%
>12 months	58%	100%	____	____			100%	100%
Lact 1 cows	58%	63%	____	____			100%	100%
Lact 2 cows	57%	64%	____	____			100%	100%
Lact 3 cows	60%	60%	____	____			90%	90%
Lact 4 +	47%	47%	____	____			75%	75%

Complete IDs can be developed rather rapidly, even if starting with an unidentified herd. The tagging of cows, recording of sires used, and tattooing calves at birth would result in 100% calf ID in only one year. A five-year objective of 100% ID through 1st lactation animals could easily be accomplished.

The selection of sires used in the herd is a practical, economical, and accurate method of herd improvement. Each 100 lb plus Predicted Difference Male (PDM) will equate to approximately 87 lb 1st lactation gain in production of that sire's daughters compared to daughters of 0 PDM bulls. To stay abreast of genetic trend, the sires used each breeding season must average 100 lb PDM above the previous year. A minimum genetic objective would be to use sires at least 400 lb +PDM above the average of AI sires available. For example, Holstein sires summarized January 1982 and available through AI organizations averages +1170 PDM, +33 PDF, and +138$. A practical genetic objective for 1982 would be to set minimum sire limits of +1570 PDM. PDM could be altered if a particular advantage in other traits could justify lowering the PDM limits. However, in most cases there are sufficient numbers of bulls available to select from that type or BF traits can be improved as well as PDM.

Sire Slection Profile

Age groups	Mid-States Average		Your herd		12 mo. 5 yr objectives	
	PDM	PDF	PDM	PDF	PDM	PDM
Service sires	—	—	—	—	—	—
Sires of replacements < 1 yr	—	—	—	—	—	—
Sires of replacements > 1 yr	—	—	—	—	—	—
Sires of lact 1 cows	—	—	—	—	—	—
Sires of lact 2+ cows	—	—	—	—	—	—

Culling

Culling of cows is the surest way to eliminate problems, but not necessarily the most profitable way. The cause for cull cows must be determined and corrected. High production comes only when maximum culling can be done for production reasons. Culling high producing cows for mastitits or reproduction problems will not increase the profitability of a herd. To be profitable, herds must be managed in such a way that culling for reasons other than low production are at a minimum. By doing so selection pressure for milk production level can continually increase.

Typically, less than 25% of cows are removed from herds because of low production. Reproduction problems and udder disorders eliminate more than half of cows sold.

Culling Profile and Objectives

Measurable criteria	Mid-States average	Your herd no.	12 month % objectives	5 year Objectives
Reproductive problems	28%			
Mastitis or udder	24%			
Low production	22%			
Conformation defects	6%			
Dairy, accidents, etc.	20%			

Herd Replacements

Raising herd replacements must be an integral part of herd improvmenet. Commercial dairymen that purchase all herd replacements seldom advance above breed average in production. They fail to take advantage of genetic improvement discussed earlier. In most herds, approximately 25% of the herd wil be of higher genetic-producing ability if breeding objectives were met.

Replacements should be fed and managed to meet growth standards for the respective breeds (table 3). Freshening should be at approximately 24 months of age unless market demands for milk would make earlier or later calving more profitable.

Some common problems in raising replacements are:
- Cows in improper condition at time of calving.
- Cows not separated from herd at time of calving.
- Person not at hand during or immediately following calving.
- Colostrum not fed quickly enough after birth.
- Separate calf housing not available.
- Improper feeding of animals.
- Delayed breeding so that 1st calving occurs after 24 months of age.

TABLE 3. RECOMMENDED AGES AND BODY WEIGHT FOR BREEDING DAIRY HEIFERS

Breed	Breeding age	Body weight
Ayrshire	13-15	600 - 700
Brown Swiss	15-18	750 - 900
Guernsey	13-15	550 - 650
Holstein	14-17	750 - 900
Jersey	13-15	500 - 600
Milking Shorthorn	14-17	750 - 850

SUMMARY

Plans for dairy herd improvement are only as effective as the use made of them. The meeting of objectives such as those outlined here will not be accomplished without day-to-day management to produce change toward those objectives. Oklahoma dairymen cooperating in a demonstration herd project have made increases well above the annual objectives set here. DHI Records are intended for dairy herd improvement. Use them for that objective.

16
GENETIC INDEXES USED IN DAIRY COW AND SIRE SELECTION

Jack D. Stout

The primary objective of any breeding program should be
to produce cows with the greatest possible genetic capability to make a profit. The fulfillment of this objective
requires cows that can produce large amounts of milk efficiently and with a minimum of care, cows that can compete at
the feed bunk, and cows that will remain strong and healthy
throughout the rigors of a long and productive life. This
goal can be attained through a balanced breeding and
replacement selection program that improves the genetic
merit of a herd for economically important traits.

Figure 1 is a good example of a performance pedigree
that would meet the goals of most dairymen. My goal in this

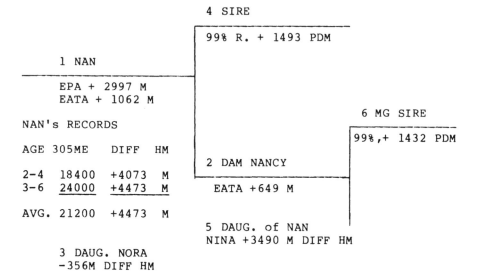

```
                        4 SIRE
                        ────────────────────
                        99% R. + 1493 PDM
        1 NAN
        ────────────────────
        EPA + 2997 M
        EATA + 1062 M
                                            6 MG SIRE
NAN's RECORDS                               ────────────────────
                                            99%,+ 1432 PDM
AGE 305ME     DIFF   HM
                        2 DAM NANCY
2-4   18400   +4073   M  ────────────────────
3-6   24000   +4473   M    EATA +649 M

AVG. 21200    +4473   M
                        5 DAUG. of NAN
                        NINA +3490 M DIFF HM
        3 DAUG. NORA
        -356M DIFF HM
```

Figure 1. Pedigree of NAN

paper is to provide an understanding of the information listed in such a pedigree and of how to use that information. For simplicity, only milk is listed; however, the same information is also available for Fat, Fat %, Dollars, and Type.

The sire and dam contribute equally to the genetic makeup of each offspring (figure 2). However, the sire has a much greater influence on the genetic merit of a herd because of the number of offspring they may have and the accuracy with which we can predict a sire's true genetic merit. With artificial insemination (AI), the sire can be selected from among the outstanding sires in the entire country, whereas the cows selected must constitute the majority of those in a single herd.

Figure 3 shows how the use of superior sires for several consecutive generations can improve the genotypic makeup of a herd. In each generation, one-half of the remaining portion of the original genotypes is replaced by the genes from the superior sire. Therefore, after only three generations of breeding to superior sires, seven-eighths of the herd's genetic makeup comes from the sire and only one-eighth of the original genotypes remain. This is the basis for the enrollment ID program of many of the dairy breeds.

To assure that superior herd replacements are purchased or the best replacement heifers are developed, dairymen must have the knowledge and understanding of genetic measurements or indexes available for sire and cow evaluation. Measurements used in sire evaluation are Pedigree Index (PI) for young bulls and Predicted Difference (PD) and Repeatability (R) being used for bulls with drug information. Cows are evaluated on their own record, PI, Estimated Producing Ability (EPA), and Estimated Average Transmitting Ability (EATA). Young animals are evaluated by PI or Pedigree Estimate of Breeding Value (PEBV).

SIRE EVALUATION

Predicted Difference

Since there is no measure of a sire's individual performance, his evaluation is based on the performance of his daughters. The USDA provides estimates of genetic transmitting ability of dairy sires in the form of Predicted Differences (PD). The base, or zero point, for PDs is the merit of the average sire for the cows included in the Modified Contemporary Comparison (MCC) Sire Summary, Fall 1974. First calving dates for cows summarized at that time ranged from 1960 to 1974 and averaged about 1968. A zero PD bull then is an average sire for cows calving about 1968, depending on breed. The exact date is unimportant, but the fact

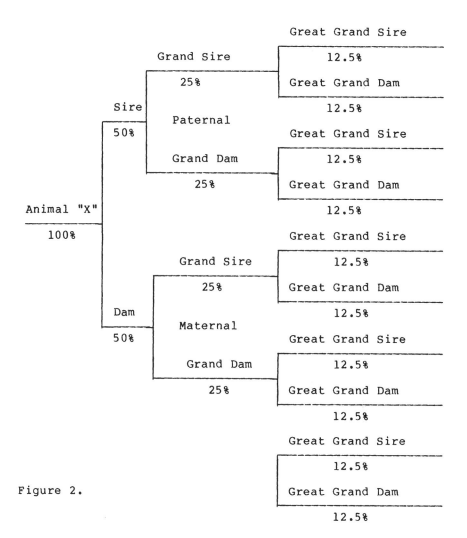

Figure 2.

148

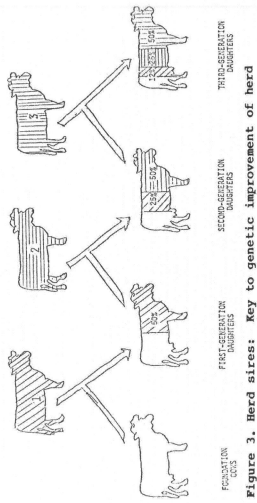

FOUNDATION COWS   FIRST-GENERATION DAUGHTERS   SECOND-GENERATION DAUGHTERS   THIRD-GENERATION DAUGHTERS

Figure 3. Herd sires: Key to genetic improvement of herd

that all MCC sire summaries to date have been adjusted to the same base is important. PDs computed with that base are correctly referred to as PD74.

The procedure for computing PD74 accounts for genetic differences both within and across time. the formula can be writen as:

$$PD74 = R(D - MCA + SMC) + (1 - R)GA$$

where:

R is the repeatibility of the progeny test.

D is the weighted average production for daughters.

MCA is the weighted average production for modified contemporaries.

SMC is the weighted average sire merit for modified contemporaries.

GA is the genetic group average or the average merit of bulls with similar pedigrees.

A bull is expected to have the same PD74, regardless of when he is evaluated. For a later evaluation, D-MCA would be expected to be smaller, because the modified contemporaries are sired by genetically superior bulls compared to earlier contemporaries (but this should be offset by the increase in SMC). Thus, PDs for a given bull are expected to be the same whenever he is summarized. Also, PDs for different bulls are directly comparable even though the bulls were summarized at different times. (This would not hold true for PD$ as the dollar value of milk changes from year to year). Since differences in SMC between sires are accounted for, sires are fairly compared even though they may have been evaluated in herds that have used sires of greatly different genetic merit. Accounting for competition is a feature of PD74 and is necessary to evaluate all bulls relative to the common base.

Expected genetic differences among the sires themselves also are considered in PD74. These differences are determined from a bull's pedigree through the use of GA. GA has, in part, accounted for a genetic trend in sires evaluated because younger bulls tend to have higher pedigree values than older bulls. However, SMC is an essential element in accounting for changes in genetic merit of contemporaries across time.

Key features of PD74 are the consideration of competition and the expected merit of the sire as suggested by the pedigree. These features are necessary to capitalize on the benefits of the common genetic base. Under PD74, sires can be compared directly and fairly regardless of the date of summary, level of competition, or ancestry.

Practical Use of Sire PD

In choosing sires, the only important consideration is the difference between sires. Unless your herd has the same genetic base as PD74 (which is unlikely), it is difficult to know how much improvement to expect in your herd from the use of any certain bull. You can, however, expect to realize differences in milk and fat production between daughters of different bulls used in your herd by the amount of the differences between the bulls in PD Milk and PD Fat, respectively. Daughters of a +2000 lb PDM bull can be expected to produce 500 lb more milk in your herd on a mature equivalent (ME) basis than daughters of a +1500 lb PDM bull. Therefore, use PDs to compare the difference between bulls rather than as a measure of their absolute level.

Research by Barr and Allaire, Ohio (table 1) indicates that as a bull's PDM increases by 100 lb, the percentage of daughters that outproduce herdmates increases by 2%. The level of PDM used in your herd must increase each year to

TABLE 1. EXPECTED PERCENT OF DAUGHTERS EXCEEDING HERDMATES*

Predicted difference	Ayrshire Jersey Guernsey	Brown Swiss Holstein
+1400	76	72
+1000	69	66
+ 800	66	63
+ 600	62	59
+ 400	58	56
+ 200	54	53
0	50	50
- 200	46	47
- 400	42	43
- 600	38	40
- 800	34	37
-1000	31	34

* Barr and Allaire, Ohio.

keep ahead of genetic trend, which increases approximately 100 lb per year. As an example, the averge of all Holstein sires offered through AI studs in 1974 was +368 PDM, +9 PDF, +39 PD$ compared to January 1982 average of +1170 PDM, +33 PDF, and +138 PD$.

Repeatability

Repeatability (R) is a measure of the reliability of the estimate of the transmitting ability of PD. The repeatability of a sire's summary varies in magnitude depending on the following information available:

No. of daughters
No. of herds
Distribution of daughters across herds
Records per daughter

Days in milk of records in process
No. of herdmates
No. average R of sires of herdmates
The addition of more daughters in one herd increases the R very little. The additional representation of more herds in the sire's summary makes a greater increase in R. This should be expected because more dairymen are involved, more feeding programs, milking systems, facilities, types of management, and (especially) more sets of herdmates. Sire summary R will vary from 15% to 99%. At low levels, we have some indication of the sire's true transmitting ability, while at the higher levels we may be reasonably certain that the summary has accurately evaluated the sire. Most AI organizations will sufficiently sample a bull so that his early proof will have R values of 50% or higher.

Practical Use of Repeatability

An important use of R is to set "confidence limits" on an estimate of transmitting ability. Sixty percent and 80% confidence intervals for PDM at given R levels are listed in table 2. The confidence limits show the range within which a bull's true transmitting ability lies, with a specified probability. Since these confidence intervals are not dependent on PD, the same confidence interval applies to all bulls at a given level of R regardless of PD. Therefore, for Holstein and Brown Swiss bulls at 90% R, 60% would have true transmitting abilities within +146 lb of their PDM and 80% would have true transmitting abilities within +223 lb of their PDM. This is true no matter what level their PDM may be.

COW EVALUATION

Cow's Own Record

To properly evaluate the producing ability of a cow, we need to adjust her performance records for the things that we know affect her performance and mask her real producing ability. For example, how good is a 16,000 lb cow? Before answering, more questions need to be asked. For example, what breed was the cow? How old was the cow? How did she produce in relation to herdmates? And what was the length of lactation? Performance records are adjusted for factors influencing performance in dairy record progress as shown in table 3.

Herdmate Comparison

Most Dairy Herd Improvement (DHI) reports list the Herdmate Comparison as Difference from Herdmates (DIFF HM). This value represents a comparison of the individual cow's record with average of Herdmates. Herdmates are identified

TABLE 2. SIXTY PERCENT AND 80% CONFIDENCE INTERVALS FOR PREDICTED DIFFERENCES AT GIVEN REPEATABILITY LEVELS FOR BROWN SWISS AND HOLSTEINS

Repeatability	60% confidence interval	80% confidence interval
(%)	(lb)	(lb)
20	±414	±630
30	±387	±589
40	±358	±546
50	±327	±498
60	±293	±446
70	±253	±386
80	±207	±315
90	±146	±223
Ayrshire, Guernsey, and Jersey		
20	±301	±458
30	±281	±429
40	±261	±397
50	±238	±362
60	±213	±324
70	±184	±281
80	±150	±229
90	±106	±162

Source: Hoard's Dairyman 113:993, 1968.

as other cows within the same herd of the same breed that calve in the same year and same season; before comparisons are made all records are adjusted to 305 days in length, two times per day milking, and the average age of freshening (305D-2x-ME).

Herdmate Differences are especially valuable and meaningful because they minimize environmental effects, e.g., season and management. When purchasing cows, it is more important to look at DIFF HM than individual level of milk production. Herdmate Difference evaluates the cow's response to that management.

Estimate Producing Ability

The EPA is the best estimate of a cow's ability to produce under the condition of her previous environment and is based only on the cow's own records. EPA summarizes the past performance of the cow and estimates future performance based on the appropriate weighing factor for the number of lactations completed. Table 4 gives the weighing factors for cows with 1 to 10 lactations.

TABLE 3. EVALUATING THE COW'S ABILITY TO PRODUCE*

What affects a cow's production?	What can we do about it?
1. Length of lactation 2. Times milked per day 3. Age	Standardize the records to a 305-2x-M. E. basis[1]
4. Breed 5. Herd 6. Year 7. Season	Compare the cow's production to that of her herdmates[2]
8. Temporary Conditions (affect only one lactation) Calving interval Days dry Sickness or injury Chance and other	Average her records[3]
9. Permanent conditions (affect all of her lactation) 10. Genetic merit	The average difference from herdmates weighted for the repeatability of that number of records is her Estimated Producing Ability (EPA)[4]

*Eastwood, Iowa

[1] The lactation records are standardized to 305 days in length, twice a day milking, to an average month of freshening, and to a mature equivalent or prime of life basis so that these three won't hinder a comparison between cows.

[2] The herdmates of a cow are all other cows of that breed that calved in that herd during that year and season.

[3] If a cow has only one record it is impossible to "average" and this is the reason we don't know as much about first lactation cows as we do about a cow with several records.

[4] EPA = N÷N+1 (difference from herdmates) where N = number of records.

The formula for calculating repeatability of milk records is N÷N+1 where N = the number of lactations. Thus, the repeatability of one record is 1÷2 = .5, three records = 3 ÷ 4 = .75 etc.

EPA estimates are listed on DHIA - 203 Pedigree and Performance Evaluation and DHIA - 204 Herd Ranking and Summary. "NAN" in figure 1 has an average DIFF HM of 4473 with an EPA of 2997 lbs.

The calculations would be:

EPA = DIFF HM N÷N+1 = 4473 x .67 = 2997

EPA of two lactations averaging 4473 lb above Herdmates would be +2997 lb (4472 x .67 = 2997).

TABLE 4. EFFECT OF INCREASING LACTATION NUMBERS ON ACCURACY OF PREDICTING FUTURE MILK PRODUCTION RECORDS

Number of records	1	2	3	4	5	6	7	8	9	10
Weighing factor	.50	.67	.75	.80	.83	.86	.88	.89	.90	.91

Cow Index

Estimate Average Transmitting Ability. The Cow Index (CI) is the information on a cow equivalent to the PD of a bull. Cow Indexes are called by various names depending on the center doing the computing. The Cow Index (CI) as figured by USDA or Breed Association, or the ETA or EATA as calculated by DHI data processing centers, is a measure of the cow's ability to transmit. Estimated Average Transmitting Ability (EATA) is the term used as calculated by the Mid-States Center, which uses five sources of information if all are available: (1) the cow, (2) her paternal half-sibs (other daughters by the same sire), (3) her daughters, (4) her dam, and (5) maternal grand sire.

Information from each source is weighted in accordance with how much the information contributes to the EATA. The Cow Index calculated by USDA or Breed Associations only uses the cow's own records plus information on her Sire and Maternal Grand Sire. Adding the other two sources increases accuracy from 5% to 7%, but would be too massive a calcultion for USDA to run for all cows. Proper weighting factors for each pedigree sources are listed in table 5.

EVALUATION OF YOUNG ANIMALS

Pedigree Estimate of Breeding Value

The Pedigree Estimate of Breeding Value for heifer evaluation (EBV) provided by most of the dairy record processing centers considers that both sire and dam contribute

TABLE 5. WEIGHTS TO BE APPLIED TO INFORMATION ON RELATIVES TO A COW FAMILY WHEN ESTIMATING THE TRANSMITTING ABILITY OF ONE OF THE FAMILY MEMBERS

INFORMATION	1 Cow's record	2 Dam's record	3 Dau.	4 Paternal ½ sisters	5 Maternal ½ sisters	Repeat- ability
Full family pedigree, A, B, C-3 Dau., D-100 PS E-3MS	.085	.040	.100	.315	.045	49%
Same as full family except only 1 dau. at C.	.090	.045	.035	.340	.050	46%
Same as full family except only 10 PS at D	.095	.040	.115	.140	.045	43%
Same as full family except only 1 PS at D	.100	.035	.120	.020	.040	39%
Partial family. Dam record mission	.090		.105	.310	.060	46%
Partial family. No daughters	.095	.045		.355	.055	44%
Partial family. No paternal ½ sisters	.100	.035	.125		.045	38%
Partial family. No maternal ½ sisters	.085	.045	.105	.315		48%
Cows record + 100 PS	.105			.345		38%
2 records on cow	.180					36%
1 cow record only	.125					25%

equally to the offspring. The estimates of these contributions are the PD of the Sire and Cow Index (CI) of the Dam.

The EBV formula would be:

Pedigree EBV = Sire's PD + Dam's CI

Referring to figure 1, a daughter of cow "NAN" would have an EBV of 2187.

The calculations would be as follows:

EBV = Sire's PD + Dam's CI
$$= +1493 + 694$$
$$=+2187$$

This estimate looks rather high, but we must remember that for Sires (PD) and Dams (CI) we are dealing with transmitting abilities not breeding values. An animal transmits a sample half of its genes, therefore the breeding value has been divided by 2 to obtain transmitting ability. In this example, the best estimate of the future CI would be +1094.

A study of the 300 Holstein heifers entering the Oklahoma State University herd from 1968 through 1978 provides good indication that EBVs have validity (table 6). The heifers were divided into high, medium, and low groups based on Predicted Pedigree EBV. Following the completion of first lactation, the EATA of each group was averaged. The Difference in predicted EBV between the high and medium group was 742 lb and 736 lb between the medium and low groups. Actual differences between the groups were 703 and 710 lb, respectively.

Pedigree Index

When selecting heifers or cows from non-DHI tested herds, the Pedigree Index (PI) would be the best method to use. The PI considers the Sire and Maternal Grand Sire (MGS) of the animal. Since a high proportion of dairymen use artifical insemination (AI), Sire PD's are the most common information available. The Pedigree Index is calculated using one-half of the Sire PD plus one-fourth of the Maternal Grand Sire PD.

In figure 1, the Pedigree Index of cow "NAN," assuming that only the PDs of the two Sires were available, woud be +1104M. The calculation would be:

PI = 1/2 Sire PD + 1/2 MGSPD
$$= 1/2 (1493) + 1/4 (1432)$$
$$= 746 + 358$$
$$= 1104$$

PI is the estimate of one-half the Breeding Value or PD.

SUMMARY

We have discussed many indexes to use in evaluating sires and cows for herd improvement. As you use these indexes, be sure to select the one with the most complete

and current information. For example, there would be no
need to use a Pedigree Index on a cow after she had
completed DHI records and an EATA. For maximum genetic
improvement, sires must be selected with PD levels well
above breed average. Dairymen must continually breed for
improved production, or genetic trend will pass them by.

TABLE 6. OKLAHOMA STATE UNIVERSITY HOLSTEIN BREEDING VALUE
STUDY 1968-1978

Group	No. cows	Average EBV	Difference between groups
High 1/3	98	1142	
Medium 1/3	102	400	742
Low 1/3	98	-336	736

Actual Average 305-2x ME Difference from Herdmates

Group	No. cows	Average EBV	Difference
High 1/3	98	668	703
Medium 1/3	102	-35	710
Low 1/3	98	-745	

FORM AND FUNCTION OF DAIRY CATTLE

Jack D. Stout

Milk production and reproduction are the primary functions of dairy cows. Cows that will perform these functions profitably and be pleasing to the eye--have a beautiful form--are the answer to a dairyman's prayers. Beautiful "show-ring-winning-form" and profitable "on-farm-function" in the early days seemed to be at opposite ends of genetic realities. Today the most successful dairymen combine form and function by breeding for production--then culling for type and production.

Dairy cattle traits, both those for production and for type, are sufficiently heritable so that progress can be accomplished through selection. Table 1 shows the heritability estimates for various dairy cattle traits. Type traits

TABLE 1. HERITABILITY OF DAIRY CATTLE TRAITS

Trait	Heritability	Trait	Heritability
Production			
Milk	.25	Fat	.25
Fat %	.50	Protein %	.50
Type (score card)			
Final score	.31	General appearance	.29
Dairy character	.19	Body capacity	.27
Mammary system	.22		
Descriptive type			
Stature	.51	Feet	.11
Head	.10	Fore udder	.21
Front end	.12	Rear udder	.21
Back	.23	Udder support	.21
Rump	.25	Udder quality	.00
Hind legs	.15	Teat placement	.31

Source: J. M. White. Holstein Science Report (1976).

of Final Score (FS), General Appearance (GA), Dairy Character (DC), and Mammary System (MS) were the basis for breed classification and show-ring judging for approximately 20 years. The Unified Score Card was developed and adopted by the Purebred Dairy Cattle Association in 1946. Descriptive Type Classification was developed around 1968, which was a step toward making type classification fit a more functional design of the cow.

To put the dairy cattle geneticist's dilemma in proper perspective we need to consider the correlation between type traits and milk production (table 2). These correlations were developed using progeny-test averages for Predicted Difference (PD) Milk and PD Type from 435 Holstein bulls having an average of 393 officially classified daughters. All correlations between milk and type were negative except DC. The correlation between udder support and PDM was nonsignificant. These results do not imply that the breeder should not select for both type and production. They do indicate, however, that he would have fewer bulls to choose from. Less than 20% of the bulls listed in the various AI Sire Directories are sufficiently plus in all categories PDM, PDF, PDT, and PDS. Even though type is negatively correlated with production, it must not be ignored in the breeding program. To do so would allow weaknesses in conformation to develop in some of the functionally important traits.

TABLE 2. CORRELATIONS BETWEEN PD MILK AND TYPE

Trait	Correlation	Trait	Correlation
Final score	-.23	Back	-.16
General appearance	-.24	Rump	-.23
Dairy character	.41	Hind legs	-.15
Body capacity	-.22	Feet	-.16
Mammary system	-.24	Fore udder	-.36
		Rear udder	-.14
Stature	-.11	Udder support	-.08
Head	-.10	Udder quality	-.13
Front end	-.19	Teats	-.09

Source: J. M. White. Holstein Science Report (1976).

High "type" cattle are of economic importance to some breeders due to their show-ring activities and the sale of breeding stock. But for type to be important to all, it must influence profitability. Length of herd life and total milk yield for that period was correlated to the Descriptive Type Traits of the Holstein classification program by Vinson, et. al., VPI. A portion of the results of that study are listed in table 3.

161

TABLE 3. WHAT DO INDIVIDUAL TYPE TRAITS CONTRIBUTE TO
LIFETIME PRODUCTIVITY?

| | | | Contribution to: | |
Type trait	Code	Description	Herd life	Lifetime milk
Stature	1	Upstanding	-15	-562
	2	Intermediate	8	49
	3	Low set	7	511
Front end	1	Smooth, wide, strong	-8	-465
	2	Medium	9	346
	3	Coarse	-30	-1089
	4	Narrow, weak	29	1210
Rump	1	Long, wide, nearly level	-29	-1166
	2	Medium	6	234
	3	Pins higher than hips	-8	-245
	4	Narrow	17	653
	5	Sloping	14	527
Hind legs	1	Straight, clean, squarely placed	5	96
	2	Acceptable	2	31
	3	Sickled or close to hock	-6	-152
	4	Bones too light	22	477
	5	Too straight	-23	-452
Udder support	1	Strong, ligament clear halving	66	2366
	2	Lacks defined halving	26	844
	3	Floor too low	-22	-437
	4	Tilted	4	723
	5	Broken	-74	-3494
Rear udder	1	Firmly attached, high, wide	51	2670
	2	Intermediate	22	871
	3	Low	1	112
	4	Narrow and ponched	1	-412
	5	Broken	-76	-3241

The study covered a period of seven years using 35,000
cows with the average cow staying in the herd 1010 days and
producing 38,995 lb of milk. Results were that the desirable type traits were not always the most economical for the
dairymen in terms of length of herd life or total yield.
For example, cows that were upstanding in stature (Code 1)
but otherwise breed-average for type remained in the herd 15
fewer days and produced 562 lb less milk than did the average cow. Similar reduction in economic returns were found
in each of the nonudder traits. Improvement of udder traits
did result in longer herd life and increased milk yield.
Code 1 udder-support cows, for example, with a strong ligament and clear halving, remained in the herd 66 days longer
and produced 2366 lb more milk than did the average cow.

TABLE 4. CONTRIBUTIONS OF UDDER CODES TO FINAL SCORE
(POINTS), HERD LIFE (PERCENT CHANGE) AND
LIFETIME MILK YIELD (PERCENT CHANGE)

			Contribution to:		
Trait	Code	Description	Final score	Herd life	Lifetime milk
			%		%
Udder support	1	Strong ligament, clear halving	.3	6.6	6.1
	2	Lacks defined halving	-.4	2.6	2.2
	3	Floor too low	-1.4	-2.2	-1.1
	4	Tilted	-1.3	.4	1.9
	5	Broken ligament, weak floor	-4.8	-7.3	-9.0
Rear udder	1	Firmly attached, high and wide	1.8	5.1	6.9
	2	Intermediate height and width	.6	2.2	2.2
	3	Low	-.6	.1	.3
	4	Narrow & pinched	-1.2	.1	-1.1
	5	Loose and/or broken	-3.4	-7.5	-8.3
Fore udder	1	Moderate length, firmly attached	1.5	4.2	3.6
	2	Moderate length, slightly bulgy	.4	3.4	2.3
	3	Short	-.8	.7	.6
	4	Bulgy or loose	-.8	-1.3	-1.5
	5	Broken and/or very faulty	-4.1	-6.9	-5.0

Source: W. E. Vinson. Hoard's Dairyman, July 25, 1980.

Table 4 presents data showing the gain in herd life and production by moving udder traits toward the "desirable" Code 1. The effect of change on udder traits through selective mating was an encouraging part of Vinson's data. When cows with broken fore udders (Code 5) were mated to Code 1 sires (sires with high PD for Code 1), 55% of the resulting daughters were coded "desirable," in fore udder. Even though breed average for fore udder was 69% "desirable," this was a substantial one-generation improvement. Data such as this was a boon to the AI organizations' special mating programs.

Table 5 lists how a change of one of each of the descriptive-type traits will affect the final type score, herd life, and lifetime milk yield. In general, the effects of individual type traits on lifetime productivity in the herd are rather small. This indicates very strongly that attention given to improving even the most important type components should not be at the expense of milk yield for cows, or of PD milk for sires.

TABLE 5. EFFECT ON HERD AVERAGE FINAL SCORE (POINTS), HERD LIFE (PERCENT) AND LIFETIME MILK (PERCENT) OF CHANGING ALL COWS IN HERD TO CODE 1 FOR EACH TRAIT

Trait	Effect on:		
	Final score	Herd life	Lifetime milk
		%	%
Udder support	.4	1.7	.6
Rear udder	1.5	3.2	2.2
Fore udder	1.4	2.0	1.0
Teats	.5	1.9	1.1
Hind legs	.9	.6	.3
Feet	.2	-.9	-.9
Rump	1.0	-3.2	-1.5
Back	.4	-.7	-.2
Front end	.6	-1.3	-.7
Head	.7	.8	.3
Stature	.8	-1.2	-.4
Miscellaneous*	.1	.0	.0

*Effect of changing all cows to code 1, no miscellaneous code.
Source: W.E. Vinson. Hoard's Dairyman, July 25, 1980.

Most breed associations and AI organizations have adopted the Linear Scoring System for type evaluation. Each has made some minor adaptations to fit its particular use, but the basic principles are the same. The Linear Scoring System is simply the use of numerical scores to evaluate traits from one biological extreme to another, i.e., tall to low set, sharp to thick, strong to weak, wide to narrow, sickled to posty, low pins to high pins, etc. Several numerical systems are currently being used (1-9, 50-99, 1-50); however, the concept remains the same.

The Linear Scoring System is now an integral part of herd improvement programs sponsored by any of the industry organizations. It is the most accurate means of type evaluation developed to date. Using a numerical scale to evaluate the degree of trait, not judging what is best, allows for developing sire PD for a particular trait the same as for milk. Genetic profiles or trait summaries have been developed by each organization that fit their particular scale. However, these scales are easily converted. Table 6 is a Genetic Trait Summary from one of the leading AI organizations. On their scale (50 is breed average) bulls can easily be ranked and compared to each other and breed average by trait.

TABLE 6. S-W-D VALIANT GENETIC TRAIT SUMMARY, JULY 1981

30 Daughters	27 Herds	60% Rpt.
Fore udder	50	
Rear udder height	52	High
Rear udder width	53	& wide
Udder center support	51	Strong support
Udder depth	52	
Teat placement	52	Good
Teat size	49	
Stature	58	Extreme scale
Strength and capacity	55	& strength
Width of rump	53	Wide
Pelvic angle	46	High at pins
Rear leg set	50	Correct

Source: ABS Autumn 1981 Holstein Sires.

Rear legs, rump, and udder are conformation areas that have the most influence on productivity (table 3). These areas also need more than one measurement to accurately describe the condition, such as height and width, or width, height and strength. Linear Scoring divided the areas so that each measurement could be made separately. Four of the Linear traits are now considered two-way traits. That is, a score in the intermediate range is the most desirable. These traits are: Rear legs (side view), Pelvic Angle, Teat Placement (rear view) and Udder Depth (related to point of hock). Legs, as an example, need some set to be correct. The intermediate range is desired when measuring for a biological extreme of posty (too straight) to extremely sickled. A corresponding effect on milk yield (table 3) would be -452 lb for too straight to +96 lb for straight, squarely placed for a total change of +548 lb milk. Udder traits are much more pronounced but more difficult to separate out of the data.

The 13 Basic Form Traits of the Linear Scoring System
with artist sketch follows:

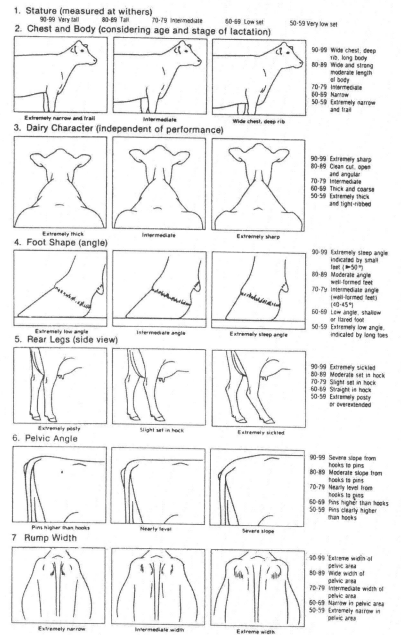

1. Stature (measured at withers)
   90-99 Very tall   80-89 Tall   70-79 Intermediate   60-69 Low set   50-59 Very low set

2. Chest and Body (considering age and stage of lactation)

90-99 Wide chest, deep rib, long body
80-89 Wide and strong moderate length of body
70-79 Intermediate
60-69 Narrow
50-59 Extremely narrow and frail

Extremely narrow and frail   Intermediate   Wide chest, deep rib

3. Dairy Character (independent of performance)

90-99 Extremely sharp
80-89 Clean cut, open and angular
70-79 Intermediate
60-69 Thick and coarse
50-59 Extremely thick and tight-ribbed

Extremely thick   Intermediate   Extremely sharp

4. Foot Shape (angle)

90-99 Extremely steep angle indicated by small feet (>50°)
80-89 Moderate angle well-formed feet
70-79 Intermediate angle (well-formed feet) (40-45°)
60-69 Low angle, shallow or flared foot
50-59 Extremely low angle, indicated by long toes

Extremely low angle   Intermediate angle   Extremely steep angle

5. Rear Legs (side view)

90-99 Extremely sickled
80-89 Moderate set in hock
70-79 Slight set in hock
60-69 Straight in hock
50-59 Extremely posty or overextended

Extremely posty   Slight set in hock   Extremely sickled

6. Pelvic Angle

90-99 Severe slope from hooks to pins
80-89 Moderate slope from hooks to pins
70-79 Nearly level from hooks to pins
60-69 Pins higher than hooks
50-59 Pins clearly higher than hooks

Pins higher than hooks   Nearly level   Severe slope

7 Rump Width

90-99 Extreme width of pelvic area
80-89 Wide width of pelvic area
70-79 Intermediate width of pelvic area
60-69 Narrow in pelvic area
50-59 Extremely narrow in pelvic area

Extremely narrow   Intermediate width   Extreme width

Figure 1. Appraisal standards and scoring ranges for 13 functional type traits

166

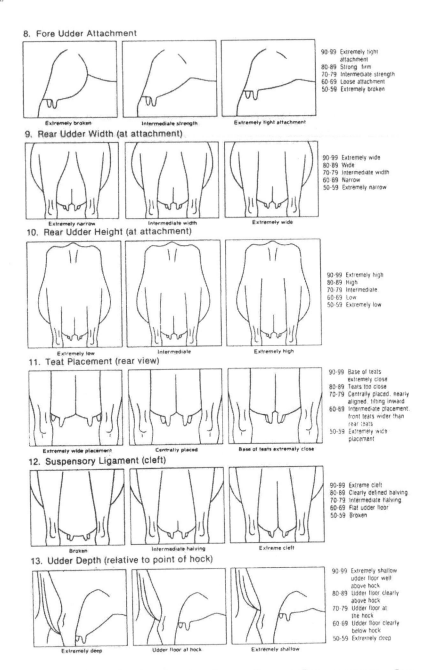

8. Fore Udder Attachment

90-99 Extremely tight attachment
80-89 Strong firm
70-79 Intermediate strength
60-69 Loose attachment
50-59 Extremely broken

Extremely broken / Intermediate strength / Extremely tight attachment

9. Rear Udder Width (at attachment)

90-99 Extremely wide
80-89 Wide
70-79 Intermediate width
60-69 Narrow
50-59 Extremely narrow

Extremely narrow / Intermediate width / Extremely wide

10. Rear Udder Height (at attachment)

90-99 Extremely high
80-89 High
70-79 Intermediate
60-69 Low
50-59 Extremely low

Extremely low / Intermediate / Extremely high

11. Teat Placement (rear view)

90-99 Base of teats extremely close
80-89 Teats too close
70-79 Centrally placed, nearly aligned, tilting inward
60-69 Intermediate placement, front teats wider than rear teats
50-59 Extremely wide placement

Extremely wide placement / Centrally placed / Base of teats extremely close

12. Suspensory Ligament (cleft)

90-99 Extreme cleft
80-89 Clearly defined halving
70-79 Intermediate halving
60-69 Flat udder floor
50-59 Broken

Broken / Intermediate halving / Extreme cleft

13. Udder Depth (relative to point of hock)

90-99 Extremely shallow udder floor well above hock
80-89 Udder floor clearly above hock
70-79 Udder floor at the hock
60-69 Udder floor clearly below hock
50-59 Extremely deep

Extremely deep / Udder floor at hock / Extremely shallow

Figure 1. **Appraisal standards and scoring ranges for 13 functional type traits (con't)**

The American Jersey Cattle Club was the first breed association to adopt the Linear Scoring System. They have just published (July 25, 1982) the first correlations of Linear Scoring to milk production. Their study compared the score and official production records of Elite Cows to all cows that had been scored. Elite cows are those that have Cow Index Scores for production ranking them in the top 2% of the breed.

Table 7 indicates to me that Form and Function of dairy cattle have now been coordinated. Elite cows of the Jersey breed had a higher Performance Index--226.2 compared to 85.5 for all cows. (The Total Performance Index is based on a ratio of 3:1:1 for milk, fat %, and type, respectively.) Elite cows also had scores that were higher than breed average for 10 of the 13 traits. The -1.0 in Fore Udder and -1.7 in Udder Depth indicate that the higher-capacity udders produce more milk. However, the current scoring system lets us know that degree.

TABLE 7. APPRAISAL AVERAGES OF UNIFORM FUNCTIONAL-TYPE TRAITS FOR ELITE COWS AND FOR ALL JERSEYS IN 1980-1981

	Elite cow	All cows	Difference	Cow index of elite cows
Number of cows with UFTTs	926	40,287		
Number of cows final scores	1,044	40,287		
Cow index milk	1,254			
Cow index fat	49			
Cow index fat %	-0.14			
Cow index dollars	171			
Cow performance index	226.2	85.5		
Stature	77.1	75.2	1.9	0.6
Chest and body	82.6	81.2	1.4	0.4
Dairy character	86.9	82.5	4.4	1.0
Foot shape (angle)	68.0	68.0	-0.3	-0.2
Rear legs (side view)	80.7	80.3	0.4	0.3
Pelvic angle	78.5	78.0	0.5	0.2
Rump width	79.8	78.1	1.7	0.3
Fore udder attachment	78.6	79.6	-1.0	-0.2
Rear udder width	81.2	79.4	1.8	0.3
Rear udder height	83.2	81.9	1.3	0.2
Teat placement (rear view)	68.3	68.1	0.2	0.0
Suspensory ligament	81.6	79.8	1.8	0.3
Udder depth	78.8	80.5	-1.7	-0.4

Source: B. Norman, Jersey Journal, July 25, 1982.

The past 80 years of development and study of methods measuring the dairy cow's Form has been quite interesting. The show-ring had the most influence until the AI industry developed a Linear Scoring System that could be adapted to computer mating. This system was used on all cows, grade and registered, which gave a much larger data bank. The increase in information proved that sires must be selected for minimum levels of milk as the basic criteria, then for functional traits to improve conformation. Linear Scoring is now accepted by all breed associations with slight modification of miscellaneous points and the addition of a final classification score to fit show-ring approval.

An Oklahoma-developed cow, Ball's Pollyboots Demand, a two-generation Gold Medal Sire (GMS) mating by American Breeders Service (ABS), that developed into the top Index Cow and highest selling cow of the breed--represents what all of the industry wants from improved breeding systems. Form and Function must go together and complement each other. Linear Scoring makes this happen. One AI organization is evaluating approximately a half-million cows per year. The repeat user spending dairy-produced dollars to buy the service is proof of dairyman acceptance and monetary gain from the program.

Sires should be selected on an Index system according to the ratio of type value to the herd's milk income. Most breed associations have a Type Production Index (TPI) as a formula 3:1:1 milk, fat %, and type ratio. In practice most commercial breeders are using a ratio closer to 6:1:1.

Part 4

PHYSIOLOGY,
REPRODUCTION, AND MANAGEMENT

18
HORMONAL REGULATION
OF THE ESTROUS CYCLE

Roy L. Ax

INTRODUCTION

All reproductive events are regulated by hormones. In simple terms, if the organs of reproduction correspond to the "plumbing," the endocrine system can be called the "wiring." A delicate balance exists between the nervous system and the endocrine system. We are entering an era in which artificial control of the estrous cycle with hormones promises to become more commonplace. Thus, hormones can be considered valuable management tools. If producers are to use the tools effectively, we must develop a better understanding of the complex hormonal interrelationships between the hypothalamus, pituitary, and ovary.

HORMONES DEFINED

A classical definition for a hormone is that it is a substance produced in one tissue that is transported to another tissue to exert a specific effect. (Some of the confusion about hormone actions should be clarified in the next section). Hormones have many chemical classifications; some of the most common reproductive hormones are briefly described here. Gonadotropin-releasing hormone (GnRH) is composed of amino acids and is thus a polypeptide in nature. The follicle-stimulating hormone (FSH) and luteinizing hormone (LH) are glycoproteins. This means they are composed mostly of protein, with some carbohydrate attached to the protein. Estrogen and progesterone are steroids that are synthesized from cholesterol. Prostaglandins are produced from a fatty acid--arachidonic acid. The diversity of the composition of hormones leads to the variation in their biological functions. Most hormone concentrations are in billionths or trillionths of a gram per milliliter of blood.

HOW DO HORMONES WORK?

The fact that a hormone is produced by a tissue does not necessarily imply that it will exert a physiological effect somewhere else. The ability of one tissue to respond to a particular hormone rests in whether that tissue possesses a receptor to the particular hormone. A receptor functions as the lock, and the hormone functions as the key that fits the lock. Therefore, as an example, if a tissue is going to respond to estrogen, its cells must possess estrogen receptors. After a hormone is bound to its receptor, a cellular response is initiated in the target tissue. A target tissue may possess receptors for several different hormones, and exposure to the various hormones can modulate the final response.

THE HYPOTHALAMUS

The hypothalamus is located at the base of the brain. It contains nerve endings to integrate sensory information and sorts out hormonal signals as well. The major reproductive hormone of the hypothalamus is gonadotropin-releasing hormone (GnRH) that is sometimes called luteinizing-hormone-releasing hormone (LHRH). For purposes of this discussion we will use GnRH nomenclature. GnRH is transported in blood vessels to the pituitary gland to regulate secretion of FSH and LH from the pituitary.

THE PITUITARY

The pituitary is positioned underneath the hypothalamus directly above the roof of the mouth. The major reproductive hormones produced in the anterior lobe of the pituitary are called gonadotropins, which means to stimulate the gonads. Follicle-stimulating hormone (FSH) and luteinizing hormone (LH) are the two gonadotropins that regulate the ovary. They are secreted by the pituitary and transported in the circulation to the ovary where they interact with their respective receptors to affect ovarian functions. The main action of FSH is to initiate growth of follicles on the ovary. Continued follicle growth depends on the presence of both FSH and LH. The major effect of LH is to promote ovulation, but there is increasing evidence that FSH can exert a major influence to facilitate ovulation.

THE OVARY

The ovary has two biological functions: (1) to provide the eggs (ova) for the female genetic contribution to the next generation, and (2) to produce hormones to coordinate behavioral changes with ovulation and prepare the reproductive tract for pregnancy.

Estrogen is the hormone produced by follicles as they develop on the ovary. As the predominant follicle or follicles approach ovulatory size, the increased amounts of estrogen are transported to the hypothalamus to cause behavioral heat. The pituitary also responds to the elevated estrogen by releasing a surge of LH which leads to ovulation. Thus, estrogen coordinates behavioral acceptance of a male when the egg will be released into the female tract. This is Mother Nature's attempt to ensure that the probability of fertilization occurring is maximized.

After ovulation has occurred, the tissue that a moment ago was a follicle starts a dramatic change into becoming a corpus luteum (yellow body). The corpus luteum produces progesterone to prepare the female tract for a possible ensuing pregnancy. The corpus luteum forms, regardless of whether or not mating occurs in farm animals. If the corpus luteum remains functional due to pregnancy occurring, the sustained production of progesterone prevents cyclicity. Until the corpus luteum regresses, the typical pattern of cyclic hormonal changes is absent.

THE FOLLICLE

In a simple sense, the follicle is the dwelling of the ovum. Ovulation of a follicle is the exception rather than the rule because over 99% of all potential oocytes are never shed from the ovary. This loss is called atresia. Atresia can occur at any time during follicle growth. When animals are injected with gonadotropins to induce superovulation, some follicles that would have undergone atresia are rescued. This supports the hypothesis that continued follicle growth is dependent upon continued exposure to gonadotropins and the presence of gonadotropin receptors in the follicle.

OVULATION

Once the LH surge has been elicited from the pituitary, the follicle starts to undergo a series of changes to prepare for impending ovulation. The cells lining the inside of the follicle begin to luteinize and secrete progesterone as the major steroid rather than estrogen. The oocyte commences its maturational steps to get it in the proper meiotic configuration for chromosome pairing with the meiotic contribution from a sperm.

Enzymes are activated to degrade the follicle wall and permit the egg to pass into the oviduct. Biochemical studies have pointed to FSH being responsible for stimulating production of those enzymes. FSH also promotes the spreading apart of cells that are tightly surrounding the egg, which then leads to some of the subsequent maturational changes in the egg. Prostaglandins are required for normal

174

ovulation. Substances known to inhibit prostaglandin formation prevent ovulation. Due to the enzyme, steroid, and
prostaglandin effects, a hypothesis was formulated comparing
the ovulatory process to an inflammatory reaction.

THE CORPUS LUTEUM

The scar tissue remaining after ovulation becomes the
corpus luteum; this endocrine tissue has been studied extensively. Low amounts of LH from the pituitary are essential
for establishment and continued function of the corpus
luteum. In all livestock species, the corpus luteum functions to maintain early pregnancy by secreting progesterone. Progesterone prevents cyclicity. The placenta of the
developing fetus eventually sustains the pregnancy by producing progesterone in the bovine, equine, and ovine. In the
porcine, the corpus luteum is required to support the entire
gestational period.
If cyclicity is to resume, the corpus luteum must regress and cease progesterone production. In pregnant animals, an embryonic signal leads to maintenance of the
corpus luteum, (discussed in the next section). In nonpreg-
nant animals, the uterus recognizes the absence of an embryo
and secretes prostaglandins. Those prostaglandins are transported to the corpus luteum and cause it to regress. Thus,
the inhibitory effects of progesterone are removed, new follicles start to develop, and heat occurs in a few days.

PREGNANCY RECOGNITION

The corpus luteum continues to function to provide a
signal from the embryo to the dam. These signals are hormones identified in the human as chorionic gonadotropin
(hCG) and in the mare as pregnant mare serum gonadotropin
(PMSG). These two gonadotropins can be used to regulate the
estrous cycle of animals. PMSG is biologically similar to
FSH, and hCG exerts an action similar to LH. Thus, PMSG can
be used to induce follicle growth and hCG will promote ovulation of these follicles. However, both PMSG and hCG are
recognized as foreign proteins by livestock, and the animals
build up antibody resistance to them, if they are used too
frequently.
An ideal pregnancy test would be the identification of
the embryonic signal in the bovine, ovine, and porcine. Experiments have shown that embryonic extracts will maintain a
corpus luteum in an animal that has not been bred. However,
even with sophisticated biochemical tests, specific signals
from embryos in the dam's circulation have not been detected. The livestock industry could benefit significantly from
pregnancy tests of these types if they are ever developed
successfully. Since an embryonic signal would have to be
apparent to prevent a subsequent heat in the dam, a preg-

nancy test would also pinpoint which animals would be returning to heat within a few days.

HORMONAL REGULATION OF THE CYCLE

The preceding sections indicate that follicle growth, ovulation, and corpus luteum formation are a dynamic process of sequential steps in an intricate balance. Administration of a hormone to mimic the effect of that hormone in the animal can be used to regulate the cycle. If hormone administration is to produce the desired result, it must be given at a time that is physiologically compatible with the cycle. The common hormones that have been used experimentally or commercially are progesterone-like drugs (progestins), GnRH, and prostaglandins.

Progestins

These compounds were the first to be experimentally employed to regulate the cycle. Progestins have been injected, fed, implanted, or administered via vaginal sponges. Regardless of what stage of the cycle an animal is in when the progestin commences, cyclic fluctuations in other hormones are arrested. As long as progestin is administered, cyclicity ceases. Removal of the source of the progestin results in renewed follicle growth, and estrus, within a few days. Field trial data suggest that fertility at the first estrus after progestin withdrawal is lowered. Thus, it is usually recommended that breeding be done at the second estrus, since the cycles of the animals will still be in close synchrony.

Prostaglandins

These compounds have largely replaced the use of progestins because (1) only one or two injections are required, and (2) fertility is not affected by use of prostaglandins. Prostaglandins are only effective if the animal possesses a functional corpus luteum. Contrary to some opinions, prostaglandins are not a heat-inducing drug. Rather, they cause a corpus luteum to regress, and the animal secretes her own gonadotropins to regulate the ovary and cause a physiological heat. Success has occurred regularly by breeding animals at a predetermined time after prostaglandin injection. Greater success in conception rates can occur if animals are watched for estrual behavior after receiving prostaglandins and are bred in relation to standing heat. Care must be exercised with prostaglandins, because injections into an animal with a functional corpus luteum sustaining a pregnancy could induce an abortion.

GnRH

GnRH is composed of 10 amino acids. It can now be chemically synthesized in a laboratory, and this has permitted chemists to develop some powerful analogs. There are no noticeable ill effects from administering GnRH; its action is to promote a release of gonadotropins from the pituitary. Maximum gonadotropin output occurs approximately 2 to 4 hours after GnRH injection.

A common use for GnRH is to initiate cyclicity in animals with anestrous. GnRH is the most widely used therapy for treating cystic ovarian degeneration. Cystic ovaries usually result from inadequate gonadotropin production. Thus, GnRH triggers release of gonadotropins to restore ovarian function.

A new use for GnRH is for injection after prostaglandin administration; the interval to the gonadotropin surge, and hence, ovulation, can be coordinated more closely. This reduces the variation in time between prostaglandin injection and standing heat that is ordinarily seen among animals.

We have an ongoing study at the University of Wisconsin to evaluate the efficacy of GnRH injections at the time of insemination in dairy cattle. The heifers receiving GnRH have shown no advantages over heifers receiving the saline control. In lactating cows, administration of GnRH 14 days postpartum or at the first artificial insemination has improved first-service conception rates by 15% to 19%. In cows presented for third service (and thus classified as "repeat" breeders in commercial herds) conception rates were about 30% higher for cows that received the GnRH. The physiological effect elicited by GnRH has yet to be experimentally established. We have postulated that gonadotropins produced in response to GnRH cause a corpus luteum to form that may have otherwise been deficient and led to early embryonic death. GnRH could also promote what would have been a delayed ovulation to occur sooner or have a direct effect on the ovary. The lactational stress imposed on a dairy cow may make her unique to respond to GnRh in this manner. Experiments with other farm animals are needed to determine if similar effects result.

SUMMARY

The reproductive cycle is regulated by fluctuations in different hormones. The cycle can be regulated by administering hormones to mimic the effect that would occur in the animal. Therefore, producers have endocrine tools to assist them in managing their animals. For maximum success the producers must understand how the hormones work biologically and realize that they are powerful drugs. We will see an increasing frequency of producers regulating the reproductive cycle to maximize reproductive efficiency.

REFERENCES

Britt, J. H. 1979. Prospects for controlling reproductive processes in cattle, sheep, and swine from recent findings in reproduction. J. Dairy Sci. 62:651-665.

Britt, J. H., N. M. Cox and J. S. Stevenson. 1981. Advances in reproduction in dairy cattle. J. Dairy Sci. 64:1378-1402.

Foote, R. H. 1978. General principles and basic techniques involved in synchronization of estrus in cattle. Proc. 7th Tech. Conf. on Artif. Insem. and Reprod., Nat'l Assoc. Anim. Breeders, pp 74-86.

Hansel, W. and S. E. Echternkamp. 1972. Control of ovarian function in domestic animals. Amer. Zool. 12:225-243.

Jones, R. E. (Ed.) 1978. The Vertebrate Ovary. Comparative Biology and Evolution. Plenum Press, New York.

Lee, C. N., R. L. Ax, J. A. Pennington, W. F. Hoffman and M. D. Brown. 1981. Reproductive parameters of cows and heifers injected with GnRH. 76th Ann. Mtng. of the Amer. Dairy Sci. Assoc. Abstract P228.

Maurice, E., R. L. Ax and M. D. Brown. 1982 Gonadotropin releasing hormone leads to improved fertility in "repeat breeder" cows. 77th Ann. Mtng. of the Amer. Dairy Sci. Assoc. Abstract P233.

Nalbandov, A. V. 1976. Reproductive Physiology of Mammals and Birds. W. H. Freeman and Co., San Francisco.

19
NECESSARY COMPONENTS FOR A SUCCESSFUL HEAT DETECTION PROGRAM

Roy L. Ax

INTRODUCTION

Accurate detection of estrus (heat) is essential for dairy producers to maximize the benefits derived from using artificial insemination. In general, as herd size increases, the reproductive efficiency decreases. A large part of the reproductive inefficiency can be attributed to a poorer job of heat detection--approximately 50% of all physiological heats are undetected by livestock producers. Because the "weak link" in many herds can be corrected by successful heat detection, this paper reviews the necessary components for optimal results.

IDENTIFICATION

Animals must be identified so that they have a permanent place in herd record books. As herd size increases, it becomes more difficult to rely on memory. The common methods are (1) ear tags, (2) neck chains, or (3) freeze-brands. An advantage of ear tags or neck chains is that information such as sire, dam, and birth date can also be coded. The producer should compare systems used by other producers and then decide on a method that will appear to work best. The method selected should use large-sized symbols that can be easily recognized from a distance.

RECORDS

A good identification program complements a good record-keeping system. Heat-detection records are but a part of the reproductive records that should be part of the overall herd records. Scrawling on a piece of paper towel or the palm of the hand cannot be classified as a valid record system.

Many companies offer a 21-day reproduction calendar that is designed to project anticipated heat dates. Producers should watch heifers and cows prior to the projected return to heat because of the variation in estrous cycle lengths. If an animal is bred, she still requires watching for a possible return to heat--because only 50% conceive to any particular service. In today's economy you cannot wait for a pregnancy check on an animal that has been bred before deciding conception did not occur in the first place.

Another good record-keeping system is a breeding wheel, which encompasses a total reproductive profile for the herd. A particular cow registers on the wheel after she calves. Producers who use a breeding wheel regularly gain two major advantages: (1) they do not have to ruffle through a loose-leaf notebook trying to find a particular record to do a calculation, and (2) they tend to subconsciously learn the reproductive status of the entire herd by glancing at the wheel each time an observation is recorded. This permits the producer to get a quick impression of which cows should be returning to heat after calving so those cows can be watched more closely.

Computer record-keeping systems are now on the market for a reproductive program. Producers should shop around before investing in such a system. If a dairyman has a good record-keeping system now, a computer may be an asset. However, if current records are poor, a computer would not be recommended. The software should have a provision for keeping track of several previous heats, calculate days between heats, and project the next expected heat starting 15 to 16 days after the last recorded heat or breeding date. The most important consideration for a computer system is that the information derived can be used in the daily operations on the farm. Also, check to see if the program can be modified.

SIGNS OF HEAT

Signs of heat vary among cows and for individual cows. One heat may be very intense for up to 18 hours, and a subsequent heat may be very weak for only a few hours. Producers should be aware of the signs associated with heat and cows should be managed so that such signs can be observed. The ideal sign would be that observed when a cow stands to be mounted by a herdmate; however, such observations are impossible when cows are confined to stanchions or tie-stalls. Other indications include: ruffled hair over the rump, restlessness, a pink and swollen vulva, a clear mucus discharge from the vulva, a cow standing up when others are lying down, nudging and sniffing other cows, raising of the tail when contacted by others, and decreased appetite and milk production. A "silent" heat exists when an animal is physiologically in estrus, but has no outward behavioral manifestations.

Ovulation in cattle occurs 10 to 12 hours after standing heat. With all other farm animals, ovulation occurs sometime during heat. Thus, if heat activity is observed,

181

the producer is reasonably certain that the ovum has not
been released from the ovary. This is one reason why artificial insemination has succeeded so well in cattle.

WHEN TO WATCH FOR HEAT

If cows are observed continuously for 24 hours, the
producers' accuracy of heat detection can compare with that
obtained when using a bull. However, most producers cannot
devote that much of their labor to watching for heat; thus
there must be a compromise between heat detection efficiency
and other chores.

Most mounting activity occurs at night, if cows are
allowed to interact continuously throughout the day. Table
1 shows the distribution of mounting activity if cows are
allowed to interact continually.

TABLE 1. MOUNTING ACTIVITY THROUGHOUT THE DAY

Time	% of all mounts
6 a.m. - noon	22
Noon - 6 p.m.	10
6 p.m. - midnight	25
Midnight - 6 a.m.	43

Note that 68% of all mounts were recorded from 6 p.m.
to 6 a.m. This implies that producers would observe only
about one-third of these heat signs by watching during day-
light hours. For best results, animals should be observed
as early, and as late, in the day as possible. In managing
the cows, observing for estrual activity should be part of a
daily routine. Cows should not be fed while being watched
for heat.

Several studies have indicated that the frequency of
mounts escalates if more than one animal is coming into heat
simultaneously. They tend to be more aggressive and to
stand for each other. These behavioral cues are very marked
and can best be observed if the cows are exposed to each
other. Because of the definite behavioral patterns associated with heat, some dairymen feel that free-stall housing
systems lend themselves to improved heat detection. However, the cows still must be watched, and if conditions
underfoot are slippery, no cow will attempt mounts with
herdmates.

HEAT DETECTION AIDS

In the dairy industry, we have seen enormous technology
and automation developments within our feeding and milking

system. Unfortunately, we have not seen parallel changes in the reproductive management program. Practical heat detection aids are orange chalk on the tail base, or mount detectors glued on the back. Various methods have been developed to render bulls functionally sterile. Dairy bulls are often dangerous because of their size and temperament. Masculinized heifers or cows have an advantage in that they have no penis to spread disease and are not so dangerous, although their effectiveness is comparable to that of an altered bull.

The theories behind some new developments on the market are basically sound and have scientific experimental documentation. They include pedometers to measure walking activity; vaginal probes to record changes in electrolytes in mucus; and dogs trained to smell variations between estrual and nonestrual secretions in the female tract or milk. Milking machine companies are interested in the potential for measuring the decrease in milk output associated with estrus, or with some diseases. Finally, although hormones provide the ideal profile of cyclic changes in a cow, the time and expertise required to run the assays preclude using them to predict estrus. On the other hand, hormone levels are an excellent way to determine if a cow was bred when actually in heat. Hormone determinations can give an overview of the heat-detection accuracy in a herd and can provide a base for other educational work.

WHEN TO BREED

Much has been written on the subject of when to breed a cow after she has been detected in standing heat. The general "a.m.-p.m." rule still applies. That is, if a cow is observed in heat in the morning, breed her that evening. If she is detected in heat in the evening, breed her the next morning. Precise recommendations as to when to breed a cow cannot be provided because of the great variation among cycle lengths and estrual intensities. The following diagram illustrates the general scheme of events that occurs around estrus:

	Ovulation ↓			
Event	Standing heat			Life of egg
Time (hrs.)	18	12	12	
		Life of sperm		

Best time to breed ↑

Eggs have a maximum fertile life of 12 hours. Sperm live 20 to 24 hours once deposited in the female tract. About 4 to 6 hours are required for sperm to undergo capacitation before they can fertilize the egg. Therefore, the ideal situation is to have sperm deposited 4 to 6 hours prior to ovulation. A thumb rule has been adapted that it is best to breed as close as possible to the end of standing heat to ensure that the functional fertile lives of egg and sperm overlap as much as possible. An important point to remember is that cows still in heat 12 hours after breeding should be rebred because almost all the sperm in the tract will be dead when ovulation actually does occur.

ECONOMICS OF ACCURATE HEAT DETECTION

Reproductive inefficiency is not a direct out-of-pocket expense to dairymen, but suboptimal reproductive performance costs the dairy industry an estimated one billion dollars annually in the U.S. Problems with heat detection undoubtedly contribute heavily to this loss. If we assume that $2.00 to $3.00 per day are lost for an open cow, then a missed heat costs $42.00 to $63.00 per cow. If producers had to pay that out as cash, they would be more efficient in their daily heat-detection chores. Observing a cow in standing heat and breeding on time is like putting money in the bank--it's an investment that pays off.

184

REFERENCES

DeSilva, A. W. M. V., G. W. Anderson, F. C. Gwazdauskas, M. L. McGilliard and J. A. Lineweaver. 1981. Interrelationships with estrous behavior and conception in dairy cattle. J. Dairy Sci. 64:2409.

Kiddy, C. A. and D. S. Mitchell. 1981. Estrus-related odors in cows: Time of occurrence. J. Dairy Sci. 64:267.

Kiser, T. E., J. H. Britt and H. D. Ritchie. 1977. Testosterone treatment of cows for use in detection of estrus. J. Anim. Sci. 44:1030.

Pelissier, C. L. 1978. Fertility problems under large herd management. In: C. J. Wilcox et al. (Ed.) Large Dairy Herd Management Symposium. pp 201-218. Univ. Presses of Florida, Gainesville.

Schams, D., E. Schallenberger, B. Hoffmann and H. Karg. 1977. The oestrus cycle of the cow: Hormonal parameters and time relationships concerning oestrus, ovulation, and electrical resistance of the vaginal mucus. Acta Endocr. 86:180.

Williams, W. F., D. R. Yver and T. S. Gross. 1981. Comparison of estrus detection techniques in dairy heifers. J. Dairy Sci. 64:1738.

20

POSTPARTUM REPRODUCTIVE MANAGEMENT
OF THE DIARY COW

Roy L. Ax

INTRODUCTION

Calving interval is the best overall measure of reproductive efficiency. A shorter calving interval leads to
higher milk production per day of life in the herd and more
replacements born in a cow's lifetime. The ideal calving
interval is 12 months. DHI records indicate that the U.S.
national average is 13.5 to 13.8 months; thus, there is room
for improvement.
Dairymen should set realistic goals for shortening the
calving interval in their individual herds--a good rule of
thumb would be to have average days open related to average
pounds of milk produced per day at the peak of lactation.
To shorten the calving interval, producers also should know
the physiological changes that occur around calving. Better
management prior to calving and at calving are keys to
getting cows bred back sooner.

THE DRY PERIOD

Producers should be concerned about rebreeding cows in
the postpartum period--before the calf is born! A sound
program for managing the dry cow leads to (1) higher milk
production when lactation begins, and (2) greater success at
getting cows bred back as soon as possible. Cows too fat at
calving (fat-cow syndrome) have a higher incidence of
dystocia and retained placenta. Retained placenta can lead
to other disorders such as metritis and pyometra. All of
these factors can promote the onset of ovarian cystic
degeneration and interfere with cyclicity.
Condition of cows at calving also affects the percentage of observed heats after calving. For cows in good
condition at calving, 90% will show at least one heat within
60 days postpartum. For cows in moderate or thin condition,
90% will be observed in standing estrus approximately 90
days and 120 days postpartum, respectively. Therefore, a
producer can evaluate heat detection records and get an
estimate of his herd's condition at parturition.

CALVING TIME

Cows should calve in a clean, dry box stall. Delivery of a calf into unsanitary conditions puts that calf at an immediate disadvantage for survival. Those same unsanitary conditions can lead to microorganisms entering the cow's reproductive tract and causing infection. This delays rebreeding and costs the producer antibiotic and/or hormone therapy to clear up the infection.

What about calving ease? This subject receives much attention and is the reason that some producers continue to use a small jumper bull or beef bull to breed the heifers. In addition to nutrition during the dry period, three factors are known to affect calving ease. Those three are sire used, sex of the calf, and age of the cow. Table 1 shows the likelihood of calving difficulty in relation to these various factors.

TABLE 1. FACTORS AFFECTING CALVING EASE

Calf number	Calf sex	Chances of difficulty
1	Bull	1 out of 3
	Heifer	1 out of 4
2	Bull	1 out of 5
	Heifer	1 out of 20
3+	Bull	1 out of 10
	Heifer	1 out of 20

Obviously, first-calf heifers have the highest probability of requiring assistance. All of the artificial insemination organizations have excellent bulls that can contribute to calving ease. Producers can obtain high predicted differences for milk and type while gaining a calf that can have a strong genetic impact on the future of the herd.

RESUMPTION OF CYCLICITY

When does a cow normally start to cycle after calving? If a cow has a normal calving, the first ovulation ordinarily occurs 15 days after calving, the second ovulation is at 32 days, and third ovulation is at 53 days. However, at those ovulations, 75%, 50%, and 33% of the respective heats are "silent" heats. That is, the cow may have a normal physiological heat and be capable of being bred--but she never shows outward signs of estrus. A good program for heat detection is essential to maximize efficiency of heat detection. Cows should be watched at least twice per day, preferably as early and late in the day as possible. Also, use of a hormone-treated masculinized heifer or cow for heat detection can be an asset.

WHEN SHOULD COWS BE REBRED?

When should cows be rebred? This question is often asked, but there is no definite recommendation. In a cow that has had a normal delivery, the uterus is usually involuted completely at 35 to 45 days postpartum. A veterinarian can palpate to determine if involution is progressing satisfactorily. The rule of thumb in the field has been to wait 60 days before rebreeding; however, many cows have an ovulation at 53 days postpartum. By waiting 60 days, the dairyman is one week "off" the normal biology of the cow. In a study just completed at the University of Wisconsin, we waited 60 days before rebreeding in a trial using 300 cows over a two-year period. The average time to first breeding was 74 to 76 days, or 21 to 23 days after the ovulation at 53 days. Thus, waiting 60 days to rebreed is not a good practice for sound reproductive management.

Are higher-producing cows harder to breed? The majority of research evidence suggests that reproductive efficiency decreases with increased production. Physiologically, there are no differences between high and low producers; they exhibit similar hormonal changes and cycles. The major problem is that higher producers have a higher incidence of "silent" heats, making it harder to catch them in heat. The stresses of higher milk production presumably cause the manifestation of "silent" heats. Some cows will show signs of heat 40 to 60 days postpartum, then peak in production and not show another heat until 100 days or more. These cows present a challenge to dairymen, but DHI records substantiate that these cows can be bred earlier, successfully. Many excellent research projects have proven that the fetus places no nutritional demand on the cow until six months of gestation, thus it can no longer be argued that the developing fetus places strain on the cow and reduces milk production.

One final consideration for early breeding of higher-producing cows is that they may be harder to dry off before the next calving. Current research is showing that in many cases a 40 day dry period is as good as the conventional dry period. I have already mentioned in a previous paper the role of good management of the dry cow in terms of the importance to normal cycling after calving.

DECIDING ON INFUSIONS

The dairy cow has an amzing ability to restore her physiological balance. A discharge from the vulva is normal for 7 to 10 days postcalving. However, if pus is seen in this discharge, or foul odor develops, an infection may be indicated. If a placenta does not pass within 6 to 8 hours after delivery of the calf, it is considered to be retained and should be treated accordingly.

Some producers are tempted to infuse all postpartum cows. Reports from the field suggest that an increasing number of dairymen are regularly infusing cows after artificial insemination. This costs the dairyman both time and money, thus he should consult with a veterinarian about infusions and infuse only cows that require the therapy. If a low-grade infection is suspected as contributing to "repeat" breeders, then infusions should be made after cows have been presented for three or more services. Such infusions should be performed 48 hours after AI--which is well after sperm have been transported through the female tract and about two days before the ovum will enter the uterus from the oviduct. When infusions are performed too often, the reproductive management program probably needs attention in some other areas.

OVARIAN CYSTIC DEGENERATION

Cystic ovaries are a common reproductive problem in all dairy herds. Twice-weekly palpations of cows by a Michigan veterinarian revealed that up to 40% of cows in a herd may have cystic ovaries in the early postpartum period.

In a study just completed at the University of Wisconsin, we surveyed 2,000 lactation records from 777 cows in 16 commercial herds. The overall incidence of cysts diagnosed by palpation was 30%! Cows not treated for the cystic condition had an average calving interval of 416 days. Their herdmates receiving treatment had a calving interval of 390 days. The difference was statistically highly significant. This condition represents a significant economic loss to the dairy industry because of its contribution to lengthened calving intervals. The most effective treatment for cysts is hormone therapy rather than manual rupture by palpation. In 1980, agricultural economists at Cornell completed a cost-benefit computer analysis of treating cystic ovaries. They emphasized that a dairyman cannot afford not to treat a cyst after it has been diagnosed. This point must be stressed in the field.

A cow that becomes cystic is likely to have the problem again 30% of the time. There is some genetic contribution to cysts; certain cow families appear to have the problem more than others. However, management plays an important part in whether or not a cow is likely to become cystic. We recently completed a survey of 40 years of University of Wisconsin dairy herd reproductive records to find factors contributing to the probability that a cow would become cystic. Cows were defined as cystic if they required veterinary treatment for the condition. In general, stresses imposed on a cow at the time of calving markedly increased her chances of becoming cystic. Holsteins were cystic in 5.2% of all lactations. Birth of twin calves increased the incidence to 12%. Metritic cows were cystic 10.4% of the time. Retained placenta resulted in a 6.6% frequency of cysts. With increased parity, cysts increased. The more frequently a cow required treatment for milk fever in a single lactation, the higher the likelihood to develop a cyst--with four treatments cows were found cystic 20% of the time. The lowest incidence of cysts occurred when cows calved in January (2.2%), whereas the highest incidence was February (8.1%). Taken collectively these data reiterate the importance of reducing stress at calving to ensure optimum reproductive performance.

SUMMARY

Postpartum reproductive management of dairy cows requires conscientious planning, observations and good records. Accurate records are a real asset from year to year to evaluate the overall reproductive program. Calving interval is the best comprehensive measure of efficiency. Many factors affect the calving interval. The shorter the calving interval, the greater milk production per cow per day of life in the herd. To shorten the calving interval, realistic goals should be set. Constant striving for increased reproductive efficiency leads to increased profitability in the dairy enterprise.

190

REFERENCES

Berger, P. J., R. D. Shanks, A. E. Freeman and R. C. Laben. 1981. Genetic aspects of milk yield and reproductive performance. J. Dairy Sci. 64:114.

Callahan, C. J., R. E. Erb, A. H. Surve and R. D. Randel. 1971. Variables influencing ovarian cycles in postpartum dairy cows. J. Anim. Sci. 33:1053.

Erb, H. N., S. W. Martin, N. Ison and S. Swaminathan. 1981. Interrelationships between production and reproductive diseases in Holstein cows. Conditional relationships between production and disease. J. Dairy Sci. 64:272.

Garverick, H. A. and C. J. Bierschwal. 1978. Ovarian cysts in dairy cattle. In: C. J. Wilcox et al. (Ed.) Large Dairy Herd Management Symposium. pp 606-621. Univ. Presses of Florida, Gainesville.

Hardie, A. R. and R. L. Ax. 1981. A 40-year survey of cystic ovaries in dairy cows. 76th Ann. Mtng. of the Amer. Dairy Sci. Assoc. (Abstr. P194).

Morrow, D. A., S. J. Roberts, K. McEntee and H. F. Gary. 1966. Postpartum ovarian activity and uterine involution in dairy cattle. J. Am. Vet. Med. Assoc. 149:1596.

Peralta, R., R. L. Ax and W. G. Elford. 1982. Incidences of cystic ovaries in Wisconsin dairy herds. 77th Ann. Mtng. of the Amer. Dairy Sci. Assoc. (Abstr. P240).

Schneider, F., J. A. Shelford, R. G. Peterson and L. J. Fisher. 1981. Effects of early and late breeding of dairy cows on reproduction and production in current and subsequent lactation. J. Dairy Sci. 64:1996.

Wagner, W. C. and W. Hansel. 1969. Reproductive physiology of the postpartum cow. I. Clinical and histological findings. J. Reprod. Fert. 18:493.

Whitmore, H. L., W. J. Tyler and L. E. Casida. 1974. Effects of early postpartum breeding in dairy cattle. J. Anim. Sci. 38:339.

21

EMBRYO MANIPULATIONS, TEST-TUBE FERTILIZATION, AND GENE TRANSFER: LOOKING INTO THE CRYSTAL BALL

Roy L. Ax, G. D. Ball,
R. W. Lenz, N. L. First

INTRODUCTION

The food-producing animals in the U.S. today are superior to their predecessors of 20 years ago. The quality of animals has improved due to better management and genetic selection. The use of artificial insemination has contributed significantly to the genetic improvement of our farm animals. However, the potential for further genetic improvement appears to be unlimited. New frontiers for researchers, and ultimately producers, are upon us. The purpose of this presentation is to outline some of the research currently receiving attention and to point out the potential advantages to the animal breeding industry.

EMBRYO MANIPULATIONS

Embryo Transfer

Many articles have appeared recently that deal with the topic of embryo transfer. Basically, donor animals are either given injections of gonadotropic hormones to induce multiple ovulations or allowed to have a normal estrous cycle with recovery of a single embryo for transfer. The majority of people doing embryo transfers recover the embryos from donors using nonsurgical methods. For the actual transfers of embryos to recipients, embryos are either deposited into the reproductive tract (much like artificial insemination) or are placed directly into the uterus through an incision made in the flank. The main uncertainty with embryo transfers is to predict how many ovulations will occur and what percentage of multiple-ovulated eggs will be fertilized. When embryos are transferred to suitable recipients, 60% to 65% become pregnant. In its current form, embryo transfer will not have a major impact upon altering the genetic base. Granted, many more offspring are produced in a dam's reproductive life than would ordinarily be the case, but a single female would never be able to contribute to the industry to the extent that a male does by artificial insemination. We have learned many facts about reproductive biology with the advent of embryo transfer. Hormonal synchronization of the estrous cycle was elucidated. Requirements for transfer methodology were established. Culture media necessary for embryo survival were developed. All of these procedures will be used extensively to take embryos to the new frontiers: sexing, freezing, and cloning.

Sexing of Embryos

If sperm could be assayed to determine whether they contained a Y or X chromosome, it would be easy to alter the sex ratio of offspring. Geneticists tell us that if we could alter the sex ratio to be significantly female we could apply rigid selection pressure and make good genetic progress. Most efforts to identify the genetic content of sperm have either failed or have not yielded consistent, repeatable success. A method has been developed to sort X and Y-bearing sperm with a fluorescent, activated cell sorter; however, the sperm are dead after this procedure. If a alternative method of sexing is developed, the cell-sorter procedure could provide a validation. Sexing of embryos is another option. Keep in mind that this involves flushing embryos out of the reproductive tract, doing a biopsy on a few cells to determine the sex, and then transferring the "sexed" embryo to a recipient. This procedure requires the skills of highly specialized individuals. Furthermore, because of the embryo manipulations required, the industry would not see the sex ratio altered dramatically. A final technical problem is that the longer embryos remain in culture, the lower their chances for optimum fertility when they are transferred to recipients. If freezing and cloning of embryos become routine, then sexing would play an important part. One embryo from a clone could be sexed so that the genotype of the rest of the clones would be known. The rest of the clones could be frozen to ensure maximum survival and transferred to recipients at an appropriate time.

Freezing of Embryos

We all are familiar with freezing of semen; however, freezing embryos poses a different problem because they are composed of masses of cells rather than being a single cell as is a sperm. When freezing and thawing embryos, external cells are exposed to the environment before internal cells. This can cause cell-to-cell junctions to weaken and lead to disaggregation of the cell mass. The end result can be embryonic death. Extremely sophisticated thermal devices have been engineered that show promise for better control of the temperature for embryo survival. The advent of microwave technology offers a novel approach to thaw the embryo from the inside out! Preservation of whole organs such as kidneys and skin has provided researchers with some hints that have already led to significant improvements for frozen storage of embryos. Today if embryos are frozen, thawed, and transferred to recipients, the success rate is 50% of that with fresh embryos.

Cloning of Embryos

From a technical viewpoint, identical twins are clones. They were derived from a single cell, but the embryo split early in development to yield two carbon copies. Embryonic cloning involves use of chemicals and, more commonly, microsurgery to divide the embryo. Greatest success has been obtained by splitting morulae to yield two blastocysts. Current efforts involve splitting 2-cell or 4-cell embryos. By the 8-cell stage, success is markedly reduced. This is presumably because the embryo has started to differentiate at the 8-cell step of development. In theory, it would be fantastic to be able to split a 4-cell embryo into four identical individuals. Each of those could be cultured to the 4-cell stage, split again, and so forth. However, every time an embryo is split, the cells in the next generation are one-half the size of the starting cell. There appears to be a critical ratio of nucleus to cytoplasm in a cell that will have to be studied before this theory can become functional on a commercial scale. The genetic advantage of cloning has already been considered. If ten identical daughters produced one lactation record in ten herds, the remainder of the identical clones (presumably frozen) could be marketed with approximately 80% repeatability for potential milk records. Producers and distributors would both benefit from this knowledge. In the AI industry, if twin, triplet, or quadruplet bulls could be produced by cloning, the carbon copies could be saved in frozen storage. For the outstanding bulls remaining after proofs are computed, their identical genotype would still be in storage for the AI company. An inventory of embryos would take up considerably less space than identical bulls waiting in another barn someplace. This may keep the cost down on some superior sires, but their genes would persist in the industry much longer. With mammals, it will probably be quite a while before a cell from the skin or some other organ will be used to generate that identical individual. For the time being we will have to resort to embryonic cloning and progeny-test one of the embryos.

Nuclear Transfer

Nuclear transfer probably holds more promise for obtaining identical offspring than does cloning. The method involves using microsurgery to collect cells from the inner-cell mass of the trophoblast. Approximately 200 cells can be obtained. Each nucleus is then transferred to the inside of an unfertilized egg that has had its own nucleus or chromosomes removed. All of the cells that develop successfully are identical individuals. The potential of having many identical individuals was already addressed.

Nuclear Fusion

Nuclear fusion implies the union of nuclei from two gametes, which are either eggs or sperm. The combinations are as follows: uniting nuclei from two eggs--all offspring are daughters, uniting nuclei from an egg and sperm--one-half of offspring are daughters, one-fourth are males, and one-fourth bear YY chromosomes, which is lethal (for surviving embryos, twice as many daughters as sons would result from fusing nuclei from sperm), uniting nuclei from an egg and a sperm as happens at fertilization. You can quickly see the possibilities of this procedure. Two outstanding bulls (or cows) could be mated to each other by uniting gamete nuclei into an unfertilized egg and chemically activating the egg to commence division. This procedure has been successfully performed in laboratory species, but it remains to be done in cattle. A point brought out by geneticists is that this process would enable "selfing" of an individual animal. If an embryo developed to term after "selfing" by this method, we would be fairly certain that not many deleterious genes would be present. They would be eliminated in the high percentage of embryos that would not develop.

IN VITRO FERTILIZATION

In vitro (test tube) fertilization has received widespread attention and interest since the first human resulting from this procedure was born in England. Initial attempts in livestock were largely unsuccessful because scientists tried to recover eggs near the time of ovulation. Our team has worked out details for successful in vitro fertilization in cattle. Eggs are aspirated from immature follicles on the ovary and placed into culture with hormones to prepare them for penetration by sperm. In the past year we had 60% success at fertilizing almost 700 eggs with this procedure. The future for this method is extremely promising. If eggs were obtained from immature follicles with a laparascope, the donor female could live her normal reproductive life. Eggs could be collected from ovaries in a pregnant cow! In vitro fertilization permits an individual sire to be used to fertilize each egg. Thus, a cow could be progeny-tested as we do for bulls. Also, far fewer sperm are required as compared to AI. A bull that now yields 600 straws from a single ejaculate could be used for 36,000 eggs in vitro from that same ejaculate! In the future a valuable bull may have some semen packaged in smaller units to be saved specifically for in vitro fertilization. This would greatly extend his reproductive potential. The technical problem facing us now is to be able to culture 2-cell embryos to the 8-cell stage. Dr. Rob Hoppe at the University of Wisconsin routinely gets 8-cell cattle embryos to develop to blastocysts in culture. At this stage they can be transferred to the uterus of a recipient. Our 2-cell embryos would require surgery to place them into the oviduct. If we can culture embryos from 2-cell to 8-cell, we know they can go further. Unfortunately, the ideal culture conditions for 2-cell embryos have not been worked out, because previously it was difficult to obtain an abundant source.

GENE TRANSFER

Recombinant DNA technology is receiving widespread attention and research funding. Introduction of mammalian genes for various proteins, e.g., insulin, has resulted in bacteria manufacturing insulin as a by-product. This avoids the costly and time-consuming process of having to purify that protein from an animal-tissue source. Furthermore, human insulin is not recognized as a foreign substance as is pig or cattle insulin, so patients do not tend to develop immunity. A similar example from cattle stems from studies in the lab of Dr. Dale Bauman at Cornell University. His group found that injections of growth hormone into lactating cows promoted a 15% increase in milk output without increased feed. What if an additional gene or genes for growth hormone would be introduced into an embryo? The mature cow could produce more of the hormone and would not have to receive injections. Mouse embryos successfully received a gene for rabbit hemoglobin and were able to manufacture the hemoglobin. The most important genetic consideration for gene transfers is that once the gene is successfully introduced, it is transmitted to every succeeding generation. This potential opens a whole realm of possibilities for introducing superior production and type traits as well as disease resistance.

SUMMARY

We have all seen tremendous changes in livestock improvement. What took 20 years of careful selection may someday be achieved in one single microsurgical manipulation of an embryo! The animal breeding principles used for AI will be applied to yield more extensive selection pressure on females. Genetic technology poses an enormous challenge with potential that cannot escape our attention or imagination.

REFERENCES

Ball, G. D., R. L. Ax and N. L. First. 1981. Mucopoly-saccharide synthesis accompanies expansion of bovine cumulus-oocyte complex in vitro. In: V. B. Mahesh, T. G. Muldoon, B. B. Saxena and W. A. Sadler (Ed.) Functional Correlates of Hormone Receptors in Reproduction. pp 561-563. Elsevier-North Holland, New York.

Ball, G. D., L. M. Liebfried, R. W. Lenz, R. L. Ax, B. D. Bavister and N. L. First. 1982. Bovine fertilization in vitro: A temperature sensitive process. 74th Ann. Mtng. of the Amer. Soc. of Anim. Sci. (Abstr.).

Betteridge, K. J. 1981. An historical look at embryo transfer. J. Reprod. Fert. 62:1.

Brackett, B. G., Y. K. Oh, J. F. Evans and W. J. Donawick. 1980. Fertilization and early development of cow ova. Biol. Reprod. 23:189.

Greve, T. 1980. Bovine egg transplantation. Superovulation, nonsurgical recoveries and transfers. Nord. Vet-med. 32:513.

Gordon, J. W. and F. H. Ruddle. 1981. Integration and stable germ line transmission of genes injected into mouse pronuclei. Science 214:1244.

Hoppe, P. C. and K. Illmensee. 1977. Microsurgically produced homozygous-diploid uniparental mice. Proc. Nat. Acad. Sci. U.S. 74:5657.

Illmensee, K. and P. C. Hoppe. 1981. Nuclear transplantation in Mus musculus: Developmental potential of nuclei from preimplantation embryos. Cell 23:9.

Marx, J. L. 1981. Tracking genes in developing mice. Science 215:44.

McDaniel, B. T. and B. G. Cassell. 1981. Effects of embryo transfer on genetic change in dairy cattle. J. Dairy Sci. 64:2488.

Ozil, J. P., Y. Heymon and J. P. Renard. 1981. Production of monozygotic twins in cows by micromanipulation and cervical transfer. Proc. Soc. for the Study of Fertility, Fourth Anglo-French Meeting, Cambridge (Abstr. No. 6).

Powell, R. L. 1981. Possible effects of embryo transfer on evaluation of cows and bulls. J. Dairy Sci. 64:2476.

Wagner, T. E., P. C. Hoppe, J. D. Jollick, D. R. Scholl, R. L. Hodinka and J. B. Gault. 1981. Microinjection of a rabbit B-globin gene into zygotes and its subsequent expression in adult mice and their offspring. Proc. Nat. Acad. Sci. U.S. 78:6376.

Willadsen, S. M., H. Lenk-Jensen, C. B. Fehilly and R. Newcomb. 1981. The production of monozygotic twins of preselected parentage by micromanipulation of nonsurgically collected cow embryos. Theriogenology 15:23.

Williams, T. J., R. P. Elsden and G. E. Seidel, Jr. 1982. Identical twin bovine pregnancies derived from bisected embryos. Theriogenology 17:114.

22

IMPROVING REPRODUCTIVE PERFORMANCE
IN DAIRY HERDS

James N. Wiltbank, D. W. Richards,
H. C. Scott

Reproductive performance in many dairy herds is low because many cows are not detected in heat or do not conceive at first service. King and coworkers (1976) used a closed circuit television system and noted that all cows had shown estrus by 60 days postcalving, but the herdsman detected only 64% of these cows. Zemanis, et al., (1969) reported that 90% of the cows reported as anestrus had cyclic ovaries typical of cycling cows. Other data reported by these workers indicated that 43% showed anestrus with 13% failing to show estrus prior to breeding and 31% failing to return to estrus following breeding even though they were not pregnant. Developing a reliable method of detecting estrus in dairy cows would appear to be important if reproductive programs are to be successful.

Workers in California reported that the conception rate on almost 4,000 cows in nine dairies varied from 32% to 47% (Kendrick, 1978). The three main causes of variation in fertility are (1) bull fertility, (2) the technicians, and (3) uterine infection.

This paper reports two studies made to improve reproductive performance in dairy herds.

The first study was conducted in the Brigham Young University Dairy herd. The length of the interval between estrous periods (interestrous interval) was determined in this dairy herd before and after new heat checking procedures were initiated. The effectiveness of visual heat detection was compared to Kamar and chalking the tail head methods. In the second study, a management system for improving reproduction was developed and put into practice on a cooperating farm.

Only 29% of the interestrous intervals were between 16 and 24 days prior to the study, while during this study this proportion increased to 42%. Note also that 21% of the intervals were over 60 days prior to the study as compared to only 1% during the study (table 1).

Two hundred eighty-nine cows were observed visually for estrus four times per day (A.M., P.M., noon, and midnight) in Trial I. Seventy-one percent of the cows showing cyclic ovarian changes were detected in estrus one or more times (table 2). Only 54% were seen in estrus at the A.M. or P.M. check. Twelve percent of the cows were seen in heat only at midnight.

TABLE 1. INTERESTROUS INTERVALS AT BRIGHAM YOUNG UNIVERSITY DAIRY

Length of intervals (Days)	Before study[a] (%)	During study[b] (%)
15 & under	12	32
16 - 24	29	42
25 - 35	14	15
36 - 48	16	8
49 - 59	8	2
60 & over	21	1

[a] Total cycles = 4,652.
[b] Total cycles =   462.

TABLE 2. ESTRUS DETECTION (289 OVULATIONS), TRIAL 1

Time of estrus detection	False positives[a] (%)	Detected only at this time (%)	Seen in estrus at this time (%)
A.M.	1	17	29
P.M.	1	15	27
Noon	0	3	16
Midnight	1	12	25
A.M., P.M.	2	33	56
A.M., P.M., midnight	3	45	69
A.M., P.M., midnight, noon	3	48	71

[a] Declared in estrus but ovarian changes did not occur.

Results from visual detections in Trial 2 were even lower (table 3). Only 54% of the cows showing cyclic ovaries were detected in estrus when checked three times a day for estrus.

Visual detection for estrus was compared with chalk on the tail head and Kamar heat detectors. The overall acuracy was increased when the chalk or Kamar methods were utilized (table 4). The overall accuracy was increased to 87% when using the chalk. However, neither the chalking or Kamar method is a "cure all". Each cow in the study was placed on a roster and days of the week were placed at the top (see Weekly Cow Roster in program outline that follows).

TABLE 3. ESTRUS DETECTION (83 OVULATIONS), TRIAL 2

Time of estrus detection	False positives[a] (%)	Dectected only at this time (%)	Seen in estrus at this time (%)
A.M.	1	16	22
P.M.	0	22	28
Midnight	0	6	18
A.M., P.M.	1	37	48
A.M., P.M. Midnight	1	43	54

[a] Declared in estrus but ovarian changes did not occur.

Each cow was checked each day and marked present (P) or absent (A). If this was not done, the systems did not work with either the Kamar or chalk method.

A system then was developed for improving reproduction in a 270 cow herd. Prior to using this system, 53% of the cows were open for fewer than 109 days; there were averages of 71 days in milk to first service, 2.83 services per conception, and 2.36 services per cow (table 5). The system was implemented in the following way. Each cow had an ID number and an individual record card. Cows in the breeding herd were grouped as much as possible. Stage of the reproductive cycle was identified by using marking chalk on the tail head. All cows were examined rectally 25 to 32 days postpartum (post-partum check); 55-62 days postpartum (prebreeding check) and 38-43 days postbreeding for pregnancy. Cows that had uterine or ovarian problems were examined as frequently as needed to perform necessary treatments. Each week a cow roster was made and was utilized by the herdsman. The two AI technicians were retrained (the program outline follows).

Following implementation of this program for 7 months, services per conception were 1.83. Conception per cow was 2.18. Sixty-seven percent of the cows were open for fewer (table 5) than 109 days. Very few problems in reproduction have been noted in the past few months. The program works because each cow is observed each day and a check is made each week to make certain this occurs.

TABLE 4. COMPARISON OF METHODS USED TO DETECT ESTRUS

Method of detection	No. of ovulations	False positives[a] (%)	Ovula-tions detected (%)	Overall accuracy[b] (%)
Trial 1 Chalk	108	10	86	78
Trial 2 Chalk	67	6	93	87
Trial 1 Visual (A.M., P.M., noon, mid.)	289	3	71	68
Trial 2 Visual (A.M., P.M., mid.)	83	1	54	53
Kamar	42	16	90	76

[a] Declared in estrus but ovarian changes did not occur.

[b] Overall accuracy = $\dfrac{\text{number of ovulations detected - false positives}}{\text{total number of ovulations}}$

TABLE 5. CHANGES IN REPRODUCTION IN DAIRY HERD USING REPRO-DUCTIVE PLAN

	Prior to use	Following 7 months use
Days open (%)		
Less than 109	53	67
110 - 149	23	21
151 or over	23	11
Days in milk 1st breeding (%)		
60 or less	25	20
71 - 81	56	70
over 81	18	11
average	71	66
Services per cow	2-36	1.83
Services per cow	2-37	2.18

REFERENCES

Kendrick, J. W. 1978. The cause and prevention of uterine disease. Report to California Milk Advisory Board.

King, G.J., J. F. Hurnik and A.A. Robertson. 1976. Estrous behavior and ovarian function in dairy cows under intensive systems of management. VIII International Congress on Animal Reproduction and Artificial Insemination. p 149.

Zemanis, R., M. L. Fahning and R. H. Shultz. 1969. Anestrus, the practitioners dilemma. Vet. Scope. 14:15.

DAIRY CATTLE REPRODUCTION PROGRAM

Success depends on:
Manager and herdsman treating each cow as an individual each day

To accomplish this, the following are needed:
- Good ID system (double tag, brand)
- Record card or sheet for each cow
- Work sheet each week (look at card of each open cow each week)
- Weekly report (for manager)
- Weekly cow roster (for herdsman)
- Monthly report (for management)
- Retraining for technicians

RECORD SYSTEM

Cow Card or sheet
  Cow No.
  Birth date
  Date into breeding herd
  Weight into breeding herd
  Breeding dates
  Treatments

Red Tab---Pregnancy diagnosis
Blue Tab--Pregnancy confirmed
  Calving date
  Any problems
  Postpartum check
  Breeding check
  Treatments
  Breeding dates

Red Tab---Pregnancy check
Blue Tab--Pregnancy recheck

GUIDELINES FOR WORKING COWS

- Postpartum Check (P.P.)--27-34 days postcalving
- Breeding Check (Br. Ch.)--55-62 days postpartum
- Pregnancy Check (Preg)--38-45 days postbreeding
- Heifers Breeding Herd--650 lb

(To Make 7-28-82 Worksheet)

P.P. - Calved 6-24 to 7-1
Prebreeding check - Calved 5-27 to 6-3
Pregnancy check - Bred 6-13 to 6-20

Drugs and Materials Needed
Prostoglandin
Lutylase
Estrumate

GNRH
Lysol & pump
Infusion
Marking chalk

WORK SHEET 7-28-82

Cow No.	Status	Work	Finding	
1	Preg	R preg 4-28	Pregnant	Blue tab
2	P.P. 15 day	Yellow	———	
3	P.P. 27 day	Yellow	Recheck P.P.	
4	P.P. 36 day	P.P. recheck cystic ovary	Treated GNRH	Treat lut Wed.
5	P.P. 56 day	Breeding check Lut OK	CL. Lut.	Heat T. or Wed.
6	B.H. Bred 7-24	Blue		
7	B.H. Bred 6-24	Blue		
8	B.H. Bred 7-10	Orange		
9	B.H. Bred 6-15	Preg check	Pregnant	Red tab
10	B.H. Calved 5-15	No heat Lut	C.L. Lut	Heat T. Wed.
11	P.P. 30 days	P.P. yellow	PYO Lut	Heat infuse, Lysol

COW ROSTER

Cow No.	Paint	S	M	T	W	Th	Fri	Sat
2	Y							
3	Y							
4	Y							
5	L O K							
6	B			H	T			
7	B				H			
8	O							
9	L O K			H	H			

WEEKLY REPORT

Cows					Heifers				
Total No.									
Pregnant									
Prebreeding herd									
In breeding herd									
Sold									
Open									
Mastitis									
Poor production									
Other									
Not bred in breeding herd									
1-20									
21-40									
over 40									

Bred					Bred				
1X	2X	3X	4X	Total	1X	2X	3X	4X	Total
1-23									
24 or over									
Total									
Bred-preg.									

Interestrous interval

1-16		Heifers
16-24		
25-45		
over 45		

Pregnancy check

No. checked	
No. pregnant	

No. calved
Retained placenta
No. P.P. checks
Pyometra
No. cows milking herd

Monthly Report for Management

Cows	Beginning	End
Total No.		
Pregnant		
0-60 days postpartum		
Cows in breeding herd		
60-120 days		
over 120 days		
Cows not bred		
over 80 days		
Cows being milked		

Occurrences During Month

Cows calved	
Retained placenta	
Postpartum exam	
Pyometra	
P.P. rechecks	

Cows	Heifers
Checked for pregnancy	
Pregnant	
Problem animals (bred 4 times or more; not bred for 30 days)	
This month	
Last month	
Interestrous intervals correct length (17-24 days)	
This month	
Last month	
Still return	
Passed over	
Bred IX	
2X	
3X	
4X or more	
Sold	
Open	
Mastitis	
Poor production	
Other	

23
REPRODUCTION MANAGEMENT FOR DAIRY CATTLE

Samuel B. Guss

The idea of good reproduction performance developed in dairy cattle management before total herd health programming evolved. As dairymen increased herd size and adopted artificial insemination services, it became important to ensure a maximum number of cows safely in calf, with a calving interval of 12 to 14 months.

There are three basic requirements for a successful reproduction program for a dairy herd: a dedicated dairyman, good records, and a competent veterinarian.

RECORDS

It is important to record heat occurrences and heat period length and interval. Dairymen have learned that heat detection works best when it becomes a near obsession for everybody who works in a dairy herd. It is a job that requires as much attention during nighttime hours as during the day. Many good record systems have been developed that help the herdsman and veterinarian keep track of the reproductive performance of their cows from calving through calving.

THE VETERINARIAN

The veterinarian's role in herd reproduction programs demands regular 2-week or monthly visits and good records to guide him in judicious use of therapy.

When veterinarians are paid on an hourly basis for their visits, the work is better organized and more efficiently performed for both the dairyman and the veterinarian.

More important than the dairyman's dedication and the veterinarian's competence is the absolute necessity for good communication and cooperation to bring out the best in them both.

FRESH COWS

Following calving, retained placental membranes should be removed within 48 hours. Cows with atonic or inflammed uterus (metritis) should receive systemic treatment. Placentas that resist removal at 48 hours should remain 24 hours longer before attempts are made to remove them.

In many herd situations, routine antibiotic treatment is practiced on day 2 or 3, following calving. Many pills and boluses, regardless of their therapeutic value, are often manufactured to withstand long storage and may not break down and dissolve in the uterus. These could become a source of sterile pus and retained corpus luteum (CL), and could prevent conception. Antibiotic solutions are much more satisfactory if they contain a mild foaming agent (an ounce or two of hydrogen peroxide (H_2O_2) per liter of antibiotic infusion solution).

All cows in the herd should be safely in calf by the time they have freshened for 70 days. Beginning in the first month of lactation, effort should be made to detect and record heat. Veterinarians feel that rectal palpation of the genital tracts before the 30th day after freshening is helpful for detecting infections or other abnormalities that interfere with conception.

Cows that have a corpus luteum (CL) present on examination after their first heat are often given a 25 mg dose of prostaglandins F_2a to remove the CL and bring them into heat in 3 to 5 days.

Use of human chorionic gonadotropin (HCG) is now very popular for encouraging slow-ovulating cows to ovulate within 72 hours after being brought into heat with prostaglandins and injected with HCG while in heat. Playing around with hormone injections is not for herdsmen or veterinarians. It may be useful for "problem cows" but "discretion" is the word to remember here.

Injections of chorionic gonadotropin are being used to encourage ovulation in cows that may have abnormally long heat cycles. It is also used for early treatment of cystic degeneration of ova. Chorionic gonadotropin and PGF_2a are very popular now. Both are expensive and both have powerful effects. They should be used with discretion. One can throw a lot of money away on a "hormone kick" in a cattle herd.

Cows should be bred at the first normal standing heat 40 days after they have calved. If this heat takes place 18 to 23 days following a previous recorded heat, normal conception may be expected.

A recent report reveals that it costs the owner of a good dairy cow $45 for every heat that cow fails to conceive. So, the financial incentive is there. All that is needed is proper heat detection, AI technique, and insemination time.

One of the so-called "labor savers" or "short cuts" to heat detection is the use of a teaser bull kept near the cows that are intended to be bred. Where such males are able to copulate, they serve as a reservoir for infection and a very efficient way of spreading infection from cow to cow. With the "gomer" bull, the old danger returns that threatens the lives of cows and people working in dairy herds. Keeping bulls for heat detection is foolish--it's better to have people trained to detect heat to keep records.

AGE TO BREED HEIFERS

Heifers should be fed well enough to weigh 750 to 800 lb for Holsteins and Swiss, 650 to 700 lb for Guernseys and Ayrshires and 500 to 600 lb for Jerseys, by the time they are 14 mo of age.

In the southern U.S., heifers on pasture should be dewormed with an organic phosphate wormer (levamisole), then bred for the first time as soon as they come into heat.

If heifers are healthy, active, and grown out well, nothing is gained by postponing breeding. Heifers held back from breeding because of slower growth become more difficult to impregnate the longer insemination is postponed.

Some dairymen breed their undersized heifers to a bull of a smaller breed or to an Angus bull. This is uneconomical and expensive. Most AI organizations have good records on average birth weights of progeny of the sires offered. These bulls are selected by dairymen who are concerned about their heifers having very large calves. The practice is a wise one.

GENITAL EXAMINATIONS

The practice of retaining a veterinarian to examine the genital tracts of all open females in the herd on a regular basis will save time and money. At each visit, the veterinarian should examine all open cows without breeding dates, cows bred 35 to 40 days (check for pregnancy), repeat breeder cows that have had 3 services, and all cows for genital health 2 weeks following calving.

Normal cows should be listed and bred on their next heat. However, many veterinarians are injecting these animals with 25 mg of costly prostaglandins (PF_2a) and have them inseminated 72 hours later whether or not they show heat. If the ovary contains a well-developed ova and the animal is approaching heat (according to the record of her previous heat) the use of prostaglandins may not be necessary.

HEAT DETECTION

After insemination, it is extremely important to detect every heat shown in the herd. Experienced veterinarians are able to accurately diagnose pregnancy at 30 to 35 days.

One of the serious failures of young "eager-beaver" veterinarians is haste and lack of gentleness. Until a veterinarian has attained the expertise to safely detect pregnancy at 30 days, he should limit initial examinations to 45 days and strive for earlier detection as accuracy and confidence improve.

The Milk Progesterone Assay Test has been a great boon to early detection of pregnancy. Milk samples are collected from cows bred 24 to 28 days, sent to a state or regional laboratory, and reported back to the dairyman with reasonable accuracy. The "not pregnant" results are about 100% accurate. They enable both the dairyman and the veterinarian to concentrate on those cows.

Good records or the use of a breeding wheel will alert the dairyman to possible heat.

BREEDING MANAGEMENT AIDS

A single breeding record for each cow, which contains all the necessary information involving breeding, can be kept in a drawer file with color-coded tabs on the cards to alert the dairyman for such things as: date to breed, date to examine for pregnancy the first time, date to dry off, and calving date.

The breeding wheel, a very useful gadget to help the dairyman keep track of breeding, has several helpful variations, but it requires daily attention. Spending $150 for a gadget is well worth it if it helps the dairy herdsman do his job.

DRYING-OFF COWS

Forty-year-old research has documented the value of the 60-day dry period before dairy cows are expected to calve. Cows that are producing under 40 lb of milk daily can simply be kept from the milking routine to dry them off safely. After at least five days without milking, cows should be stripped out and routinely "dry treated" against mastitis. Some cows are being treated without waiting a few days and stripping them out, but I do not like to put medication into a gland that is already full of coagulated milk.

DRY COW MANAGEMENT

Dry cows that often receive less than good care should be maintained as a group on pasture or in a large enough space so that they will receive ample exercise.

It is imperative that dry cows receive long-fiber forage to restore rumen capacity, which may have shrunk from eating chopped forage and large amounts of ground feed. During the last months of late lactation and the first month of the dry period, cows should be given special attention to be sure that they are in the best condition for all cows at every stage of the annual breeding cycle: lean, muscular, alert, and not fat or emaciated.

SUMMARY

Reproductive management of dairy herds has changed tremendously since the time when a bull was turned out with the cows to take care of that chore.

In today's modern dairy herd, observation, records, and artificial insemination eliminate bulls running with the herd.

DHIA records, the breeding record file, AI technology, and the herd veterinarian cooperatively have achieved breeding success far surpassing that experienced before the new aids were first used.

However, reproductive management can put a dairyman out of the business if everybody involved fails to communicate, cooperate, and competently and conscientiously do his part of the job.

Reproductive management is always dynamic--never static. The herd manager and the veterinarian have to manage so that the job is done better than it was done even a few months ago, so that everybody does his part.

REFERENCES

Seguin, B. E. Reproduction Herd Health Program for Dairy Herds. Compodium on Continuing Education, 3:5445.

Williamson, N. B. 1980. Reproduction Performances and Recording Systems. Proceedings--AABP.

24

MANAGING OUR DAIRY FARMS

Doyle Waybright

Mason Dixon Farms proclaims itself to be a "family farm" but not like one that most people envision. The size and scope of our dairy farm requires a more intense management as compared to the average family farm of a father, his son, and maybe an outside nonfamily employee or two. Our management covers everything: raising and harvesting the crops for forage; feeding, caring, and milking the dairy herd; processing and packaging the end product--grade A fluid milk. Three different families and three generations are involved in the ownership and management of such an operation. Twenty-five family members are employed along with twenty-five nonfamily people--for the most part on a full-time basis. The business is incorporated for the primary purpose of continuous ownership from one generation to the next. Ownership and full-time employment by family members are requirements for being a part of the management and decision-making team.

We have long felt that involving family members in the decision-making is the main root for our stability and success. This has been the strength of the well-known family farm. The survival of the family farm in the U.S. is being questioned because many family operators are content to be average. Mason Dixon Farms is ahead of the pack because of the constant challenge to us to improve our farm. Taking risks with our innovativeness has become commonplace in our family life. Our failures far out-number the successes, but each failure has been used as a stepping stone toward the final achievement. Pride and self-respect keep us looking forward and optimistic about the future.

With the tilling of 2,500 acres, milking 950 cows, and raising 900 heifer replacements, the chores are divided up according to each member's likes and ability. Two separate divisions of the farm have come about naturally: 1) the shop and field crew and 2) those working with the dairy herd. Family members manage and work in a given area with employees accepting responsibility for specific chores.

The management of our farm takes the same continual effort as milking our cows or plowing our fields.

Part 5

ENVIRONMENT,
BUILDINGS, AND EQUIPMENT

25

MEASURING AN ANIMAL'S ENVIRONMENT

Stanley E. Curtis

ASSESSING ANIMAL ENVIRONMENTS

Animal environments are characterized according to problems suspected of being associated with environmental stress on the animals and the effectiveness of control measures. Most of the elements of animal environments can be measured but, interpreting the results in terms of animal well-being, facility operation, and production economics often remains a dilemma because of the interaction of environmental factors. The effect of one stressful factor on an animal quite often depends on the nature of the rest of the environment.

ANIMAL ENVIRONMENT PROBLEMS

Troubleshooters often must engage in trial-and-error to identify animal-environment problems, but there are several points to be kept in mind:
- The environment results from all external conditions that the animal experiences, so all elements must be considered. Those which cannot be measured or controlled readily might influence animal health and performance nonetheless, so even they must be considered so far as is possible.
- The environmental complex acts as a whole on the animal, so interactions must be kept in mind. The combined effects of two or more environmental components may be difficult to evaluate, but they must be considered.
- Time and space affect the environmental factors. Environmental variables should be measured where the animal experiences them--in its micro--environment--taking into account the lack of spatial uniformity that occurs in all facets of the surroundings. Most important are vertical stratification of various parameters of the thermal environment and horizontal and vertical

variation in airflow due to design or mode of operation of the ventilation system.
- Environmental elements change with time at a given place. Because weather and facility occupancy vary with time, control requirements do also. Thus, environmental assessments and control schemes must take daily and seasonal environmental cycles into account. Most animals adjust to environmental cycles readily as long as extremes of the excursions are not unduly stressful.
- The rate of environmental change is critical. Abrupt environmental changes tend to be more stressful than those occurring over a longer period. For example, preconditioning young animals to a cool environment before moving them from a warm, closed house to a cool, open one during cold weather reduces the stress. It is sometimes difficult to identify a single index of environmental stress or even an adequate multiple index or combination of indices. For example, daily temperature range per se may be of little consequence as long as the day's maximum and minimum values do not exceed or fall below respective trigger levels. Likewise, a certain rate of temperature change might be stressful if it occurs at extremes of temperature, but not within more moderate ranges. With modern statistical techniques, it is possible to develop many environmental indices, with relative ease but it is a very difficult job to determine their respective significances in terms of the environment's impingements on the animal.
- Animals modify their own environments by giving off heat, water vapor, urine and feces, disease causing microbes, and others. The animals' own processes help determine the nature of their microenvironment. Changes in age or number of animals in a facility alter these impacts and, therefore, the control measures required.
- Anthropomorphism is a common pitfall in the assessment of animal environments. A comfortable environment for a human is not necessary for an animal. Animals send signals of discomfort or uneasiness to alert caretakers. These behavioral indications are always useful signs that the environment could be improved. In some cases, the way animals behave is the only clue that stress is present.

MEASURING ENVIRONMENTAL FACTORS

A well-planned and organized approach to environmental-measurement programs is important. The means employed will depend on the amount of accuracy, the extent of detail required and the effort devoted to the actual work of measurement. Insights into sampling theory, as well as descriptions of some of the instruments and techniques that have proved especially applicable to measuring outdoor and indoor animal environments, are provided in the following sections.

Sampling Theory

Environmental assessments are based on interpretations of measurements of pertinent variables. Hence, the observations must represent the situation faithfully. To ensure this, environmental sampling programs must be planned carefully. The main reasons for this have been alluded to already: most environmental elements vary over time and space. Some of these variations are regular and predictable, others are not. The times and the places the environment is measured determine whether the resultant information reflects the character of the environment well enough to be of use. Of course, it is possible to gather more data than necessary, too.

Time considerations. To estimate the average impingement of an environmental factor over a period of hours or days, or to learn about excursions of these values over time, a continuous sample is needed. In short, continuous sampling from more or less permanent instrument stations, usually coupled with recording equipment, provides the data needed for detailed analyses of animal environments.
The most important consideration in regard to time is the length of the observation period. It should be a multiple of a well-established environmental cycle. In studying an animal facility in a temperate climate, for instance, seasonal periodicities must be accounted for to appreciate the facility's nature all year long. On the other hand, a particular problem may be limited to one season, in which case the observation period would be a multiple of the day, to account for diurnal cycles in meteorologic phenomena.
Of course, there is considerable variation among years and even among days within seasons. For example the number of cycles--the number of years or days--that should be observed to give meaningful results depends partly on the nature of this variation already known to occur in the facility's locale. Furthermore, interpretation of any results should include an historical perspective. For example, if observed values for air temperature are at the lower end of the acceptable range during a winter known to be relatively mild by local standards, the problem of a too-cold animal micro-environment might well be encountered during a more nearly normal winter.

The frequency of observations needed within a sampling period is another important decision. In general, measurements should be made as frequently as feasible, especially if automatic recording equipment is being used. Then, after observations have been completed, key periods of environmental extremes or change can be evaluated in detail, while less interesting periods can be ignored. Also, when observations are made too infrequently, errors can be made in characterizing both the ranges and the average values of the variables.

Time-averaging can obscure extremes of environmental factors that may trigger animal responses, hence it must be done only when warranted. For example, effective environmental temperature in an outdoor environment might range from -10° to 20°C one day, from 0° to 10°C another. Average temperature might be around 5°C on both days but the nature of the animals' thermoregulatory reactions would be different on these days, and thus averaging the environmental data over a day could lead to misleading impressions.

Nonlinearity in interactions among environmental factors causes additional difficulties so far as the time-averaging of environmental measurements is concerned. In other words, effects of combined factors must be calculated carefully, even when their relations for steady-state conditions are well-known. Take as an example the case of the wind-chill index (table 1). The average of the wind-chill indices for the three sets of conditions in the table is 1.6×10^3 Kcal m^{-2} hr^{-1}, while the wind-chill index for the average of the three conditions (namely, temperature -34°C and wind speed 9 m sec^{-1}) is 2.1×10^3 kcal m^{-2} hr^{-1}.

TABLE 1. AN EXAMPLE OF NONLINEAR INTERACTION AMONG
ENVIRONMENTAL VARIABLES: WIND-CHILL INDEX

	Condition		
	A	B	C
Temperature (°C)	-18	-34	-51
Wind speed (m sec^{-1})	18	9	0
Wind-chill index (kcal m^{-2} hr^{-1})	1.8 x 10	2.1 x 10	.8 x 10

Time lags between environmental occurrences and animal responses also must be recognized--taking them into account could improve the probability of defining a connection between animal and environment.

Another approach is discontinuous sampling, sometimes called spot- or grab-sampling. This usually involves more portable equipment and is aimed at gaining information over short periods, such as hours. Discontinuous sampling is best suited to determining extremes in an environmental element when the basic nature of the variation of the factor is known. For example, air temperature might be measured only in the early afternoon on relatively hot days to gain information on upper values of air temperature, or concentrations of air pollutants might be measured in closed animal houses on relatively cold days when ventilation rate is relatively low.

Discontinuous sampling has several advantages. it requires less time and often less equipment. Further, fewer data are generated, so data-processing equipment requirements are less than when continuous sampling is practiced. Finally, equipment portability often facilitates economical use of a single instrument at many locations in a facility to learn more about environmental variation over space than might be feasible otherwise.

Space considerations. Two prime considerations should determine where environmental measurements are to be taken. In the first place, environmental factors generally should be measured in the immediate surroundings of the animals. For example, most elements of the environment vary consistently with distance above the floor, but most animals reside at discrete heights within a facility. Thus, the height at which the environment is measured is crucial if the measured values are to reflect the conditions to which the animals are being exposed.

In animal facilities it is sometimes tempting to sample the environment in an alleyway or some other place where the instruments will be relatively safe from animal damage. For the most part, these temptations should be resisted. The animals affect their own surroundings so greatly that most variables differ even from animal microenvironments to nearby areas where the animals are not permitted.

Second, the environmental and animal features known to affect the variable under study should be clearly in mind when the measurement sites are chosen. Major items include heat sources: air inlets and outlets; orientation of the facility to winds, the sun, and other structures; location of mechanical services, such as feed-delivery systems; and animal size and population density.

There also is the substantial problem of the effect of the measurement instrument itself, and its protective hardware, on the environment. Some equipment stations obstruct airflow, for example, and the air samplers may be drawing air from different heights and extracting components from the air around the sampler, thereby modifying it.

Instrument Choice

Dozens of instruments for measuring environmental factors are available on the commercial market today. The choice of instrument or set of instruments to be used in a given environmental-measurement program is based on several considerations: (1) the kind of information needed, (2) the relative efficiencies of the various instruments and their reliabilities under field conditions, (3) ease of use, cost, and availability, and (4) personal choice--often based on past experiences--is an important point.

These instruments have accompanying instructions. If the manufacturer's written advice proves inadequate, get in touch with a technical representative of the manufacturing firm directly; most will assist customers by mail or telephone with problems in specific applications of their product.

In the sections that follow, some instruments commonly used in measuring environmental factors in animal facilities are described in brief detail.

MEASURING AIR TEMPERATURE

Liquid-In-Glass Thermometers

A variety of commercially available mercury- or alcohol-in-glass thermometers are used widely to measure air temperature. As temperature rises, the liquid expands to occupy more of the capillary tube in which it is held. Of course, as in all thermometry, the temperature registered is actually that of the thermometer, not necessarily that of the environment, so factors apart from the temperature of the surrounding air that affect thermometer temperature lead to errors. Chief among these are solar and thermal radiations and air movement. Precautions must be taken to shield the thermometer against them. Furthermore, the thermometer may be placed in a spot where the air is stagnant or otherwise unrepresentative of the general area.

Lag time. When environmental temperature changes, the value required for a thermometer to reduce the difference between registered temperature and actual air temperature to 36.8% of the original difference is called the lag time of the thermometer. Lag time for most mercury-in-glass thermometers is around 1 min, while that for alcohol-in-glass is roughly 1.5 min. Some electronic thermometers have much shorter lag times. Especially in discontinuous sampling, lag time is an important consideration because a measurement can be made more quickly with the shorter lag time.

Maximum-minimum thermometer. A thermometer designed to register the maximum and minimum temperatures experienced during the measurement period often has been used in animal facilities. It consists of a U-tube with bulbs at both ends. One side of the U serves as the scale for maximum temperature, the other for minimum. The bulb on the maximum side serves as a safety reservoir and is partially filled with a liquid such as creosote solution. That on the other side is completely filled with the liquid. Between these two portions of liquid, in the bottom of the U-tube, is mer-

cury. As the thermometer becomes warmer, the liquid in the
filled bulb expands, pushing the mercury up on the maximum-
scale side (the opposite side). Atop the mercury on both
sides is an iron index (a sliver of iron), and as the mercury moves up the maximum side it pushes the index ahead of
it. As the thermometer cools, the mercury retracts, but the
index remains at its highest point due to friction with the
inside of the U-tube that can be overcome by the mercury,
but not by the other liquid.   Of course, as the
mercury column retracts upon cooling, it pushes the minimum-
temperature side's index ahead of it, so this one registers
the lowest temperature of the measurement period on the max-
imum-temperature scale, which is upside-down. When maximum
and minimum temperature have been observed, the thermometer
is reset by replacing the indices atop respective mercury
meniscuses by means of a magnet.

Although maximum and minimum temperature may be all the
information needed in some situations, and despite the fact
it is relatively inexpensive and straightforward in design,
this instrument has serious drawbacks. Chief among them are
the tendencies for the mercury column to become separated or
broken and for the indices to become permanently lodged in
the U-tube.

Bimetallic-strip thermometer. The sensor of a bimetal-
lic-trip thermometer comprises a sheet of each of two metals
having dissimilar coefficients of linear thermal expansion,
which are joined along their faces. When such a strip's
temperature changes, it bends. Lag time is relatively
short--around 10 seconds.

Thermograph. The bending movement noted above for the
bimetallic strip is usually magnified by shaping the strip
appropriately--and in a thermograph the movement is recorded
by affixing a pen to its free end and applying this pen to a
piece of graph paper on a drum rotated by a clockworks. By
this relatively inexpensive means, a permanent record of
temperature variation is made. Further, it requires no
electrical supply. Of course, a thermograph is so large
that it affects the microenvironment it is used in and, because of its construction, is prone to error due to radiation and stagnant air pockets.

Calibration of a thermograph is a critical matter. The
recording element is very sensitive to physical shock, which
often occurs when the instrument is moved from place to
place, thus, after the instrument is moved, it is absolutely
necessary to calibrate a thermograph by adjusting the recording pen several days in a row, preferably near the times
of the daily high and low temperatures. An artificially
ventilated psychrometer is commonly used as the standard instrument for calibration of thermographs.

Electric thermometers

There are two general kinds of electric thermometers. One kind depends on the principle that as the temperature of a substance changes, so does its electrical resistance. The other kind--thermocouple thermometry--depends on the principle that when wires of two specific metals are joined at both ends, and when these two junctions are kept at two temperatures, an electromotive force is generated in that circuit. Voltage in such a circuit, when measured with a potentiometer, is directly proportional to the temperature difference between two junctions.

Thermistor. Certain semiconducting materials have negative temperature coefficients of electrical resistance: as temperature rises, resistance decreases. Resistance in the circuit to which a current has been applied is measured by an indicating unit. These sensing elements are called thermistors--parts of the electric thermometers most applicable to air thermometry in animal environments. Thermistor probes, indicating units, and recording units are commercially available in a wide range of models. At one time, thermistor systems were notoriously unstable, but nowadays stable, calibrated probes are on the market and in recent years they increasingly have become the sensors of choice for routine assessment of animal environments.

Thermocouples. A variety of combinations of dissimilar wires are used for the two sides of the circuit in thermocouple thermometry. Copper and constantan are frequently chosen. One of the thermocouple junctions (the reference junction) is held at a constant or known temperature so changes in electromotive force measured reflect changes in the temperature of the measurement junction, which is placed in the environment to be monitored. The voltage generated can be used to drive a millivolt recorder or registered on a millivoltmeter.

In general, use of thermocouples for air thermometry in animal facilities has some disadvantages. The needed equipment, especially the constant-temperature bath for the reference junction, is relatively cumbersome, the physical integrity of the measurement and reference junctions critical and sometimes difficult to maintain, and careful calibration of the thermocouples very important.

Metal-resistance thermometer. In metals, as temperature rises, so does electrical resistance. Small-diameter platinum wire wound on a support having a small thermal expansion coefficient is a frequent choice. A small current is introduced into the wire, and resistance of the whole circuit is measured using a Wheatstone bridge circuit. The most sensitive resistance thermometers tend to be fragile, and for this reason alone they are of limited use in animal environments. In addition, the measuring equipment is relatively expensive.

MEASURING AIR MOISTURE

The measurement of water vapor in air is called psy-chrometry or hygrometry. Several principles have been used to measure air moisture, and three have been applied widely in quantifying animal environments.

Hair Hygrometer

Hair is hygroscopic, and its length is related directly with the amount of water it contains. Further, there is a nonlinear, direct correspondence between length of hairs and the relative humidity of the air surrounding them. Hair hygrometry is most accurate when relative humidity ranges between 20% and 80%. This principle has been employed extensively in hygrometry in animal facilities, partly because hair hygrometers require no electrical supply and they are affected little by other factors.

Hygrograph. Elongation and shortening of hair bundles can be magnified by an appropriate level system and trans-formed into movement of an arm holding a pen that is applied to graph paper on a rotating drum, giving a record of changes in relative humidity. Just as for the thermograph, hygrographs must be calibrated carefully over a period of several days after they have been moved to a new location before reliable measurements can be made.

Psychrometers

Psychrometers are a class of instrument by means of which the air's moisture content or relative humidity can be estimated indirectly. They use both a wet-bulb thermometer and a dry-bulb, and their measurement principle is based on the thermodynamic relation between the air's moisture con-tent and wet-bulb temperature. (Wet-bulb temperature is af-fected by dry-bulb temperature and air pressure.) Once dry-bulb and wet-bulb temperatures of the air are known, the air's moisture content and relative humidity can be esti-mated from a psychrometric chart.

The dry-bulb thermometer can be of any type, while the wet-bulb thermometer is a similar instrument having a water-saturated wick closely surrounding the bulb or sensing the element. As the psychrometer is operated, evaporation from the wick occurs, and the temperature of the wet bulb is de-pressed. Of course, the drier the air, the greater the evaporation, and the greater the wet-bulb depression.

To give an accurate estimate of wet-bulb temperature, the thermometers must be ventilated adequately; maximum cooling of the wet bulb does not occur in still air. Also, temperature readings must not be taken until equilibrium has occurred. Other sources of error when using any psychro-meter are heat conduction down the wet-bulb thermometer (the wick is ordinarily extended up the stem), receipt of solar and thermal radiation (the latter even from the operator of the instrument), and wicks that are too thick or dirty (one way to minimize mineral crust is to wet the wick with distilled water only).

Wet-bulb temperature should always be read before that of the dry-bulb, as it will begin to rise as soon as ventilation ceases. It is also good practice to repeat the measurement several times to make sure the lowest wet-bulb temperature has been attained.

Sling psychrometer. The sling psychrometer, once the standard instrument for spot-sampling psychrometry, consists of dry- and wet-bulb liquid-in-glass thermometers in a frame that can be revolved around a handle. The thermometers are whirled—usually for a minute or more—to permit wet-bulb depression to occur. Larger sling psychrometers must be revolved at least two times per second to provide sufficient ventilation, and smaller ones five times. The movement of the operator mixes the air in the region.

Artificially ventilated psychrometer. Various models of psychrometer are now available in which air is drawn artificially past the temperature sensors. One popular model employs dry-cell batteries that supply a small fan, which pulls air at speeds up to 5 m sec^{-1} past the bulbs. Of course, this kind of instrument may draw air from as far away as 1 foot, hence it is not applicable to some microenvironmental measurements. Still, it is reliable, rugged, and portable.

Electrical Conductivity of Hygroscopic Materials

Salts such as lithium chloride are hygroscopic, and their electrical conductivity increases—thus, their resistance decreases—with increasing water content.

Electric hygrometer. Commercially available sensing units usually involve a film of lithium chloride on a nonconducting frame through which an electrical current is passed for the purpose of measuring electrical resistance changes. The logarithms of the resistance and the atmospheric humidity parameters are inversely related.

Dew cell. When a film of lithium chloride is applied to a heating-element frame, and the temperature is so high the salt is dry, the salt is also highly resistant to conducting electricity. The electrical circuitry of a dew-cell apparatus is designed so that when the salt is a conductor, the element is being heated, but the heating stops when the salt becomes warm enough to become dry. Thus, the lithium-chloride film is kept more or less at the same temperature and dry at all times. This equilibrium salt temperature is related directly with the air's dew-point temperature.

Electric hygrometers and dew cells make it possible to monitor air humidity continuously, but in animal environments they often become dirty, and this can lead to errors. Further, they remain operational for periods of only a few months under the best of conditions.

MEASURING AIR MOVEMENT

Drafts, stagnant spaces, and inadequate removal of moisture or noxious gases in animal houses are among the common symptoms of improper design or operation of a ventilation system. Air speed and distribution throughout an animal facility must be known if ventilation problems are to be remedied.

Anemometers

Several kinds of instruments to measure air speed are available. Each is best suited to a particular application in animal-environment measurement.

Pitot-tube measurement. When air moves into or across the mouth of an open tube, the air pressure in the tube changes; it increases in the former case, decreases in the latter. Such pressure changes are proportional to air speed and serve as the basis of pitot-tube anemometry.

The most common pitot-tube anemometer used today is the Velometer, a rugged and portable instrument that gives a direct reading of air speed and comes supplied with a variety of probes for different velocity ranges. This instrument is most adaptable to measuring air velocity at inlets, in areas of strong drafts, and in air ducts.

Because of its very nature, the pitot-tube is extremely directional; it is sensitive to air movement in one direction. Large errors can result when a probe is not properly oriented.

Hot-wire anemometer. Several modes of hot-wire anemometer are on the market. Some are directional and therefore applicable to measuring air speed in ducts and at inlets; others are more nearly omnidirectional. They operate on the principle that as air passes across a fine platinum or nickel wire that has been heated electrically, the wire tends to cool by an amount proportional to air speed. This instrument is designed so current flowing through the wire automatically changes so as to keep wire temperature nearly constant. The amount of current required to achieve this is thus related directly with air velocity.

Hot-wire anemometers are relatively sensitive and thus especially applicable to situations--such as animal microenvironments--where velocity can be as low as .5 cm sec^{-1}. Another advantage is that some designs are very portable and fairy rugged, and so small they do not interfere much with the environment. They also respond very quickly to changes in air speed. However, the sensing wire can be affected over time by atmospheric pollutants, and for this and other reasons frequent calibration is necessary. Further, the sensing wire itself is exceedingly fragile and subject to damage. The hot-wire anemometer cannot be used outdoors during rainy periods or in any environments where water can reach the sensing wire.

Rotating-vane anemometer. There are various kinds of anemometers employing a lightweight propeller with eight or more blades. All are highly directiónal, and for applications where air direction changes greatly they are commonly attached to a vane device that keeps them aimed into the main flow of the air.

A variety of physical and electronic metering devices are used in conjunction with these anemometers. They are most useful at high air speeds, and thus are most applicable in outdoor settings. Indoor models are relatively fragile and subject to damage by corrosive gasses.

Cup anemometer. The device used to measure air velocity outdoors consists of several cups, each connected to a rod radiating from a rotor. A revolution of the assembly is directly related to air velocity. Cup anemometers are not suitable for measuring low air speeds. They are most applicable to estimating the average speed of the wind during periods of at least several hours.

Kata thermometer. A simple instrument was developed over a century ago to estimate the cooling power of the air. The kata thermometer is capable of measuring very low air velocities. It is simply an alcohol-in-glass thermometer with the stem marked at the 37.5° and 35°C levels.

The thermometer is warmed in a water bath to around 40°C, removed from the water and wiped dry, and placed in the environment to be measured. The time required for temperature to fall from 37.5° to 35°C is determined using a stopwatch, and this value substituted into a formula (which also includes air temperature and an individual instrument-calibration factor supplied by the manufacturer) for calculating air velocity.

Airflow-pattern measurement

Certain visible particles suspended in the air can be used as an aid to tracing how the air is distributed in an animal house. The source of the particulate matter is simply placed at the air inlet or in the microenvironment to be studied, and the course of the pollutant followed visually through the space of interest. Much cigar and pipe smoke has been used quite effectively for this purpose in the past. More reliable sources of larger amounts of visible tracers are now on the market.

In the absence of an anemometer, low and moderate air speeds can be estimated to a first approximation by briefly interrupting the tracer's flow and monitoring the movement of the turbulence so induced with the aid of a measuring tape and a stopwatch.

"Smoke"vials. When a solution of titanium tetrachloride is exposed to air, it gives off copious whitish fumes. Small glass vials of this solution areideal for use in studying airflow patterns in microenvironments or parts of a room. A vial is broken open and simply can be held in the location to be observed. An open vial is commonly placed in a cup, which may be attached to a pole in order to reach certain parts of the room of interest. Several devices are also available to increase the evolution of fumes by passing air over the solution. Titanium tetrachloride fumes in the concentrations used are not toxic to animals.

Talcum aerosolizer. Fine talcum powder is also used as a tracer in small-scale air-distribution studies in animal houses. Special talcum-powder aerosolizers are available commercially. Talcum tends to precipitate faster than does titanium tetrachloride, and it is generally more difficult to generate an adequate aerosol of the powder. For these reasons, it is inferior to titanium tetrachloride for this purpose.

"Smoke" pellets and candles. When long-term or large-scale observations are desired, more tracer might be needed than can be supplied by titanium tetrachloride or talcum aerosols. The commercial market has a wide range of pellets and candles that, when lighted, produce very large amounts of fumes. These are generally set in the way of the incoming air. The fumes of some of these devices are toxic and therefore must not be used in occupied houses.

Static-Pressure Measurement

In negative-pressure or exhaust-ventilation systems, adequate negative-static pressure must be maintained by the fans. Static pressure inside an animal house is usually measured by means of a manometer sloped 1:10 to increase accuracy. One end of the manometer is open to the outside atmosphere, the other to the inside, and the pressure difference is registered by the manometer.

Every animal house employing mechanical ventilation should be equipped with its own static-pressure manometer.

MEASURING SURFACE TEMPERATURE

The temperature of an animal's surface is an important determinant of convective and radiant exchanges of heat.

The temperatures of environmental surfaces likewise play a central role in determining the magnitude and direction of thermal-radiant flux. Accurate measurement of surface temperature is difficult, but there are two ways this can be approached.

Portable radiation thermometer. A variety of battery-powered instruments now available provide rapid measurement of surface temperature by a technique that doesnot involve contact with the surface. The portable radiation thermometer is simply aimed at the surface to be measured, and the surface's temperature is registered on the meter. Models providing different fields of view are available.

This kind of thermometer must be calibrated before every observation by means of a Leslie cube or some other temperature-calibration device. Solar radiation reflected by a surface leads to overestimation of that surface's temperature when measured with this instrument. Also, surfaces in the surroundings substantially cooler than that being measured tend to bias these estimates in the opposite direction. The instrument is reliable, reasonably rugged, and is an excellent means for spot-sampling animal and environmental surface temperatures.

Contact thermometers. Thermometers in a variety of styles involving contact with animal and environmental surfaces have been used sometimes to measure surface temperature. These include thermistor probes in hypodermic needles, contact discs (banjo probes), as well as thermocouples taped or glued to surfaces.

There are serious drawbacks to contact thermometers. First, the contact must be flawless--otherwise insulative air pockets, even very small ones, between thermometer and surface will introduce measurement errors--and this is rarely achieved on either animal or environmental surfaces.

There are other problems of introducing artifacts with animal surfaces in particular. It is a practical impossibility to achieve the necessary contact with covered areas of an animal's surface without disrupting that cover and altering surface temperature. Even on nude areas, affixing a thermometer to the skin in such a way as to ensure adequate contact alters cutaneous blood flow and surface temperature.

MEASURING SOLAR AND THERMAL RADIATION

Solar Radiation

The total direct and diffuse (sky) solar radiation received by a horizontal surface is measured by an instrument called a pyranometer. Several designs are available on the market. The standard instrument in the United States is the Eppley pyranometer.

Eppley 180° pyranometer. This instrument consists of
horizontal concentric silver rings, one black and one
white. The glass hemisphere that covers the rings transmits
only radiation with wavelengths less than 3.5 ^{e}m (solar
radiation), for which the black and white rings have different absorptivities. The temperature difference between the
two rings generates an electromotive force in a thermopile
(a series of thermocouple junctions), alternate junctions of
which are in thermal contact with respective rings.
    The Eppley 180° pyranometer measures direct and diffuse
solar radiation received by a horizontal flat surface from
the upper hemisphere. When direct sunshine is blocked by a
shade, only the sky radiation is measured.
    This instrument is fragile and must be sited careful-
ly. Its glass bulb must be cleaned daily.

Thermal Radiation

    The rate of incoming thermal radiation is usually es-
timated as the difference between the rate of total incoming
radiation having wavelengths between .1 and 100 ^{e}m, and the
rate of total incoming solar radiation as estimated by an
unshaded pyranometer. The rate of total incoming radiation
can be measured by any of several kinds of total radio-
meter.

    Beckman and Whitley total radiometer. This instrument
is ventilated to minimize errors due to variable convective
heat loss and measures the rate of incoming radiation at
wavelengths between .1 and 100 ^{e}m. It essentially consists
of three layers of plastic: the top one is painted black
and exposed to radiation from the upper hemisphere; the
middle one contains a thermopile; and the lower one is
shielded to minimize its receipt of radiation. The temperature difference between top and bottom of the middle plate
is related directly to the upper plate's rate of radiation
absorption.

MEASURING LIGHT INTENSITY

Photovoltaic Meters

    Light meter. A selenium photovoltaic cell comprises
the basis of most light-intensity meters used by photograph-
ers. Illumination of the sensitive layer of selenium by
visible radiation (wavelengths from .3 to 7 ^{e}m set up a flow
of current in an appropriate circuit. Photographic light
meters are portable and very useful in measuring light
intensity in animal environments.
    In most commercially available light meters, sensitivi-
ty over the visible spectrum is trimmed to resemble that of
the human eye; that is, it peaks at a wavelength of about
.55 ^{e}m instead of the .63 ^{e}m wavelength radiation that
drives photoperiodisms in animals.

Illuminometer. A recording hemispherical photometer called the Illuminometer is also on the market. It is particularly suited to stationary installations where light intensity is known to vary over time, such as it does outdoors.

MEASURING SOUND LEVEL

Sound-level meter. Several battery-powered models of sound-level meter are available commercially. All provide a simple, portable, and reliable means of measuring sound level in decibels. The microphones on such instruments are relatively nondirectional, and the operation of sound-level meters is straightforward. Necessary precautions include recognizing the possible presence of obstacles to sound waves; locating the microphone at the observer's side, not between observer and sound source; shielding the microphone from any moving air; and making sure interfering electromagnetic fields from other electrical equipment are accounted for.

MEASURING AIR PRESSURE

Aneroid barometer. Measurement of air pressure in conjunction with animal production is ordinarily accomplished by means of an aneroid barometer--an instrument in which the walls of an evacuated cell move as air pressure changes. The movements are transmitted to a pointer, which indicates air pressure. While not as accurate as a mercury barometer, the aneroid version is nonetheless quite useful in animal work. If moved, it must be recalibrated against a mercury barometer.

Barograph. Movements of an aneroid barometer cell's wall can be recorded on graph paper affixed to a rotating drum when a pen is linked to that wall.

MEASURING AIR POLLUTANTS

Measuring Aerial Gases and Vapors

Colorimetric indicator tubes. Several systems of the same general type are available commercially for the convenient and, when properly used, reasonable accurate measurement of aerial gases and vapors. These consist of an indicator tube and a precision piston or bellows pump operated manually to draw air. The detector tube contains a specific chemical that reacts with the gas or vapor being measured. These small detector tubes are available for all the major gases and some of the vaporous compounds commonly present in animal-house air.

When air is pulled through an indicator tube, the pollutant for which the tube's indicating gel is specific reacts with the chemical, resulting in discoloration. The extent of this change is related to the concentration of the pollutant in the air. One problem with such a measurement system is that gases associated with dust particles--for example, some of the ammonia in dusty air--are filtered out of the air before it reaches the colorimetric indicator. This tend to cause underestimation of the pollutant's concentration in the air.

The pump for this kind of system must be kept leakproof and must be calibrated. Also, the detector tubes must be handled carefully and their predicted shelf lives observed.

Measuring Aerial Dust

High-volume sampler. Several models of dust samplers in wide use draw through a filter made of cellulose paper, glass or plastic fibers, or organic membrane. Dust particles too large to pass the filter are collected on it.

In practice, a filter is dried and tared before sample collection, and the particle-laden filter is dried and weighed again at the end of a sampling period. The difference between the two is an estimate of the mass of the particles collected during the sampling period. This usually is divided by the product of sampling period and average airflow rate to give the concentration of the pollutant. Filters are commonly handled with tweezers and transported outside the laboratory in large, covered petri dishes.

Another critical factor is calibration of airflow rate. most high-volume air samplers are equipped with some sort of airflow meter, but these instruments should be calibrated frequently because errors in this estimate are perpetuated as errors in all concentration estimates.

Particle counting and sizing. Several dozen models of instruments to count airborne particles are on the market. Some are primarily collectors--based on the principle of impaction on a solid surface, impingement in a liquid medium, centrifugation, or settling--and used in conjunction with subsequent visual observation. Others are direct-reading instruments employing optics and electronics.

Impactors are most commonly used for discontinuous sampling in animal environments. Two popular instruments are the Anderson six-stage, stacked-sieve, nonviable sampler, and the four-stage cascade impactor. Both feature impaction of dust particles on pieces of glass, which are then inspected microscopically for counting. As for sizing of the particles, both instruments have several stages designed so that the polluted air is drawn through a series of jets with progressively smaller cross-sections. The result is that relatively large particles are impacted in early stages, smaller ones at later stages. The size ranges monitored by these instruments are pertinent to the site of deposition of the particles within an animal's respiratory tract.

MEASURING AERIAL MICROBES

Qualitative studies of airborne microbes in animal environments have long involved opening a petri dish of culture medium, permitting viable particles to settle out of the air onto the medium's surface. Of course, special media can be used when there is interest in particular kinds of microbes. When the aerial concentration of microbes or microbe-carrying particles must be determined, another method must be used.

All-glass impinger. One method of quantifying airborne microbes is to impinge them in an isotonic solution that can then be diluted appropriately, combined with nutrients, and cultured in preparation for counting. This method is well-adapted to situations where aerial microbic level is high, but has the disadvantage of disintegrating airborne particles containing more than one microbe so that, for instance, an airborne particle that would give rise to one colony in an animal's respiratory tract might give rise to hundreds to be counted in the culture dish.

Andersen six-stage viable sampler. The viable version of the Andersen stacked-sieve sampler holds special culture-medium plates instead of flat pieces of glass as in the nonviable model. Particles are impacted onto the solid medium's surface where colonies can grow and be counted.

The Andersen viable sampler is a very useful instrument for both counting and sizing airborne microbic particles in animal environments. Special media can be used when desirable, and both the size and the number results can be interpreted in terms of the challenge the aerial microbic particles present to the animals' respiratory tracts. On the other hand, because of its relatively high air-sampling rate, the Andersen sampler is less well-adapted to air environments in which microbic populations are very high, as in some closed animal houses during cold weather. In such case, the sampling period may have to be short as 15 seconds, and thus special care must be exercised to ensure accuracy in estimating the volume of air drawn through the instrument during the sampling period.

Andersen disposable two-stage viable sampler. A less expensive device is a disposable-plastic, two-stage sampler fashioned after the original Andersen six-stage model. Commercial petri dishes available in hospital microbiology laboratories and a variety of vacuum sources can be used with this system.

This instrument has a critical orifice providing an air-sampling rate of 1 ft^3 per min when a vacuum of at least 10 in. of mercury is maintained. When operated in this way, the colonies that grow on the upper stage have arisen from particles having an aerodynamic diameter greater than 7 ^{e}m, and hence they would not have deposited in the lungs of an animal. The particles that are impacted on the lower stage are between 1 and 7 ^{e}m in diameter, and many of these could have reached the animal's lungs.

Like the Andersen viable sampler, the disposable sampler sometimes must be operated for a short sampling period in commercial animal houses.

Also, the collection efficiency of the Andersen disposable sampler seems to be less than that of the standard version. Despite these drawbacks, the disposable model is an inexpensive, relatively accurate means of estimating the concentration of microbe-bearing particles in the air, and whether they are of such a size as to directly threaten pulmonary health.

240

REFERENCES

Anonymous. 1972. Air Sampling Instruments for Evaluation of Atmospheric Contaminants. Fourth Ed. Am. Conf. Gov. Indust. Hygienists. Cincinnati.

Curtis, S.E. 1981. Environmental Management in Animal Agriculture. Animal Environment Services, Mahomet, Illinois.

Gates, D.M. 1968. Sensing biological environments with a portable radiation thermometer. Appl. Optics 7:1803.

Hosey, A.D. and C.H. Powell (Eds.). 1967. Industrial noise--a guide to its evaluation and control. Pub. Health Serv. pub. 1572. U.S. Gov. Printing Off., Washington.

Johnstone, M.W. and P.F. Scholes. 1976. Measuring the environment. In: Control of the Animal House Environment. Vol. 7, Laboratory Animal Handbooks. Laboratory Animals, Ltd., London.

Kelly, C.F. and T.E. Bond. 1971. Bioclimatic factors and their measurement. In: A Guide to Environmental Research on Animals. Nat. Acad. Sci., Washington.

Munn, R.E. 1970. Biometerological Methods. Academic Press, New York.

Platt, R.B. and J.F. Griffiths. 1972. Environmental Measurement and Interpretation. Krieger, Huntington, NY.

Powell, C.H. and A.D. Hosey (Eds.). 1965. The industrial environment--its evaluation and control. Pub. Health Serv. Pub. 614, U.S. Gov. Printing Off., Washington.

Schuman, M.M., et al. 1970. Industrial Ventilation. Eleventh Ed. Am. Conf. Gov. Indust. Hygienists, Cincinnati.

Spencer-Gregory, H., and E. Rourke. 1957. Hygrometry. Crosby Lockwood, London.

Stern, A.C. (Ed.). 1976. Measuring, Monitoring, and Surveillance of Air Pollution. Vol. III, Air Pollution. Third Ed. Academic Press, New York.

Tanner, C.B. 1963. Basic instrumentation and measurements for plant environment and micrometerology. Soils Bull. 6, Univ. of Wisconsin, Madison.

Wolfe, H.W., et al. 1959. Sampling Microbiological Aerosols. Pub. Health Serv. Pub. 686. U.S. Gov. Printing Off., Washington.

26
MEASURING ENVIRONMENTAL STRESS IN FARM ANIMALS

Stanley E. Curtis

Relations between agricultural animals and their surroundings always have been important. Those species
recruited for domestication generally differ from their wild
cousins in that the domesticated animals are adaptable to a
wider range of environments than are their cousins (Hale,
1969). Hence, these animals we keep are more amenable to
being confined and managed by the humans they serve (Bowman,
1977).
Ecology always has been at the heart of animal production. The shelter aspect of environmental management has
been applied for a long time. shepherds kept their flocks
in folds at night thousands of years ago. Only with the
advent of widespread spacewise and time wise intensiveness
in animal agriculture have animal-environmental relations
become so important relative to other factors of production. And only with this intensiveness has major environmental modification been possible not to mention economically feasible. Now in addition to increasing the fit of
the animals to the environment, we are coming closer to
meeting the animals' needs by modifying their environments.

A DIGRESSION

"Measuring Environmental Stress in Farm Animals," calls
to mind that which we should keep in mind. Let us examine
the last five words of the title first and the first word
last.

Environmental Stress in Farm Animals

Stress is of the environment, not of the animals
(Fraser, et al., 1975; Curtis, 1981). Nevertheless, we measure stress in the animals, not in the environment. An
animal is under stress when it is required to make extreme
functional, structural, or behavioral adjustments in order
to cope with adverse aspects of its environment. Thus, an
environmental complex is stressful only if it makes extreme
demands on the animal.

In other words, an environment is not stressful in and of itself; it is stressful only if it puts an animal under stress. And because animals differ in the ways they perceive and respond to the environmental impingements, the very same environment can be stressful to one animal and not to another.

An environmental factor that contributes to the stressful nature of an environment is called a stressor. When we "measure stress in an animal" we really measure the effects of the stress: the changes the stressor causes in the animal (such as the rise in body temperature when the animal is experiencing a net gain of heat from the environment) or the responses the animal invokes in an effort to establish a normal internal state in the face of a stressor (such as the rise in breathing rate when the animal needs to increase its heat-loss rate to bring body temperature back down to the desired point).

Measuring Environmental Stress

Scientists in a wide range of disciplines have been "measuring stress in animals" with increasing frequency over the past century and a half. Almost twenty years ago, The American Physiological Society published an epic tome of some 1056 pages called Adaptation to the Environment (Dill, 1964). The means of measurement have continued to develop as the scientific inquiry in animal ecology has blossomed profusely in the intervening two decades.

Yet the measurement of stress in animals is but the first step in applying ecological knowledge to animal production. The second step is the interpretation of the values. And of the two, the second step is by far the more difficult. In particular, it is necessary to determine where stress leaves off and distress (excessive or unpleasnt stress) begins.

Interpretation of stress parameters and indices is thus the real challenge as we continue to generate more knowledge and endeavor to use more completely what is already known for the purpose of increasing the fit between agricultural animals and their environments. And so it is this interpretation step on which we shall dwell.

STRESS RESPONSES: TRADITIONAL CONCEPTS

It is the unusual moment when an animal--in the wild or on a farm--is not responding to several stressors at once. Stress is the rule, not the exception. And nature has endowed the animals with a marvelous array of reactions to these impingements.

External environment comprises all of the thousands of physical, chemical, and biological factors that surround an animal's body. Each environmental factor varies over space and time. The animal's environment is therefore exceedingly complex.

The animal must maintain a steady state in its internal environment despite fluctuating external conditions. Claude Bernard (1957) said: "All vital mechanisms, however varied they may be, have only one object, that of preserving constant the conditions of life in the internal environment." This is the concept based on negative-feedback control loops that Walter Cannon later called homeostasis. More recently it has been called homeokinesis to emphasize its dynamic, yet consistent, nature.

All sorts of external environmental elements tend to modify corresponding internal environmental elements in an animal. Ultimately, if no homeokinetic mechanism acted, the internal environment would resemble the external, and life would cease.

Homeokinetic Control Loops

The homeokinetic animal attempts to control all aspects of its internal environment via adaptive responses similar in principle to a house's temperature-control system. Neural mechanisms participate in input reception and analysis, decision-making, and effector activation. Neuroendocrine mechanisms link neural and endocrine elements and activate effectors. Endocrine mechanisms take part in neural-endocrine and endocrine-endocrine linkages, as well as effector activation, and in some cases even effector action. These processes occur in specific configurations in the animal's many specific control loops. Muscles and glands are the body's chief effectors. Effector action is usually specific for the particular remedial reaction required.

Nonspecific stress response. In addition to specific stimulus/effector activation loops, Hans Selye (1952) has developed the concept of a nonspecific initial reaction to diverse stimuli. According to this facet of Selye's general adaptation syndrome, the rate of adrenal glucocorticoid secretion increases abruptly following any insult to the body. The teleological reason for this is that glucocorticoids promote mobilization of proteins from tissues. The amino acids liberated in this way can be used either as fuel or for synthesis of other proteins, such as immunoglobulins or scar tissue, that might be crucial at the moment.

This nonspecific reaction no doubt occurs, but specific impingements sooner or later require specific counterreactions. Further, for domestic animals, the nonspecific alarm reaction seems to be superfluous, if not counterproductive, whereas insult or injury might interfere with a wild animal's getting food, and thus crucial amino acids, food-getting is ordinarily not a problem for domestic animals. Finally, all productive processes involve protein synthesis, so a high glucocorticoid secretion rate can be detrimental to food-animal performance at least in the short term.

Adaptation: Stress and Strain

An environmental adaptation refers to any functional, structural, or behavioral trait that favors an animal's survival or reproduction in a given environment, especially an extreme or adverse surrounding. Rates of life processes are the criteria used most often to assess adaptation. Adaptation can involve either an increase or a decrease in the rate of a given process.

A strain is any functional, structural, or behavioral reaction to an environmental stimulus. Strains can be adaptive or nonadaptive. Many enhance the chances of survival, but others are seemingly of little consequence.

A stress is any environmental situation--and a stressor any environmental factor--that provokes an adaptive response. A stress might be chronic (gradual and sustained) or acute (abrupt and often profound). Thus, by definition, environmental stress provokes animal strain, or in other words environmental stress provokes a stress response.

Environmental stress occurs when a given animal's environment changes so as to stimulate strain (as when environmental temperature falls below the crucial level) or when the animal itself changes in relation to a given environment (as when shearing reduces a sheep's cold tolerance).

Kinds of adaptation. There are several categories of environmental adaptation. A given animal represents one stage in a continuum of evolutionary development. An animal's heredity determines the limits of its environmental adaptability. Hence, there are genetic adaptations to environment. Genotypic changes occur naturally due to genetic mutations. Environmental stress theoretically permits mutations having adaptive utility to be realized and ultimately to become fixed in animal populations. Artificial selection pressures for productive traits reduce such natural selection pressures. But individuals selected on the basis of productive performance are at least adequately adapted to the production environment; otherwise, they would neither perform at relatively high levels nor reproduce.

There are also induced adaptations. A given stressful environmental complex provokes various responses depending on the individual animal's current adaptation status, which is determined by heredity and by its life history, as well.

Acclimation is one kind of induced adaptation. It refers to an animal's compensatory alterations due to a single stressor acting alone, usually in an experimental or artificial situation, over days or weeks. A hen in a layer house might acclimate to altered day length, for example. Acclimatization, on the other hand, refers to reactions over days or weeks to environments where many environmental factors vary at the same time. A ewe at pasture acclimatizes to seasonal variations in day-length in conjunction with variations in other environmental factors.

Finally, an animal may become habituated to certain stimuli when they occur again and again. Sensations and effector responses associated with particular environmental stimuli tend to diminish when these stimuli occur repeatedly. A pig raised near an airport becomes habituated to the roars of jet airplanes, for example.

Level of Adaptation. An animal's environmental adaptation can be analyzed at several levels of organization. At one end of the spectrum, adaptations can take the form of enzyme inductions or of changes in other modifiers of catlyzed biochemical reactions. In the middle are changes associated with adaptive responses to environmental stress in sensory, integrative, and coordinative neural functions, in neuroendocrine and endocrine functions, and in effectors' outputs. At the other end of the range, the animal's behavior often changes in response to environmental stimuli. Malcolm Gordon (1972) said: "There is certainly no logical basis for any claim that understanding the nature of life at one level of organization is more fundamental to overall understanding than comprehension at any other level."

STRESS RESPONSES: PSYCOLOGICAL COMPONENTS

It is now generally recognized that the amount of stress an animal is under depends not only on the intensity and duration of the noxious agent (the traditional concept), but on the animal's ability to modify the effects of the stressor as well (Mason, 1975; Archer, 1979).

Lack of Control

A recent study of stress effects on tumor rejection demonstrated psycological components of stress responses (Visintainer, et al., 1982). Stressors such as mild electrical shock depress an animal's ability to reject certain tumors in experimental settings. In this particular experiment, individually held rats were inoculated with a standard dose of tumor-causing cells and assigned to three treatments: control (no shock), mild shock that could be stopped by pressing a switch (escapable shock), and mild shock that stopped anytime the escapable-shock rate in the trial pressed its switch but over which this rat itself had no control (inescapable shock).

In other words, the amount of physical impingement received by animals in the two shock treatments was the same, but those in one group (escapable shock) could control the duration, while those in the other group (inescapable shock) could not. Fifty-four percent of the control rates rejected their tumors. Inescapable shock caused so much stress that tumor rejection occurred in only 27% of the rats, while 63% of those subjected to escapable shock rejected their tumors. The conclusion: the low rate of tumor rejection was due not to the shock itself, but to the animal's inability to control this stressor.

Alliesthesia Modification

Central perception ("alliesthesia") of stress intensity depends on the context within which it occurs. Alliesthesia in the form of comfort rating or pleasure rating is affected by the animal's internal state and, hence, by its external surroundings as well.

For example, a thermal stimulus can feel pleasant or unpleasant depending on the body's thermal status. Hypothermic humans find cold stimuli very unpleasant and hot very pleasant, while hyperthermic humans have the opposite perceptions (Cabanca, 1971). Similarly, gastric loading with glucose decreased the human subjects' pleasure rating of the sweet taste of sucrose in a thermoneutral environment (26°C), but this negative alliesthesia due to glucose loading was eliminated when ambient temperature was reduced to 4°C (Russek, et al., 1979).

These findings remind us that "variety is the spice of life" and suggest that "taking the bitter with the sweet" is pleasurable in the long run. Extrapolating the concept of alliesthesia modification to agricultural animals' lives, it would appear that stress of one sort often primes the animal to receive pleasure from some other aspects of its environment.

In any case, the fact that an animal's psychological state can modify its perception of stress makes it all the more difficult to interpret how a specific stressor is affecting a specific animal.

MEASURING AND INTERPRETING STRESS RESPONSES

The scientific literature stores report of hundreds of experiments purported to measure stress in food animals (Hafez, 1968; Hafez, 1975; Johnson, 1976a,b; Stephens, 1980; Craig, 1981; Curtis, 1981). It is a relatively simple task to subject experimental animals to a controlled stressor and measure a resultant change in some physiological, anatomical, or behavioral parameter. Hormonal, cardio-respiratory, and heat-production parameters have been studied most in the past. Behavioral and anatomical changes are being characterized more lately. But an objective index of stress in terms of animal health, performance, and well-being has been elusive. As Graham Perry (1973) said: "Even marked physiological changes may indicate only that an animal is successfully adapting to its environment--not necessarily that it is succumbing to adversity." And, similarly, Ian Duncan (1981) said: ". . . it should be of no surprise that chickens behave differently in different environments. This may simply demonstrate how adaptable they are." Again: at what point does stress become dis-stressful.

As for methodology, it is very difficult to study the effects of specific supposed stressors on an animal without introducing artifacts due to the stressfulness of the investigative techniques themselves (Adler, 1976). This is especially so in real or simulated production situations. What is the baseline adrenal-glucocorticoid secretion rate of an animal? Will it ever be certain beyond a reasonable doubt that the experimental manipulation necessary to obtain the needed samples or observations is not itself so stressful as to compromise the results?

Also, interpretation of the results of this kind of research is hampered by the fact that, by and large, there is not yet consensus as to the meaning of data on specific behavioral and hormonal changes in responses to stressors. What does it mean when an animal increases breathing rate by 250% in one environment compared with another? Does a 65% increase in plasma glucocorticoid concentration indicate the animal is under stress? If so, is the stress mild or severe? Scientists still do not understand how findings such as these relate to an animal's well-being, its health, and its productivity; consequently, we cannot rely on physiological or behavioral traits as valid indicators of the amount of stress an animal actually perceives, let alone how these might be related to the animal's health and productivity.

STRESS AND PRODUCTIVITY

Environmental stress generally alters animal performance (Curtis, 1981). The stress provokes the animal to react, and this reaction can influence the partition of resources among maintenance, reproductive, and productive functions in one or more of five ways:

1. The reaction may alter internal functions. Many bodily functions participate in productive processes as well as in reactions to stress. Survival responses may thus unintentionally affect productive preformance. For example, increased adrenal glucocorticoid secretion in response to stress can impair growth.

2. The reaction may divert nutrients. When an animal resonds to stress, it in effect diverts nutrients to use in higher-priority maintenance processes. Adaptive reactions are implemented even at the expense of productivity.

3. The reaction may reduce productivity directly. The animal's response sometimes partly comprises intentional reductions in productive processes. This generally frees some nutrients for maintenance uses. For example, an animal might reduce its productive rate in a hot environment in an attempt to re-establish heat balance with its surroundings.

4. The reaction may increase variability. Individual animals within a species differ from each other in functions, behaviors, and structures by what have been called "individuality differentials." Individual animals therefore differ in their responses to the same environmental stressor. In other words, two animals in the same group might successfully cope with the same stressful situation by calling different mechanisms into play. Then, if the complements of mechanisms used by the two individuals differ in the energy expenditure required to achieve them, the amount of energy diverted from productive processes will be different for the two animals. The result of this is that the amount of variation in individual performance in a group of animals tends to be related directly with the environmental adversities to which the animals are subjected.

5. The reaction may impair disease resistance. Because the animal's reaction to stress can impair disease resistance, that reaction influences the frequency and severity of disease. Of course, infection itself is a stress, so once established it in turn can influence the animal's productive performance. The mechanisms involved in the relations between environmental stress and resistance against infectious disease are just now being elucidated.

Kelley (1980) identified eight stressors: heat, cold, crowding, regrouping, weaning, limit-feeding, noise, and movement restraint. He documented the fact that all of these have been accorded a central role in stress-induced alterations of resistance against infection.

Having developed a framework for analyzing relations between adverse environments and animal productivity, it would be unrealistic to leave the impression that the link between stresses and productive processes are clear and simple. Consider two examples.

Lactating dairy cows held in a natural subtropical summer environment and provided no shelter are obviously under severe stress at mid-afternoon. They have markedly higher body temperatures and respiratory rates than their herdmates under the shade. Yet there might be no significant difference in fat-corrected milk yield between the two groups of cows (Johnson et al., 1966).

Socially and physically deprived animals often grow faster than do their counterparts in more enriched environments (Fiala et al.) So there is a risk in assuming that an animal stressed by a specific environmental complex is necessarily unfit for productive use in that environment. While one often might be justified in presuming that strain against stress reduces animal productivity, the animal can still be putting out an acceptable amount of product per unit of resource input.

Robert McDowell (1972) refers to "physiological adaptability" to environment (measured by physiological traits such as breathing rate) that is associated with survival responses, and to "performance adaptability" to environment (measured by productive-performance traits such as growth rate). These two often bear little positive relation to each other.

Thus, it is not sufficient for an animal producer to be concerned only with physiological and behavioral indices of environmental adaptability. Producers are more interested in the size of decrement, if any, in production associated with an animal's living in a particular environment. And to learn the quantitative effects of a given environment on animal performance, the productive traits themselves must be measured. After all, knowing a hen's breathing rate tells one little or nothing about her rate of lay.

There has been unfortunate ambiguity on this point among researchers and producers alike. An animal exhibiting marked strain has generally been assumed to be having markedly depressed performance. This is not necessarily so. Indeed, visible strain signifies that the animal is attempting to compensate for an environmental impingement. These attempts might succeed, and they might interfere with production only slightly or not at all.

The marvelous homeokinetic phenomena they possess make for resilient beasts and birds on our farms and permit profitable performance in a wide range of circumstances. The response flexibility that animals demontrate in the face of myriad stressors seem more remarkable than those instances when defensive reactions are inadequate and the environmental complex drastically reduces health or performance.

REFERENCES

Adler, H. C. 1976. Ethology in animal production. Livestock Prod. Sci. 3:303.

Archer, J. 1979. Animals Under Stress. Edward Arnold, London.

Bernard, C. 1957. An Introduction to the Study of Experimental Medicine. Dover, New York.

Bowman, J. C. 1977. Animals for Man. Edward Arnold, London.

Cabanac, M. 1971. Physiological role of pleasure. Science 173:1103.

Cannon, W. B. 1932. The Wisdom of the Body. Norton, New York.

Craig, J. V. 1981. Domestic Animal Behavior. Prentice-Hall, Englewood cliffs.

Curtis, S. E. 1981. Environmental Management in Animal Agriculture. Animal Environment Services, Mahomet, Illinois.

Dill, D. B. (Ed.) 1964. Handbook of Physiology. Section 4: Adaptation to the Environment. American Physiological Society, Washington.

Duncan, I. J. H. 1981. Animal rights-animal welfare: a scientist's assessment. Poul. Sci. 60:489.

Fiala, B., F. M. Snow, and W. T. Greenough. 1977. "Impoverished" rates weigh more than "enriched" rats because they eat more. Devel. Phychobiol. 10:537.

Fraser, D., J. S. D. Ritchie, and A. F. Fraser. 1975. The term "stress" in a veterinary context. Brit. Vet. J. 131:653.

Gordon, M. S. 1972. Animal Physiology: Principles and Adaptations. (Second ed.) Macmillan, New York.

Hafez, E. S. E. (Ed.) 1968. Adaptation of Domestic Animals. Lea and Febiger, Philadelphia.

Hafez, E. S. E. (Ed.) 1975. The Behavior of Domestic Animals (Third ed.) Williams and Wilkins, Baltimore.

Hale, E. B. 1969. Domestication and the evolution of behavior. In: E. S. E. Hafez (Ed.). The Behavior of Domestic Animals. William and Wilkins, Baltimore.

Johnson, H. D. (Ed.) 1976a. Progress in Animal Biometerology, Volume 1, Part I. Swets and Zeitlinger, Amsterdam.

Johnson, H. D. (Ed.) 1976b. Progress in Animal Biometerology, Volume 1, Part II. Swets and Zeitlinger, Amsterdam.

Johnston, J. E., J. Rainey, C. Breidenstein, and A. J. Gidry. 1966. Effects of ration fiber level on feed intake and milk production of dairy cattle under hot conditions. Proc. Fourth Int. Biometerological Cong., New Brunswick.

Kelley, K. W. 1980. Stress and immune function: A bibliographic review. Ann. Vet. Res. 11:445.

Mason, J. W. 1975. Emotion as reflected in patterns of endocrine integration. In: L. Levi (Ed.) Emotions-- Their Parameters and Measurement. Raven, new York.

McDowell, R. E. 1972. Improvement of Livestock Production in Warm Climates. Freeman, San Francisco.

Perry, G. 1973. Can the physiologist measure stress? New Scientist 60 (18 October):175.

Russek, M. M. Fantino, and M. Cabanac. 1979. Effect of environmental temperature on pleasure ratings of odor and testes. Physiol. Behav. 22:251.

Selye, H. 1952. The Story of the Adaptation Syndrome. Acata, Montreal.

Stephens, D. B. 1980. Stress and its measurement in domestic animals: a review of behavioral and physiological studies under field and laboratory situations. Adv. Vet. Sci. Comp. Med. 24:179.

Visintainer, M. A., J. R. Volpicelli, and M. E. P. Seligman. 1982. Tumor rejection in rats after inescapable or escapable shock. Science 216:437.

27
LIVESTOCK PSYCHOLOGY AND HANDLING-FACILITY DESIGN

Temple Grandin

Handling your cattle and sheep will be much easier if you learn a little livestock psychology. Many people do not realize that cattle and sheep have panoramic vision and they can see all around themselves without turning their heads (Prince, 1977; McFarlane, 1976). Sheep with heavy fleeces would have a more restricted visual field depending on the amount of wool on their head and neck. Both cattle and sheep depend heavily on their vision and are easily motivated by fear (Kilgour, 1971). Livestock are sensitive to harsh contrasts of light and dark around loading chutes, scales, and work areas. "Illumination should be even and there should be no sudden discontinuity in the floor level or texture" (Lynch and Alexander, 1973).

Solid shades should be used over the working, loading, and scale areas (Grandin, 1981). Slatted shades are fine for areas where the animals live and feel familiar. However, when the animals come into the handling areas they are often nervous. The zebra stripe pattern cast by the slatted shades constructed from snow fence or corrugated sheets suspended on cables will cause balking. The pattern of alternating light and dark has the same effect as building a cattle guard in the middle of the facility. Contrasts of light and dark have such a deterrent effect on cattle that in Oregon lines are painted across the highway to take the place of expensive steel cattle guards.

Shadowy stripes will cause balking problems with sheep. A single-file chute for sorting sheep should be oriented so that the sun does not form a shadow down the middle of the chute. The worst possible situation for sheep is to have half the floor of the chute in the shade and the other half in the sunlight. In shearing sheds and sheep holding areas, the wooden slats on the floor should face so that the sheep walk across the slats instead of in the same direction as the slats (Hutson, 1981). If you get down on your hands and knees and look at the floor, the floor appears more solid if you move across the slats. The floor should also be constructed to prevent sunlight from shining up through the slats.

256

A single shadow that falls across a scale or loading chute can disrupt handling. The lead animal will often balk and refuse to cross the shadow. If you are having problems with animals balking at one place, a shadow is a likely cause. Balking can also be caused by a small bright spot formed by the sun's rays coming through a hole in a roof. Patching the hole will often solve the problem. Handlers themselves should be cautious about causing shadows. Figure 1 illustrates a shadow that was formed when the handler waved at the cattle. The animals refused to approach the shadow of the waving handler cast at the entrance to the single-file chute.

Figure 1.   The handler's shadow cast on the entrance to the single-file chute caused the cattle to balk. This is just one of the many kinds of shadows which can cause balking problems in your cattle handling facility.

APPROACH LIGHT

Both cattle and sheep have a tendency to move towards the light. If you ever have to load livestock at night, it is strongly recommended that frosted lamps that do not glare in the animals face be positioned inside of the truck (Grandin, 1979). However, loading chutes and squeeze chutes should face either north or south; livestock will balk if they have to look directly into the sun. Sometimes it is difficult to persuade cattle or sheep to enter a roofed working area. Persuading the animals to enter a dark, single-file chute from an outdoor crowding pen in bright sunlight is often difficult. Cattle are more easily driven into a shaded area from an outdoor pen if they are first lined up in single file. Many people make the mistake of placing the single-file chute and squeeze chute entirely inside a building and the crowding pen outside. Balking will be reduced if the single file chute is extended 10 to 15 feet outside the building. The animals will enter more easily if they are lined up single file before they enter the dark building. The wall of the building should NEVER be placed at the junction between the single file chute and the crowding pen. Either cover up the entire squeeze chute and crowding pen area or extend the single file chute beyond the building. If you have just a shade over your working area, make sure that the shadow of the shade does not fall on the junction between the single file chute and the crowding pen.

PREVENT BALKING

Drain grates in the middle of the floor will make both sheep and cattle balk because the animals will often refuse to walk over them. A good drainage design is to slope the concrete floor in the squeeze chute area toward an open drainage ditch located outside the fences. The open drainage ditch outside the fences needs no cover and so it is easier to clean.

Animals will also balk if they see a moving or flapping object. A coat flung over a chute fence or the shiny reflection off a car bumper will cause balking. You should walk through your chutes and view them from a cow's eye level before moving or loading animals. You will be surprised at the things you may see. When cattle and sheep are being worked, the handlers should stand back away from the headgate so that approaching animals cannot see them with their wide angle vision. The installation of shields for people to hide behind can facilitate the movement of livestock (Kilgour, 1971; Freeman, 1975).

Problems with balking tend to come in bunches; when one animal balks, the tendency to balk seems to spread to the next animals in line (Grandin, 1980). When an animal is being moved through a single-file chute, the animal must never be prodded until it has a place to go. Once it has balked, it will continue balking. The handler should wait until the tailgate on the squeeze chute is open before prodding the next animal (Grandin, 1976). A plastic garbage bag attached to a broom handle is a good tool for moving cattle in pens. The cattle move away from the rustling plastic. When livestock are being moved, well-trained dogs are recommended for open areas and large pens. Once the animals are confined in the crowding pen and single-file chute, dogs should not be allowed near the fences where they still can bite at the cattle or sheep.

SOLID CHUTE SIDES

For both cattle and sheep the sides of the single-file chute, loading chute, and crowding pen should be solid. Solid sides prevent the animals from seeing people, cars, and other distractions outside the chute. A study with sheep showed that they moved more rapidly through a single-file chute that had solid sides (Hutson and Hitchcock, 1978). The principle of using solid sides is like putting blinkers on the harness horse. The blinkers prevent the horse from seeing distractions with his wide-angle vision. Cattle and sheep in a handling facility should be able to see only one pathway of escape--this is extremely important. They should be able to see other animals moving in front of them down the chute, when sheep are being sorted, the approaching animals should be able to see the previously sorted sheep through the end of the sorting chute.

Livestock will balk if a chute appears to be a dead end (Brockway, 1975; Hutson, 1980). Sliding and one-way gates in the single-file chute must be constructed so that your animals can see through them, otherwise the animals will balk (figure 2). The sides of the single-file chute and the crowding pen should be solid. The crowding-pen gate also should be solid so that animals cannot see through and will head for the entrance to the single-file chute (Rider, 1974). Mirrors could be used to attract sheep into pens and other areas that appear to be a dead end. The sheep are attracted to the image of sheep in the mirror (Franklin & Hutson, 1982).

HERD BEHAVIOR

All species of livestock will follow the leader and this instinct is strong in both cattle and sheep (Ewbank, 1961). Many people make the mistake of building the single-file chute to the squeeze too short. The chute should be long enough to take advantage of the animal's tendency to follow the leader. The minimum length for the single-file chute is 20 ft. In larger facilities 30 to 50 lineal ft is recommended.

Cattle and sheep are herd animals and, if isolated, can become agitated and stressed. This is especially a problem with Brahman-type cattle. An animal left alone in the crowding pen after the other animals have entered the single file chute, may attempt to jump the fence to rejoin its herdmates. A lone steer or cow may become agitated and charge the handler. A large portion of the serious handler

injuries occur when a steer or cow, separated from its herdmates, refuses to walk up the single file chute. When a
lone animal refuses to move, the handler should release it
from the crowding pen and bring it back with another group
of cattle.

Figure 2.   The single-file chute to the squeeze should have
            solid sides to prevent the cattle from seeing
            distractions outside the fence. Sliding gates in
            the single-file chute must be constructed from
            bars so that the cattle can see through them.
            Solid sliding or one-way gates will cause
            balking.

EFFECTS OF SLOPE AND WIND

        To prevent livestock from piling up against the back
gate in the crowding pen, the floor of the pen must be
level. A 10° slope in the crowding pen will cause the animals to pile and fall down against the crowding gate.   A
small 1/4 in. to 1/8 in. slope per foot for drainage will
not cause a handling problem. Livestock move more easily
uphill than down, but they move most easily on a flat surface (Hitchcock and Hutson, 1979).
        Research by Hutson and Mourik (1982) indicates that
sheep will move more easily when they are heading into the
wind.   Heading into the wind can stimulate sheep to start
moving along a chute.

WHY A CURVED CHUTE WORKS

A curved chute works better than a straight chute for two reasons. First it prevents the animal from seeing the truck, the squeeze chute, or people until it is almost in the truck or squeeze chute. A curved chute also takes advantage of the animal's natural tendency to circle around the handler (Grandin, 1979). When you enter a pen of cattle or sheep you have probably noticed that the animals will turn and face you, but maintain a safe distance (figure 3). As you move through the pen, the animals will keep looking at you and circle around you as you move. A curved chute takes advantage of this natural circling behavior.

Figure 3. When you walk through a pasture the cows will turn and look at you. They will circle around you as you move about the pasture. Curved chutes take advantage of the cow's circling behavior.

Cattle can be driven most efficiently if the handler is situated at a 45° to 60° angle perpendicular to the animal's shoulder (Williams, 1978) (figure 4). A well-designed, curved single-file chute has a catwalk for the handler to use along the inner radius. The handler should always work along the inner radius. The curved chute forces the handler to stand at the best angle and lets the animals circle around him. The solid sides block out visual distractions except for the handler on the catwalk.

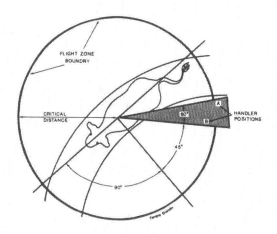

Figure 4. The shaded area shows the best position for
moving an animal. To make the cow move forward
the handler moves into Position B which is just
inside the boundary of the flight zone. The
handler should retreat to Position A if he wants
the animal to stop. The solid curved lines
indicate the location of the curved single-file
chute.

The catwalk should run alongside of the chute and NEVER
be placed overhead. The distance from the catwalk platform
to the top of the chute fence should be 42 inches. This
brings the top of the fence to belt-buckle height on the
average person.

Figures 5 and 6 illustrate curved facilities for handling cattle and sheep. Curved designs are recommended by
both Grandin (1980) and Barber (1977).

FLIGHT DISTANCE

When a person penetrates an animal's flight zone, the
animal will move away. If the handler penetrates the flight
zone too deeply, the animal will either turn back and run
past him or break and run away. Kilgour (1971) found that
when the flight zone of bulls was invaded by a mechanical
trolley, the bulls would move away and keep a constant distance between themselves and the trolley. When the trolley
got too close the bulls bolted past it. The best place for
the handler to work is on the edge of the flight zone. This
will cause the animals to move away in an orderly manner.
The animals will stop moving when the handler retreats from
the flight zone.

The size of the flight zone varies depending on the
tameness or wildness of the animal. The flight zone of
range cows may be as much as 300 ft whereas the flight zone

262

Figure 5. Cattle handling facility utilizing a curved
single-file chute, round crowding pen, and wide
curved lane. Up to 600 cattle per hour can be
moved through the dip vat with only three
people. The handlers work along the inner radius
of the single-file chute and the wide curved lane
(designed by Temple Grandin).

Figure 6. Sheep handling and sorting facility with a curved
bugle crowding pen. The inner radius is solid to
prevent the sheep from seeing the handler
standing at the sorting gates (designed by Adrian
Barber, Australia).

of feedlot cattle may be only 5 to 25 ft (Grandin, 1978). Extremely tame cattle or sheep are often difficult to drive because they no longer have a flight zone. Many people make the mistake of getting too close to the cattle when they are driving them down an alley or putting them in a crowding pen. Getting too close makes cattle feel cornered. If the cattle attempt to turn back, the handler should back up and retreat to remove himself from the animal's flight zone instead, of moving in closer.

Cattle will often rear up and get excited while waiting in the single-file chute. The most common cause of this problem is the handler leaning over the single file chute and deeply penetrating the animal's flight zone. The cattle will usually settle down if the handler backs up.

When sheep are being handled in a confined area, pile-ups can occur if their flight zone is deeply penetrated. This is why dogs should not be used in the crowding pen or the single-file chute, because a dog, in a confined area, deeply penetrates the flight zone and the sheep have no place for escape. Dogs are recommended only for open areas and larger pens where there is room for the sheep to move away. During handling, minimize yelling and screaming so as to avoid enlarging the size of the animal's flight zone.

BREED DIFFERENCES

The breed of the cattle or sheep can affect the way it reacts to handling. Cattle with Brahman blood are more excitable and may be harder to handle than the English breeds. When Brahman or Brahman-cross cattle are being handled, it is important to keep them as calm as possible and to limit use of electric prods. Brahman and Brahman-cross cattle can become excited; they are difficult to block at gates (Tulloh, 1961) and prone to ram into fences. With this type of cattle it is especially important to use substantial fencing. If thin rods are used for fencing, a wide belly rail should be installed to present a visual barrier. Angus cattle tend to be more nervous than Herefords (Tulloh, 1961). Holstein cattle tend to move slowly (Grandin, 1980). Brahman cattle tend to stay together in a more cohesive mob than English cattle.

Brahman and Brahman-cross cattle can become so disturbed that they will lie down and become immobile, especially if they have been prodded repeatedly with an electric prod (Fraser, 1960). When a Brahman or Brahman-cross animal lies down, it must be left alone for about five minutes or it may go into shock and die. This problem rarely occurs in English cattle or European cattle such as Charolais.

There are distinct differences in the way various breeds of sheep react during handling (Shupe, 1978; Whately et al., 1974). Rambouillet sheep tend to bunch tightly together and remain in a group; crossbred Finn sheep tend to turn, face the handler, and maintain visual contact. If the handler penetrates the collective flight zone of a group of Finn sheep, they will turn and run past the handler.

Cheviots and Perendales are the easiest to drive into a crowding pen; the Romney, Merino-Romney cross, and the Dorset-Romney are the most difficult. The Romney tends to follow the leader but it is easily led into blind corners. Cheviots have a strong instinct to maintain visual contact with the handler and to display more independent movements than other breeds.

DARK BOX AI CHUTE

For improved conception rates, cows should be handled gently for AI and not allowed to become agitated or overheated. The chute used for AI should not be the same chute used for branding, dehorning, or injections. The cow should not associate the AI chute with pain. Cows can be easily restrained for AI or pregnancy testing in a dark box chute that has no headgate or squeeze (Parsons & Helphinstine, 1969; Swan, 1975). Even the wildest cow can be restrained with a minimum of excitement. The dark box chute can be easily constructed from plywood or steel. It has solid sides, top, and front. When the cow is inside the box, she is inside a quiet, snug, dark enclosure. A chain is latched behind her rump to keep her in. After insemination the cow is released through a gate in either the front or the side of the dark box. If wild cows are being handled, an extra long dark box can be constructed. A tame cow that is not in

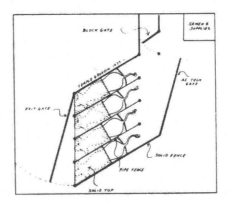

Figure 7.   Chutes for A.I. can be laid out in a herringbone design. The two outer fences and the grates should be solid. The inner partition in between the cows should be constructed from bars. Cows will stay calmer if they know they have company.

heat is used as a pacifier and is placed in the chute in front of the cow to be bred. Even a wild cow will stand quietly and place her head on the pacifier cow's rump. After breeding, the cow is allowed to exit through a side gate, while the pacifier cow remains in the chute. If a large number of cows have to be pregnancy checked or inseminated, two to six AI chutes can be laid out in a herringbone pattern (figure 7). This design is recommended by McFarlane (1976) from South Africa. The chutes are set on a 60° angle. They are built like regular dark box AI chutes except that the partitions inbetween the cows are constructed from open bars so the cows can see each other. The cows will stand more quietly if they have company. The two outer fences should be solid. If the cows are reluctant to enter the dark box, a small 6 in. by 12 in. window can be cut in the solid front gate in front of each cow.

LOADING CHUTE DESIGN

Loading chutes should be equipped with telescoping side panels and a self-aligning dock bumper. These devices will help prevent foot and leg injuries caused by an animal stepping down between the truck and the chute. The side panels will prevent animals from jumping out the gap between the chute and the truck.

A well-designed loading ramp has a level landing at the top. This provides the animals with a level surface to walk on when they first get off the truck. The landing should be at least 5 feet wide for cattle. Many animals are injured on ramps that are too steep. The slope of a permanently installed cattle ramp should not exceed 20°. The slope of a portable or adjustable chute should not exceed 25° (Grandin, 1979). Steeper ramps may be used for loading sheep but they are NOT recommended for unloading. Sheep will move up a steep ramp readily.

If you build your ramp out of concrete, stairsteps are strongly recommended. For cattle the steps should have a 3.5 to 4 in. rise and a 12 in. tread width. The surface of the steps should be rough to provide good footing. For sheep the steps should have a 2 in. rise and a 10 in. tread width.

On adjustable or wooden ramps, the cleats should be spaced 8 in. apart from the edge of one cleat to the edge of the next cleat (Mayes, 1978). The cleats should be 1 1/2 to 2 in. high for cattle and 1 in. by 1 in. for sheep.

Chutes for both loading and unloading cattle should have solid sides and a gradual curve (figure 8). If the curve is too sharp, the chute will look like a dead end when the animals are being unloaded. A curved single-file chute is most efficient for forcing cattle to enter a truck or a squeeze chute. A chute used for loading and unloading cattle should have an inside radius of 12 ft to 17 ft, the bigger radius is the best. A loading chute for cattle should be 30 in. wide and no wider. The largest bulls will fit through a 30 in. wide chute. If the chute is going to be used exclusively for calves, it should be 20 to 24 in. wide.

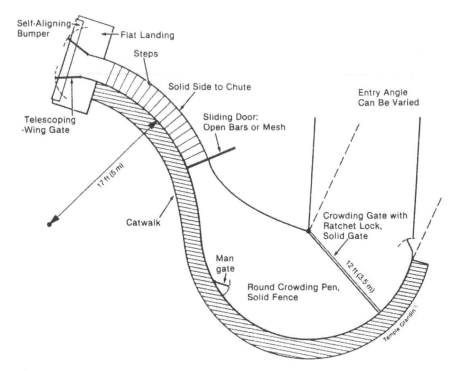

Figure 8. Curved loading chute with a round crowding pen. The sides of the chute are solid.

In auctions and meat packing plants where a chute is used to unload only, a wide straight chute should be used. This provides the animals with a clear path to freedom. These chutes can be 6 to 10 ft wide. A wide, straight chute should not be used for loading cattle.

SHEEP LOADING

Since most trucks have a 30 in. wide door, a good chute design for sheep is a 30 in. wide ramp that enables two sheep to walk up side by side. If a single-file chute is used, it should be 17 in. to 18 in. wide. The chute should be designed so that the animals walk up either in a single file or two abreast. Don't build a chute that is one and one-half animals wide--this creates jamming problems.

A wide ramp is recommended for loading sheep into shearing sheds or onto trucks that can be opened up the full width of the vehicle. In Australia sheep moved easily up ramps 8 to 10 ft wide when loading sheep onto ships for shipment to the Middle East (figure 9). The entrance ramp into a raised shearing shed should be 8 ft to 10 ft wide (Simpson, 1979).

Figure 9. A wide ramp is used to load sheep onto a ship in Australia. Once the flow of sheep was started the animals moved easily up the ramp.

REFERENCES

Barber, A. 1977. Bugle sheep yards. Fact Sheet, Dept. of Agriculture and Fisheries South Australia, Adelaide, Australia.

Brockway, B. 1975. Planning a sheep handling unit. Farm Buildings Center, Nat. Agr. Center, Kenilworth, Warickshire, England.

Ewbank, R. 1961. The behavior of cattle in crises. Vet. Rec. 73:853.

Franklin, J. R., G. D. Hutson. 1982. Experiments on attracting sheep to move along a laneway. III Visual Stimuli, Appl. Animal Ethology 8:457.

Fraser, A. F. 1960. Spontaneously occurring forms of "tonic immobility" in farm animals. Canad. J. Comp. Med 24:330.

Freeman, R. B. 1975. Functional planning of a shearing shed. Pastoral Review 85:9.

Grandin, T. 1981. Innovative cattle handling facilities. In: M.E. Ensminger (Ed.). Beef Cattle Science Handbook, 18:117. Agriservices Foundation, Clovis, Calif.

Grandin, T. 1980. Livestock behavior as related to handling facilities design. Int. J. Stud. Animal Problems 1:33 etc.

Grandin, T. 1979. Understanding animal psychology facilitates handling livestock. Vet. Med. and Small Animal Clinician 74:697.

Grandin, T. 1978. Observations of the spatial relationships between people and cattle during handling. Proc. Western Sec. Amer. Soc. of Animal Sci. 29:76.

Grandin, T. 1976. Practical pointers on handling cattle in squeeze chutes, alleys, and crowding pens. In: M.E. Ensminger (Ed.). Beef Cattle Science Handbook 13:228.

Hitchcock, D. K., G. D. Hutson. 1979. The movement of sheep on inclines. Australian J. Exp. Agr. and Animal Husbandry 19:176.

Hutson, G. D., S. C. van Mourik. 1982. Effect of artificial wind on sheep movement along indoor races. Australian J. Exp. Agr. and Animal Husbandry 22:163.

Hutson, G. D. 1981. Sheep movement on slatted floors. Australian J. Exp. Agr. and Animal Husbandry 21:474.

Hutson, G. D. 1980. The effect of previous experience on sheep movement through yards. Appl. Animal Ethology 6:233.

Hutson, G. D. and D. K Hitchock. 1978. The movement of sheep around corners. Appl. Animal Ethology 4:349.

Kilgour, R. 1971. Animal handling in works, pertinent behavior studies. 13th Meat Industry, Res. Conf. Hamilton, New Zealand. pp 9-12.

Lynch, J. J. and G. Alexander. 1973. The Pastoral Industries of Australia. pp 371. Sydney University Press, Sydney, Australia.

Mayes, H. F. 1978. Design criteria for livestock loading chutes. Technical Paper No. 78-6014, Amer. Soc. Agr. Eng. St. Joseph, Michigan.

McFarlane, I. 1976. Rationale in the design of housing and handling facilities. In: M. E. Ensminger (Ed.). Beef Cattle Science Handbook 13:223.

Parsons, R. A. and W. N. Helphinstine. 1969. Rambo AI breeding chute for beef cattle. One-Sheet-Answers, University of California Agricultural Extension Service, Davis, California.

Rider, A., A. F. Butchbaker and S. Harp. 1974. Beef working, sorting, and loading facilities. Technical Paper No. 74-4523, Amer. Soc. Agr. Eng. St. Joseph, Michigan.

Shupe, W. L. 1978. Transporting sheep to pastures and markets. Technical Paper No. 78-6008, Amer. Soc. Agr. Eng. St. Joseph, Michigan.

Simpson, I. 1979. Building a modern shearing shed. Division of Animal Industry Bulletin A3.7.1. New South Wales Dept. of Agr., Australia.

Swan, R. 1975. About AI facilities. New Mexico Stockman. Feb., pp 24-25.

Tulloh, N. M. 1961. Behavior of cattle in yards: II. A study of temperament. Animal Behavior 9:25.

Whately, J., R. Kilgour and D. C. Dalton. 1974. Behavior of hill country sheep breeds during farming routines. New Zealand Soc. Animal Production 34:28.

Williams, C. 1978. Livestock consultant, personal communication.

28

DESIGN OF CORRALS,
SQUEEZE CHUTES, AND DIP VATS

Temple Grandin

CORRALS

A corral constructed with round holding pens, diagonal sorting pens, and curved drive lanes will enable you to handle cattle more efficiently because there is a minimum of square corners for the cattle to bunch up in. The principle of the corral layout in figure 1 is that the animals are

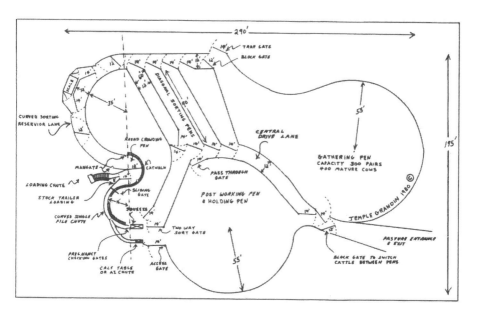

Figure 1. General purpose corral system for shipping, branding, sorting, and AI. It can handle 300 cow and calf pairs or 400 mature cows. Capacity can be increased by adding more diagonal pens and holding pen space (Grandin, 1981).

gathered into the big round pen and then directed to the curved sorting reservoir lane for sorting and handling. The curved sorting reservoir lane serves two functions: It holds cattle back into the diagonal pens, that are being sorted. It also holds cattle waiting to go to the squeeze chute, AI chute, or calf table.

Large Corral

The corral shown in figure 1 is a general-purpose system for shipping calves, working calves, sorting, pregnancy checking, and AI. It can handle 300 cow-calf pairs or 400 mature cows. It is equipped with a two-way sorting gate in front of the squeeze chute for separating the cows that are pregnant from cows that are open. Depending upon your needs, you can position either the squeeze chute, AI chute, or calf table at the sorting gate. If the cattle are watered in the large gathering pen, they will become accustomed to coming in and out of the trap gate. When you need to catch an animal, you merely shut the trap gate and direct her up the curved reservoir lane to the chutes. This is an especially handy feature for AI.

The curved sorting reservoir terminates in a round crowding pen and curved single-file chute. The crowding gate has a ratchet latch that locks automatically as the gate is advanced behind the cattle. To load low stock trailers, open a 8 ft gate that is alongside the regular loading chute. This provides you with the advantage of the round crowding pen for stock trailers. All fences in the curved single-file chute and the round crowding pen are solid. The ratchet crowd gate also should be covered with sheet metal or plywood.

Figure 1 can also be adapted for use with a prefabricated steel circle crowding pen and curved single-file chute. Since the prefabricated units have a 12 ft radius instead of the 16 ft radius shown in the drawing, you will have to move the sorting gate. If you plan to build the entire setup yourself, out of either wood or steel, keep the 16 ft radius, especially if you have large cows.

Diagonal Sorting Pens

When cows and calves are being separated, the calves are held in the diagonal pens and the central drive lane, and the cows are allowed to pass through one of the diagonal pens into the large post working pen. The diagonal pens and the central drive lane in figure 1 can hold 300 weaned calves overnight or 500 weaned calves crowded together. Each 70 by 12 ft diagonal pen holds 60 weaned calves overnight or 85 weaned calves crowded together. If the mother cows are put in the diagonal pens, each pen holds 40 cows overnight or 50 cows crowded. These capacities may vary depending on the size of your cattle.

To expand the corral system to handle more cattle, you can add more diagonal pens. Do NOT increase the length of the diagonal pens! If they are too long, the cattle will bunch up. You can increase the diagonal pen capacity to 1000 calves. It is NOT recommended to increase the size of the round gathering pen beyond the 55 ft radius shown. If the round gathering pen is too large, you may have difficulty getting the cattle into the curved reservoir lane. (Grandin, 1980[a]).

In order to increase the gathering area, you can build an additional round gathering pen at the pasture entrance. After the first 300 pairs are worked or sorted you can bring in 300 more pairs. The post working pen can be enlarged to hold cows after sorting or handling in the squeeze or AI chute.

Small Corral

The layout in figure 2 is designed for smaller ranches as a main working corral or a pasture corral on larger

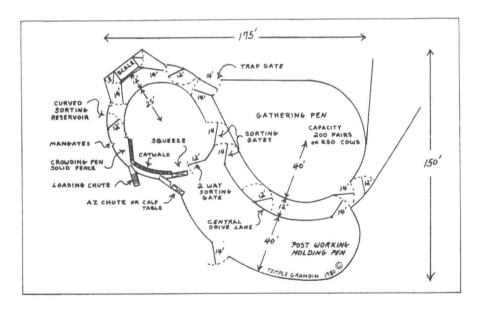

Figure 2. This is an economical corral system for a smaller operation or a pasture corral on a large ranch. It can handle 200 cow and calf pairs or 250 mature cows. It can be expanded to handle 300 cow and calf pairs (Grandin, 1980a).

operations. It is economical to build but still retains many of the features of the larger corral. It can handle

200 cow-calf pairs or 250 mature cows. By increasing the radius of the gathering pen to 55 ft and lengthening the central drive lane, it can be expanded to 300 pairs or 400 cows.
In figure 2 you can sort two ways out of the squeeze chute and three ways from the curved reservoir lane. Groups of cattle held in the curved reservoir lane can be sorted back into the post working pen, the central drive lane, or the round pen that is formed by the inner radius of the curved reservoir lane. When calves are being separated from the cows, the cows can be sorted into the post working pen and the calves into the central drive lane. For additional photos of corrals and a diagram of a minicorral, see Grandin, 1981, Beef Cattle Science Handbook 18:117.

Corral Construction Tips

Five foot high fences are usually sufficient for cattle such as Hereford and Angus. For Brahman cross and exotics a 5 1/2 ft to 6 ft fence is recommended. Solid fencing should be used in the crowding pen, single-file chute, and loading chute. If your budget permits, solid fencing should be used in the curved reservoir lane. If solid fencing is too expensive, then a wide belly rail should be installed. This is especially important if the corral is constructed from sucker rod (figure 3).
A V shaped chute can be built that will accommodate both cows and calves. It should be 16 to 18 in. wide at the bottom and 32 in. wide at the top. The 32 in. measurement is taken at the 5 ft level. If the single-file chute has straight sides it should be 26 in. wide for the cows and 18 to 20 in. wide for calves.
When a funnel-type crowding pen is built, make one side straight and the other side on a 30° angle. This design will prevent bunching and jamming. The crowding pen should be 10 to 12 ft wide (figure 4).
To prevent animals from slipping in areas paved with concrete, the concrete should be scored with deep grooves. The grooves should be 1 in. to 1 1/2 in. deep in an 8 in. diamond pattern. A diamond pattern should be used because it is easier to wash. If cattle are falling down when they exit from the squeeze chute in an existing facility, a grid constructed from bars will prevent falls. Construct the grid from 1 in. steel rods in 12 in. squares. Each intersection must be welded and the grid securely fastened to the concrete floor.
In areas with solid fence, small man-gates must be installed so that people can get away from charging cattle. The best type of man-gate is an 18 in. wide, spring-loaded steel flap. The gate opens inward towards the cattle held shut by a spring. A person can quickly escape because there is no latch to fool with. The man-gates can be constructed from 10 gauge steel with a rim of 1/2 in. rod.

Figure 3. Curved corral system shown in figure 1 constructed from steel with sucker rod fences. A wide 24 in. belly rail has been installed in the curved reservoir lane to provide a visually substantial fence. The round crowding pen and loading chute are covered with 10 ga. steel sheets.

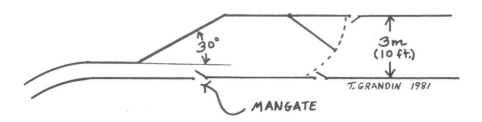

Figure 4. A funnel crowd pen should have one straight side, and the other side on a 30 degree angle. Jamming will occur if both sides are angled.

Many people have asked questions about how the corrals should be laid out. It is really very simple. In figure 1 the curved single-file chute, round crowding pen, and curved reservoir lane are laid out along the dotted line. The first step is to place a string on the site in the position of the dotted line. The radius points of the curved single-file chute, round crowding pen, and curved reservoir lane are all located along the string.

Layout steps for figure 1 (To be done in order.)
- Make a 16 ft 180° half circle for the single-file chute.
- Make a 12 ft 180° half circle for the round crowding pen.
- Make a 35 ft 180° half circle for the curved reservoir lane.
- Layout the diagonal pens on a 60° angle by placing a transit on the string.
- Layout gathering pen with a 55 ft radius. The radius point for the gathering pen is located 95 ft from the strike post of the last gate in the row of diagonal pens. The row of diagonal pen gates and the gathering pen radius point should be at a 90° angle relative to your string (dotted line).
- Layout the post working pen. The 55 ft radius point is found by measuring 55 ft from the hinge of the 14 ft sorting gate in front of the squeeze chute. The exact location of the hinge may vary depending on the length of the squeeze chute. Leave a 3 ft to 4 ft space between the end of the sorting gate and the headgate on the squeeze. This provides enough room so you can swing the sorting gate in front of the headgate without hitting the cow's head.
- After laying out the basics, finish laying out both sides of the lanes. The pasture entrance gates and the central drive lane will have to be laid out by eye.

Lay out everything in lime before building anything. This will prevent mistakes. Walk through the layout. If it looks like the drawing, then you have got it right. If there is a hill next to the site, look at the lime layout from there. If an aircraft is available, use it to check your layout.

SQUEEZE CHUTES

Herd health care is virtually impossible without a headgate or a squeeze chute for restraining animals. There are many headgates on the market and each type is especially suited for certain handling procedures. There are four basic types of headgates: scissors stanchion, full-opening stanchion, positive control, and self-catcher (Grandin, 1980b).

Scissors-stanchion headgates consist of two biparting halves that pivot at the bottom (figure 5). The full-opening stanchion consists of two biparting halves that work

Figure 5. Scissors stanchion headgate with curved neck bars. The gate opens like a pair of scissors and has pivots at the bottom. The curved neck bars provide a good combination of head control and protection against choking.

like a pair of sliding doors. A positive-control headgate locks firmly around the animal's neck. This type of gate completely restricts up and down movement. The self-catcher headgate can be set like a trap. When the animal enters, its forward movement will close the gate automatically around its neck. The advantages and disadvantages of the four types of headgates are summarized in table 1.

Self catching, scissors stanchion, and full-opening stanchion are available with either straight or curved stanchion bars. A straight-bar stanchion headgate is extremely safe and will rarely choke an animal. The disadvantage of a straight bar stanchion is that an animal can slide its head up and down unless a nose bar or other restraint is used. The straight bar stanchion is recommended if the headgate is going to be used primarily for AI or pregnancy testing.

TABLE 1—Types of Manually-Operated Headgates Compared

	Self-Catcher	Scissors Stanchion	Positive	Full-Opening Stanchion
Recommended for	Hornless cattle, gentle cattle, one-man AI	General purpose, big feedlots, wild cattle, minimum maintenance, cattle of mixed sizes (because the gate has to be readjusted to catch animals of different sizes)	Dehorning, wild cattle, horned cattle, good head control, big feedlots. Requires less strength to operate than stanchion gates.	General purpose, vet clinics, mixed cattle sizes (because the gate seldom needs adjustment). Large bulls can exit easily.
Not recommended for	Wild cattle, big feedlots, horned cattle, groups of mixed-size cattle (because the gate has to be readjusted to catch animals of different sizes)	Very large bulls (because they may have trouble exiting due to the narrow space between the two bottom pivots)	Vet clinics where the animal is held in the headgate for a prolonged time. When AI and pregnancy testing are the primary uses of the headgate.	Big wild cattle, big feedlots (because many full-opening stanchion headgates are not sturdy enough to withstand constant heavy usage)
Warnings	Mechanism requires careful maintenance. Head and shoulder injuries may result if the animals are allowed to slam into the gate.	Be careful not to catch the animal's legs or knees between the two halves of the gate or the animal may be injured.	More likely to choke than a self-catcher, scissors, or full-opening stanchion.	Mechanism requires careful maintenance to prevent jamming. Animal may trip over the lower gate track if it becomes excited.

Self-catcher, scissors-stanchion, and full-opening stanchion headgates are available in models with either a straight or curved stanchion. Refer to the text for discussion on choking hazard *versus* head control.

Source: T. Grandin (1980).

A curved-bar stanchion is a good compromise between control of the animal's head and protection from choking. It is more likely to choke than a straight-bar stanchion, but it is safer than a positive type gate. A nose bar is not needed for ear implanting or tagging, if the animal is backed up in a stanchion headgate.

The problem of choking in a curved-bar stanchion or positive type headgate can be reduced by adjusting the squeeze sides of the chute. The V shape of the chute should support the animal. The proper spacing at the bottom of the squeeze sides is 6 in. for 250 to 400 lb calves, 8 in. for 600 to 800 lb animals, and 12 in. for cows and most fed steers. The space should be 14 to 16 in. for large bulls. The measurements are taken on the inside of the chute at the floor level. The best type of chutes have two squeeze sides that fold in evenly when the squeeze is applied.

Operator Skill Important

Results of a survey conducted by the author indicated that the main cause of handling accidents in hydraulic squeeze chutes was a careless operator handling cattle too fast. The skill of the operator affected the incidence of choking and escape from the squeeze chute (tables 2 and 3). Problems such as balking and falling while exiting from the squeeze chute were largely determined by conditions such as slick floors or shadows in the handling facility. You should be able to do a better job of operating your squeeze chute than that indicated by the percentages on tables 2 and 3.

TABLE 2—Effect of Cattle Breed and Operator's Skill on the Frequency
of Handling Accidents in Hydraulic Squeeze Chutes

	Average for All 22 Groups (2150 Head)	Average for 12 Brahman Groups (1210 Head)	Average for 10 Non-Brahman Groups (940 Head)	Number of Groups with a Perfect Score	Single Worst Group Score	Observed Cause of the Worst Score
Mild choke	0.40%	0.17%	0.77%	17	3.00%	Rushing and carelessness*
Severe choke	0.30%	0.17%	0.40%	18	2.00%	Inexperienced operator
Entry balk	11.25%	10.30%	12.40%	0**	30.00%	Electric pole in front of the chute
Exit balk	15.20%	17.90%	12.00%	0	25.00%	Brahman cattle backed up after release
Partial escape	2.30%	3.90%	0.55%	10	11.00%	Long horns
Total escape	0.74%	1.00%	0.20%	14	5.00%	Carelessness
Falling	4.90%	7.30%	2.00%	5	15.00%	Slick, smooth concrete floor in front of chute
Headgate leg	2.50%	2.20%	3.00%	7	12.00%	Rushing

*All occurred in the same feedlot.
**Best balking score: entry balk 1.00%, exit balk 3.00%.

TABLE 3—Frequency of Handling Accidents in Hydraulic-
Stanchion Headgate Squeeze Chutes*

	Ear Implant Only** (1230 Head Weighing More Than 600 lb)***	Full Processing[†] (920 Head Weighing 250 to 600 lb)*** 45% Castrated
Mild Choke	0.00%	0.88%
Severe Choke	0.11%	0.30%
Entry Balk	10.95%	12.70%
Exit Balk	13.90%	15.65%
Partial Escape	2.20%	1.39%
Total Escape	1.18%	0.07%
Falling	4.68%	3.15%
Headgate Leg	0.88%	3.77%

*Bowman and Trojan hydraulic squeeze chutes.
**Cattle received an implant of growth promotant in the ear. No other treatment was given.
***The sample consisted of both Brahman-cross and English-type cattle.
[†]In places where no Brahman or Brahman-cross cattle are handled the percentages should be lower. Full processing consisted of a minimum of two brands, two injections, an ear implant, and at least one other treatment such as deworming, or pour-on insecticide.

Source: T. Grandin (1980).

Different breeds of cattle react differently to handling (Ewbank, 1968; Tulloh, 1961). More handling accidents occurred when Brahman-cross cattle were being handled. The Brahman-cross cattle had more total escapes and partial escapes. A partial ecape was recorded when the animal was caught around the middle by the headgate.

Choking in the headgate occurs when the headgate applies excessive pressure to the carotid arteries or the wind pipe (White, 1961; Fowler, 1978). Excessive squeeze pressure in a hydraulic squeeze chute can also cause choking. A hydraulic chute is safe if the pressure relief valve is set correctly. In fact it may be safer for both man and animal because the dangerous levers are eliminated. At several feedlots, some cattle died several days after going through the hydraulic squeeze chute. The animals weighed over 600 lbs and appeared to have pneumonia. An autopsy revealed that the cattle had been ruptured internally due to excessive squeeze pressure.

Cattle can also be injured if a fast-moving animal is stopped suddenly by clamping the headgate around its neck. Examination of beef carcasses revealed old healed injuries to the back and neck. Even if the animal appears normal it can sustain a spinal injury if it slams into the headgate. A skillful squeeze chute operator can slow the animal down in the squeeze before it reaches the headgate. Rubber strips can be placed on the headgate to absorb shock and help reduce injury. Old, split motorcycle tires will work well.

Too many people try to set the world speed record for working cattle and they end up injuring a lot of animals. The survey indicated that a skilled crew can actually handle more cattle per hour by handling them gently and skillfully. A four-person crew using a hydraulic stanchion-type chute in a well-designed circular cattle working facility could catch an animal and place an ear implant every 15 seconds. The crew could also brand, vaccinate (up to four injections), and ear implant an animal every 45 seconds. In 60 seconds the crew could castrate, brand, vaccinate, ear implant, and give at least one other treatment such as pour on insecticide or clip a tail (Grandin, 1980b). When the crowding pen was filled with cattle, these timed procedures could be achieved without rushing. If a crew goes faster than these times, they will be doing a sloppy job and may be injuring the cattle.

DIPPING VAT DESIGN

Building a slide to make the cattle slide into the dip vat on their rear ends is wrong. The animal should be provided with good footing as it enters the water (Grandin, 1980c).

A 9 ft downward sloping, hold-down rack (figure 6) prevents the animals from leaping to the center of the vat.

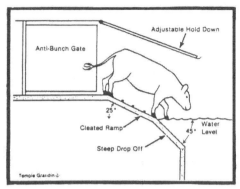

Figure 6. A grooved or cleated ramp with a nonslip surface
enables the animal to enter the vat without hesi-
tation. The steep drop off is hidden under the
water. A 9 ft. adjustable hold-down rack makes
the animal dive in and immerse its head. The
distance between the hold-down rack pivot and the
floor is 5 ft. The cleated portion of the ramp
is 6 ft. long and on a 20 to 25 degree angle.
The 45 degree angle portion is 4 ft. long
(Grandin, 1980c).

The hold-down rack forces the animals to immerse their heads
instead of being pushed under with a forked stick. It also
helps prevent the chemicals from splashing out. Splashing
can be further reduced by installing a 3 in. pipe along the
inside and top edges of the dip vat. The pipe should be 3
to 4 ft above the surface of the water.
Each animal enters the vat by walking down a 6 ft gradual declining ramp that is on a 20° to 25° angle. This ramp
has deep grooves in the concrete to provide the animal with
good footing. The grooves should be 2 in. deep and 8 in.
apart. The purpose of the ramp is to orient the animal's
center of gravity towards the water. The steep drop-off is
hidden under the water, but the ramp appears to continue on
into the water (figures 6 and 7). When the animal steps out
over the water it falls in. The ramp must have a **nonskid**
surface or the animal may become scared and attempt to back
out.
If both large and small cattle are going to be dipped,
the entrance should be equipped with antibunch gates. These
7 ft gates allow only one animal to enter the vat at a
time. The antibunch gates work on the same principle as
the trigger trap one-way gates that are used to trap wild
cows at the water hole. The opening between the ends of the
two gates is adjusted to equal the width of one animal. The
gates act as a valve to slow down incoming cattle. One of
the antibunch gates can be spring loaded. The entrance
should also be equippped with a semicircular block gate or a
sliding gate to shut off the flow of cattle. A stanchion
headgate powered by hydraulics also works well for a shut-
off gate.

Figure 7. Cattle enter the vat and immerse their heads.
They do not need to be pushed under with a forked
stick. The hold down rack is easily constructed
from 10 ga. steel and pipe.

Stairsteps are the best type of ramp for exiting from a
dip vat. A steeper ramp can be used in a dip vat than for
loading because the water supports the animal. Dip vat exit
steps can have a 6 to 7 in. rise and a 12 in. tread width.
The steps must be grooved to prevent slipping.

Drip Pen Design

A divided drip pen is recommended for dripping. When
the cattle are drying on one side, the other side can be
filled (figure 8). Figure 9 illustrates a dipping system
layout with curved lanes and angled drip pens. Each side of
the drip pen should be 30 to 40 ft long and 16 ft wide. The
drip pens should be sloped 1/4 of an inch every foot towards
the vat. The drip pen should be curbed with an 8 in. high

Figure 8. Divided drip pens with remote controlled exit gates. The pipes go to cylinders to operate the gates. The advantage of remote controlled gates is they are labor saving and they allow the handler to open the gates without entering the flight zone of the cattle.

curb. To prevent the cattle from slipping, the drip pen floor should be scored in an 8 in. diamond pattern. When the wet concrete is scored, the last pass should be made towards the vat so that the water will drain more easily.

Since often either hydraulic or air pressure is available at the site, placing cylinders on the exit gates is recommended. This will let the cattle out by remote control. The exit gates can be opened without the handler entering the flight zone of the cattle. In conventional systems, the cattle often become agitated when the handler walks up to open the gates manually. In some instances, the cattle will ram the fences and attempt to jump back into the vat when the handler approaches.

The drip pen exit gates and the dividing fence in between the two drip pens should be solid. This prevents the cattle that are confined on one side from pushing on the gate and attempting to follow the cattle that have just been released. If Brahman-cross cattle are being dipped, a one-way gate should be installed to prevent them from reentering the vat.

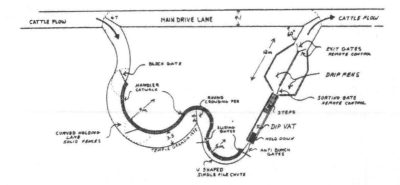

Figure 9.   Dipping vat system with curved lanes. A squeeze
chute can easily be incorporated into this layout.   A 14 ft gate positioned in front of the
squeeze chute can be used to divert animals you
do not want to dip.   English measurements for
metric measurements are: inner radius of curved
holding lane 9m (30 ft); width of holding lane
and length of crowd gate 3.5m (12 ft); and inner
radius of the single file chute 5m (16 ft)
(Grandin, 1980a).

To help keep the vat clean, the single-file chute and
the crowding pen should be installed to prevent hair and
manure from reentering the vat from the dip pens. The sump
should have a valve to divert rain water that falls on the
drip pen away from the vat. It is also wise to install a 2
ft wide curbed concrete apron alongside the vat.
    To help keep the vat clean, the single-file chute and
the crowding pen should have a concrete floor to prevent the
animals from tracking dirt into the vat. This floor should
be washed down after each dipping session.

Agitation and Aeration

Regular agitation and skimming off of floating debris
will keep a dip vat cleaner. A vat that is left standing
will quickly become stagnant and foul smelling. An easy way
to agitate a vat is to install an airline in the bottom.
The line should be a 1 in. pipe with 1/16 in. holes drilled
at 5 in. intervals. The pipe is mounted 1 to 2 in. off the
vat bottom. Two rows of holes are drilled on a 45° angle
facing downward on both sides of the pipe (Saulmon, 1972).
Connect the pipe to an air compressor capable of delivering
0.5 cu ft per minute at 40 psi per foot of pipe.
    Frequent agitation of the water with the air compressor
wil aerate the water and prevent it from becoming septic and
foul. Good results have been obtained by connecting a timer
and a' solenoid to control the release of air. A 30 sec
blast of air every 30 min almost completely eliminates bad
odors.

Filtering and Cleaning

The installation of a filtering system can double the life of the chemicals in the vat. Use of a Hydrasieve sloping screen to remove solids from the vat has reduced chemical disposal requirements by 50% and pesticide usage by 30% (Sweeten, 1976; Miller, 1975). Almost twice as many cattle can be dipped before the vat requires recharging.

A cleaning system with a Hydrasieve is very simple. It consists of a 28 in. wide Bauer Hydrasieve screen with a .2 in. screen spacing. A 3 in. centrifugal trash pump brings the water to the top of the screen. Don't use a smaller diameter pump, it will clog up. The water runs back through the screen and the solids fall off the screen into a container. The water returns to the vat by gravity. When this system is built, the suction line should be exposed. Don't bury the suction line under the slab. If you get an air leak, you will not be able to fix it. The water-return lines and other pipes may be buried.

Another type of cleaning system is a sluice box or settling basin. This system can also double the life of the vat. The advantage of the sluice box or settling basin is that it is inexpensive. The use of this system and a Hydrasieve together may enable you to quadruple the life of the chemicals in the vat (Sweeten, 1982).

The Hydrasieve removes the bigger solids, and the sluice box or settling basin removes the fine dirt. Settling systems can vary in capacity from 300 to 100 ga. The bigger systems will remove more solids but they are harder to clean. An easy sluice box system to construct consists of a metal or concrete box 10 to 18 ft long and 2 ft deep. The box contains a number of removable baffles. The highest baffles should be located at the end where the dirty water enters. The box should be narrower where the dirty water enters and becomes wider as the discharge point is reached. Two ft wide at the entrance and 4 ft wide at the discharge point will work well. The idea is to make the water move very slowly at the discharge point so the solids will be left in the bottom of the box. The water should flow through a 300 gal box at a rate of 25 to 66 gal per min (Sweeten, 1982). The sluice box or sedimentation tank must be run at least 4 hours for every 1000 cattle dipped. It is of the utmost importance to keep the sluice box clean. If it is not cleaned out regularly, it will put dirt back into the vat. After each use, the baffles should be pulled out and the box cleaned out with a shovel.

A good way to design a cleaning system is to connect the Hydrasieve and the sluice box in a tee circuit. By using two valves, the larger portion of the pump output is directed to the Hydrasieve and the smaller portion of the output is directed to the sluice box. If the flow rate is too great the sluice box will not work. Sluice boxes or settling basins must be used with care when using wettable powders because they will remove the chemical. Check with the pesticide manufacturer.

Chemical Disposal

Haphazard dumping of used chemicals around a feedlot or ranch is not recommended. Used dip can be evaporated in a shallow concrete basin (Fairbank et al, 1980). For large feedlots in the Southwest the evaporation basin should hold three times the vat volume and the water level should be less than 2 ft deep. In high rainfall areas, an evaporation basin should have cover. Check with a local engineer to determine evaporation rates for your area. Each state has its own regulations on pesticide disposal. Check with your own state.

287

REFERENCES

Ewbank, R. 1968. The behavior of animals in restraint, In: M. W. Fox (Ed.) Abnormal Behavior in Animals. W. B. Saunders, Philadelphia, PA.

Fairbank, W. C., T. Grandin, D. Addis, and E. Loomis. 1980. Dip vat design and management leaflet. 21190 Division of Agricultural Sciences, University of California, Davis, Calif.

Fowler, M. E. 1978. Restraint and handling of wild and domestic animals. Iowa State University Press, Ames, Iowa.

Grandin, T. 1980[a]. Efficient curved corrals. Angus Journal, October 1980, pp 95-97.

Grandin, T. 1980[b]. Good cattle restraining equipment is essential. Vet. Med. and Small Animal Clinician. 75:1291.

Grandin, T. 1980[c]. Safe design and management of cattle dipping vats. Technical Paper No. 80-5518. Amer. Soc. Agr. Eng. St. Joseph, Michigan.

Grandin, T. 1981. Innovative cattle handling facilities. In: M. E. Ensminger (Ed.) Beef Cattle Science Handbook.18:117. Agriservices Foundation, Clovis, Calf.

Miller, D. 1975. Evaluation of Bauer separation equipment as a dip vat filtration system. (Unpublished). USDA, Beltsville, Maryland.

Saulmon, E. E. 1972. Ticks and scabies mites...dipping vat management and treatment procedures. Veterinary Services Memorandum 556.1, USDA/APHIS, Washington, D.C.

Sweeten, J. M. 1976. Results of Hydrasieve cattle dip recycling study. Field Day on Dip Vat Management Systems for Cattle Feedyards. Texas A & M University.

Sweeten, J. M. 1982. Dipping vat sedimentation tanks compared. Beef. March, 1982, p. 104.

Tulloh, N. M. 1961. Behavior of cattle in yards: II A study of temperament. Animal Behavior 9:25.

White J. B. 1961. Letter to the editor. Vet. Record 73:935.

Part 6

FEEDS AND NUTRITION

29

MORE ECONOMICAL MILK PRODUCTION WITH COMPUTER USE IN NUTRITION AND FEEDING MANAGEMENT

Carl E. Coppock

INTRODUCTION

Since the 1950s, dairy farmers have used the computers of their Dairy Herd Improvement Cooperative to calculate, summarize, and store production and management data. The sophistication and usefulness of this information has multiplied several fold over the past 25 years. About 1970, the handheld electronic calculator made its debut, which sharply reduced the drudgery and, to some degree, the errors associated with hand calculations. Subsequently, the handheld programmable calculator (PC) was developed that provided a large increase in capability because of programs that could be recorded on magnetic cards for future use. These battery-powered devices could perform at least 50 functions from memory and 500 or more steps in program execution (Ely, 1980). These PCs were rapidly adapted to agricultural use by extension specialists, particularly those concerned with nutrition and feed formulation. The Extension Services of several universities (Chase, 1980; Linn and Spike, 1980) have been quick to offer packages of programs for the PCs including those for ration balancing, batch mixing, dry matter intake prediction, and others. The primary limitations of PCs are their limited memory and the high degree of skill needed for their successful use. Nor does their capability extend to least-cost formulation by linear programming techniques. Even so, they are used widely by those associated with the feed industry.

For over 10 years, least-cost ration balancing has been available directly to dairymen through their Extension Services (Bath et al., 1968). Earlier versions allowed access to a large main-frame computer via a mail-in service, by audio connection, and more recently by a remote terminal that is connected via phone to the main computer. At least three systems are widely used--from California, Michigan State, and Virginia Tech.

More recently, the microcomputer has become available that is designed for small business and home use and whose capability has increased several times within the past 5 years. It provides powerful computing and data handling capability, useful for many kinds of financial and management jobs. My objective in this paper is to show the application of microcomputers to nutrition and feeding management of dairy cows. I assume the microcomputer is owned by and located at the dairy.

LINEAR PROGRAMMED RATION FORMULATIONS

Linear programming (LP) is a mathematical technique used to maximize or minimize some function subject to constraints. In feed formulation, one minimizes cost, i.e., least cost (LC) of a ration--or in another mode, maximizes profit (MP) by assuming a milk-response curve to increased feed (energy) input. The recent advantage to dairymen is that now the microcomputers can handle LP formulations.

The advantages of least-cost formulation, its success, and its application have been described by Bath and coworkers (1968) in California where the system was pioneered for use by individual dairymen. Briefly, the output of such a program (assume a maximum profit type) gives the cow conditions for which the problem is specified, the forage(s) and amounts to feed, the concentrate composition and amount to feed, the forage-concentrate ratio, the optimum daily milk production, the total feed cost, and the total daily income over feed cost. In addition, the price range for those ingredients used, the opportunity price of feeds not used, and extensive ration analyses are given. My purpose here is to discuss some of the advantages and limitations of this system with emphasis on ways to resolve or minimize the limitations.

1) A least-cost ration tells one which forages to use, plus the range in price through which each forage price could move before the formulation would change to use more or less of that forage. The value of this feature is that one can include any forage that can be obtained, and even though it is not now a good buy, one can see the price where it would begin to enter the formulation. However, this valuable feature (true for both price ranges and opportunity prices) is true if only that single-ingredient price changes. But in the real world, this rarely occurs--and when two or more prices change, it is usually impossible to tell how this will affect the formulation. Nevertheless, within a given formulation, relative prices are very useful. Cropping programs and forward-purchasing contracts should be based on a consistent output over a broad spectrum of prices. The advantage of the microcomputer is that runs can be made as often as desired, with immediate results at no additional cost.

2) Many programs will not execute when very high production (90 to 100 lb/day or higher) is specified. The reason is that the energy required cannot be contained within the maximum dry matter capacity, at the minimum fiber content demanded. Yet it is this kind of cow--the cow in

negative energy balance in early lactation--that is in pre-
carious nutrient balance and should have the highest energy
density in her diet consistent with maintenance of milk fat
test. A partial solution is to use added fat (3% to 5%) or
whole cottonseed (5 lb to 8 lb), if available, to increase
energy density. In addition, some newer programs now consider the contribution of the cow's body fat, which is precisely what happens in the real world. One pound of body
fat provides energy for about 10 lb of milk. The diet of
the cow in negative energy balance should be enriched in all
nutrients, including energy.

3) Although LP formulation can consider many ingredients, it is difficult to predict the acceptance or palata-
bility of many mixtures. So, maximum acceptable amounts are
placed on early all ingredients. Many of these constraints
are only conservative "guesstimates." If these constraints
bind, but are unnecessary, cost is increased. Acceptance of
a concentrate mix is especially critical in milking parlors,
but it is of little consequence in a complete ration based
on silage that easily masks the flavor of other components.

4) If diets are formulated for several groups or production strings, each string usually needs a different formulation. Sometimes adjustment in the proportion of forages
(i.e., corn silage and alfalfa) can result in one formulation meeting requirements of nearly all production groups,
but this is an exceptional case. So to take maximum advantage of LP formulation in group-fed herds, a number of different formulations must be purchased and stored, unless
diets are mixed on the farm.

5) Nearly all feed dealers used LP-LC formulations, and
they know which major ingredients are routinely available to
them and which mixes are the most palatable. Custom mixes
are a nuisance to feed dealers; by the time grinding (and
maybe pelleting), mixing, and delivery are considered, the
custom LP mix may be more expensive than the feed dealer's
regular line. An important point is that LP-LC formulated
diets (by microcomputer or otherwise) are most relevant and
valuable to dairymen who buy commodities (individual ingredients) and who are using on-the-farm mixing. They require
formulation service; the microcomputer allows them to make
many runs with little if any additional direct cost, and
there is no delay in obtaining results.

SUPPLEMENTAL GRAIN FEEDING

As increased production makes parlor-concentrate feeding more and more inadequate, several types and systems of
supplemental-concentrate feeders are used. The simplest and
most economical way for small herds to get extra concentrate
is to offer one or more additional feeding in a bunk between
milkings. With lock-in stanchions, this becomes an especially viable system because one has the option to feed in

relation to production without social-dominance interference. The lock-in stanchions have other valuable uses
for tail-head chalking, pregnancy checking, etc. One can
retrieve some of the advantages of individual concentrate
feeding from the flat barn.

Magnetic grain feeders. This system provides supplemental grain to cows that have a magnet on their collars
that actuates a small auger in the system that delivers feed
at about one-half to 1 lb per minute. As long as the cow
has her head in the feeder, grain is slowly delivered. This
feeder allows cows with magnets to eat as much grain as they
wish within the constraints of appetite, time, and social
competition. Lower forage intake has resulted occasionally
in milk-fat depression and digestive upsets probably from
too much grain. Field observations (Gaunt, 1974; Hutjens,
1976, 1976a; Pritchard, 1973) suggests that dairymen are
well pleased with their use, but economic evaluation is
difficult.

Calan broadbent door. Another device allows cows that
carry an electronic key on their collar to open a door to
access a bunk of feed. Although the research application of
this device allows up to 50 unique identities to obtain
individual intakes on group-housed animals (Alhassan et al.,
1975), the commercial application of the door is intended to
allow all cows with the key to access the same door or
series of doors (Briggs, 1976). So the system permits free
access by keyed cows to a manger, which may contain any
available feed. An advantage of this system over that of
the magnetic collar is that grain left by one cow cannot be
obtained by a cow without a key.

COMPUTER-CONTROLLED-CONCENTRATE FEEDERS

Microcomputers (MCCF) control the most recent development in automated concentrate feeders. These are not
supplemental feeders (although they can be used that way)
but were designed to dispense all of the concentrate. At
least eight companies are marketing MCCFs (table 1). In
each case, a microcomputer controls the feed dispensed to
each cow in the herd, based on the amount keyed into the
computer by the dairymen. Each cow wears a neckstrap with a
transponder that identifies the cow to the computer when she
enters the feeding stall. The computer allocates predetermined portions of her total daily concentrate allotment and
signals the feeder to deliver it slowly to the cow while she
is in the feeding stall. When her allotment is eaten or she
leaves the stall, the feeder stops. Any of her unused
allotment is recorded by the computer and is available the
next time the cow enters a feeding stall. If she does not
consume her total allowance that day, the amount left over
is recorded and can be printed out. This can help identify

cows that are off-feed sooner than they normally would be noticed. Some feeders prorate the feed into 4 or more feeding periods through the day so a cow could not eat it all at once, even if she wanted to.

TABLE 1. COMPANIES THAT MARKET COMPUTER CONTROLLED CONCENTRATE FEEDERS[a]

1.	De Laval
2.	Surge - Babson Bros. Co.
3.	Farmtronix
4.	Germania
5.	Westfalia Systemat
6.	TeSa - Technical Industries
7.	Horn of Plenty
8.	Northco Serv-O-Trone
9.	Zero Feeding System

[a]Not intended to serve as recommendation, nor is the list exhaustive.

The number of cows that can be handled by one computer varies depending on the size of the computer and the number of feeding stations. Costs of the systems vary widely depending on the number of cows and auxiliary equipment such as computer printers, number of stalls, and the sophistication of the computer program. Estimates for the basic system (computer, stall, transponder, neckstraps, and power supply) given by manufacturers in 1981 were from $110 to $200 per cow (Hutjens, 1981). Each of the companies is developing additional equipment that can be added to the feeding module. Such information as electronic identification, automatic milk-yield recording, cow temperature sensing for health status, complete health and production records, farm accounting, and many other herd management programs will be available.

Some research studies support the use of MCCF. But any feeding system that provides additional feed energy to energy deficient cows will cause sharp increases in milk production. To decide whether one system is "better" than another, the production response must be considered in relation to the cost of the two systems. Ideally, the response should also take into account other more-difficult-to-identify factors, including reproductive performance during a long multilactation period. To date this has not been done, but Tomlinson and coworkers (1980) have compared 2-MCCF to the CALAN door and the TMR (total mixed ration) system. The study included four groups of 24 cows each, within an 84-day, continuous feeding trial. Results (table 2) suggest similar production responses from the four feeding systems. The TMR feed cost per 100 lb milk was slightly higher, but the TMR cows ate more (4+ lb of dry matter) than did the other cows. A more recent study (Schillings et al., 1982) from the same group showed significantly more milk from TMR-fed cows.

TABLE 2. EFFECT OF TOTAL MIXED RATIONS VS AUTOMATED
ELECTRONIC FEEDERS ON PERFORMANCE OF DAIRY COWS IN
EARLY LACTATION[a]

	Calan	DeLaval	TeSa	TMR
Dry matter intake (kg/cow/day)	18.60	19.30	18.60	20.90
Body weight [change (kg/day)]	40.50	40.90	47.00	39.00
Actual milk (kg/cow/day)	21.50	22.70	21.60	22.50
Milk fat (%)	4.38	4.37	4.52	4.39
FCM 4% (kg/cow/day)	22.00	23.00	22.30	23.00
Feed cost $/100 kg milk				
Corn silage	3.86	3.87	3.77	3.52
Concentrate	7.33	7.18	7.23	8.90
Total	11.19	11.02	11.00	12.42

[a]Tomlinson et al. (1980).

Microcomputer-controlled-concentrate feeding represents
a big advancement over the magnet and CALAN door feeders
because of precise control over the amount and rate of concentrate allocation permitted--plus the record of amount and
time of actual consumption by individual cows.

Apart from cost, which is high, these systems of
individual concentrate allocation, regardless of their precision, suffer from three fundamental limitations: (1) High
producers can rarely eat as much energy (concentrate) as
they require. Attempts to feed them more than about 60 to
65% of their total diet as concentrate may depress milk-fat
tests and induce digestive upsets. But even if conservative
maximums are used, the cow has some choice between forage(s)
and concentrates. In effect the cow has some freedom to set
her forage-concentrate ratio, her fiber level, and the
energy density of her consumed diet. (2) When cows are
offered two or more forages that differ greatly in their
nutrient composition (the classic extremes are corn silage
and alfalfa), cows differ greatly in selection or preference
for the forage so that the ratio of forage disappearance or
group consumption gives little indication of the consumption
by individuals. These factors severely limit the precision
of total nutrient supply. (3) One cannot predict precisely
the amount of forage consumed by individual cows.

In a study by Gray (1964), those factors that were
thought to affect forage consumption included:
- FCM production
- Amount of grain fed
- Body weight
- Daily change in body weight

- Condition ration
- Days postpartum
- Various interactions of the first six

All of these accounted for only 27% of the variation in forage consumption. Although the larger amounts of concentrate now widely fed may suppress some of this variation, there are a lot of individual differences remaining, even after accounting for all these factors. Regardless of how precisely we may be able to allocate concentrate, total nutrient allocation is sharply constrained by our inability to predict individual forage intake with precision.

Michigan State workers (Krupp and Nott, 1981) addressed another feature of computer-controlled-concentrate feeders. They advanced two hypotheses: (1) Approaching estrus is associated with a decrease in concentrate consumption, which will be indicated by the feeder. (2) Health problems will cause a decrease in concentrate intake, also shown by the feeder. If either or both of these hypotheses has validity, then the MCCF will have value in addition to its function as a concentrate feeder. Krupp and Nott (1981) used a feeder that provided a printed report for every cow each 12 hours. Based on this study in which a reduced level was defined as five percentage points below a previous stable amount, the use of such a reduction as a heat detection aid was not encouraging. And the herd manager usually diagnosed health problems in other ways, although such problems did reduce concentrate intake. But lower feed intake often did not occur until 2 or 3 half-day periods after the herd manager diagnosed the problem. Although this single study did not support these two hypotheses, more experience and study in the future may do so.

Results of research with microcomputer-controlled-concentrate feeders suggest they will be most beneficial in small- to average-sized herds, by dairymen who prefer not to feed concentrate in the parlor, whose forage is primarily hay pasture, and who would find it difficult to group cows.

CONCLUSIONS

The microcomputer offers powerful computing and data handling capability to dairymen who have acquired functional software programs and have learned to operate them. No longer are they dependent on others (including the phone company) to do their computing work. There are many valuable sets of information from linear programmed, least-cost printouts. Each set has limitations, which when understood can be resolved by innovations and improvisations. The microcomputer-controlled-concentrate feeders offer important advantages over the simple magnet feeder and key-door access feeders, but they cannot account for nutrients consumed from forage; therefore, total nutrients are dispensed imprecisely.

Microcomputers have many other applications to nutrition and feed management, including inventory control of both concentrate and forage, cropping program planning, etc. Software programs are rapidly becoming available that are documented and highly functional. The extension services are compiling lists of these programs. Rapid progress has been made within the past 3 years and much more is yet to come. The microcomputer is the most powerful tool to become available to dairymen since the milking machine.

REFERENCES

Alhassan, W. S., R. R. Hacker, and J. F. Hurnik. 1975.
Data monitoring from electronically controlled livestock feeding doors. J. Anim. Sci. 41:238 (Abstr.).

Bath, D. L., and L. F. Bennett. 1980. Development of a
dairy feeding model for maximizing income above feed
cost with access by remote computer terminals. J.
Dairy Sci. 63:1379.

Bath, D. L., S. E. Bishop, G. A. Hutton, Jr., J. C. Oliver,
and G. W. Dean. 1968. Computer-formulated least-cost
concentrate mixes for dairy cows. J. Dairy Sci.
51:1616.

Briggs, D. V. 1976. The Calan System. Personal communication. Northwood, NH.

Chase, L. E. 1980. Programmable calculators: An effective educational tool for agents. Paper at Ann. Mtg.
of ADSA (p 14), VPI.

Ely, L. O. 1980. An evaluation of the hand calculator,
mini-computer, and main-frame computer in programs for
animal science. Proc. GA Nutr. Conf., p 97.

Gaunt, S. N. 1974. Individual magnetic and electronic
activated feeders. J. Dairy Sci. 57:628. (Abstr.).

Gray, H. G. 1964. Forage consumption and efficiency of
feed utilization in lactating dairy cows. M.S. Thesis,
Cornell Univ., Ithaca.

Hutjens, M. F. 1976. Field observations with automated
magnetic grain feeders. Paper at 71th Ann. Mtg. ADSA
(Abstr.).

Hutjens, M. F. 1976a. How magnet grain feeders are working. Hoard's Dairymen. 121:555.

Hutjens, M. F. 1981. New feeders help you tailor grain
feeding. Hoard's Dairyman. 126:588, No. 11.

Krupp, I. and S. B. Nott. 1981. Computerized individual
grain feeder case studies. Paper at 76th Ann. Mtg.
ADSA (Abstr.).

Linn, J. G. and P. L. Spike. 1980. Programmable calculators and their application to feeding and management
of dairy cattle. J. Dairy Sci. 63:1390.

Pritchard, D. E. 1973. Magnets activate feeder for top
producing cows. Ohio Report 60:78.

Schillings, J. G., J. E. Tomlinson, C. D. Black, and W. H. McGee. 1982. Effect of total mixed ration vs computerized feeders on performance of dairy cows in early lactation. Paper at 77th Ann. Mtg. ADSA (Abstr.).

Tomlinson, J. E., T. P. Weisenberger, C. D. Black, and W. H. McGee. 1980. Effect of total mixed ration vs automated electronic feeders on performance of dairy cows in early lactation. Paper 75th Ann. Mtg. ADSA.

30
GREATER REPRODUCTIVE PERFORMANCE WITH NUTRITION AND FEEDING STRATAGEMS

Carl E. Coppock

INTRODUCTION

It has been observed that all nutrients needed by the reproductive organs also are needed by other tissues (Maynard et al., 1979). Deficiencies of most required nutrients, as well as some excesses and imbalances, have been shown to adversely affect reproductive performance. Extensive feed testing, careful diet formulation, and a feeding system that permits control over the consumed diet can resolve most deficiencies and excesses, except for insufficient energy consumption during the first trimester of lactation. Average milk production per cow in the U.S. has doubled within the past 25 years, but grains and other concentrates fed increased by over 2 1/2 times during this 25-year period. Genetic merit for high milk production continues to increase in the dairy cow population, without a concomitant increase in body size or capacity. Consequently, unless new formulation techniques or feeding stratagems can be devised to increase energy intake postpartum, the magnitude of energy deficit in early lactation will increase, and considerable evidence suggests that negative energy balance is detrimental to conception. It is often said that the dairy cow has great ability to maintain homeostasis--the condition of uniformity in the internal environment of the organism. But as milk production continues to increase with its greater metabolic demands, the priority of reproductive functions for required metabolites may have been superseded by demands for nutrients by the mammary gland.

Several extensive general reviews (Chase et al., 1979; Otterby and Linn, 1981) have recently been published on nutritional effects on reproduction in dairy cattle. The objective of this paper is to describe several examples of nutrients that have a direct impact on reproductive functions and to suggest some procedures to minimize these detrimental effects.

ENERGY

Heifers

In the 1950s and 60s several studies compared heifer rearing rates that were often described as low, medium, and high. In one study, these rates, respectively, were 65%, 100% and 140% of the Morrison standard. These rates refer to energy, and it is assumed that other nutrients were adequate--although, under field conditions, dairy heifers underfed in energy are often underfed in other nutrients as well. Some of the conclusions from these studies as given by Schultz of Wisconsin (1969) include:
- Within breed, heifers tend to reach puberty at about the same size (weight) regardless of feeding rate.
- Conception rate is not greatly affected by feeding level or breeding age.
- Underfeeding results in calves only slightly smaller than normal at first calving, but increases calving difficulties.
- Early bred heifers that are fed well have calves about as large as those of heifers bred later, but they also have more calving difficulties. (Today, of course, we have ease-of-calving ratings that allow earlier breeding with some reduced risk of dystocia.)
- Energy intake exceeding present standards and accompanied by delayed breeding is undesirable. Longevity may be adversely affected by overfeeding.
- Holstein heifers of high genetic worth that were fed 114% of Morrison's standard and bred the first heat period after 10 mo of age weighed less and produced less the first two lactations but eventually caught up in size and milk production with those heifers that were older at first breeding.

A recent Israeli study (Hoard's Dairyman, 1978) suggests that earlier breeding is feasible by careful selection of sires to reduce dystocia. Twenty-seven Holstein heifers fed to gain from 1.65 to 1.87 lb/day (compared to our more customary 1.5 lb/day) were bred at their first heat (about 9 mo of age). Compared to conventional 15 mo breeding, their results were very encouraging. Texas DHIA heifers calve at an average age of 27 mo. The conventional recommendation for breeding Holstein heifers at 14 to 15 mo (800 lb) and to calve at about 24 mo can be shortened by 2 to 4 months if heifers are well fed and sires are selected for smaller-than-average calves for the breed. This reduction in calving age would reduce the cost of rearing heifers more than any other known procedure.

Older Cows - Prepartum

The lactation cycle begins with the preceding dry period. The energy required for the last two months of pregnancy is rather low and averages about 30% of the energy required for about 10 lbs of milk. A modest quality grass hay would provide this energy requirement and might be deficient only in phosphorus, sodium, and a couple of trace elements. The problem of nutrition during the dry period today is largely one of excess energy consumption, which may lead to the fat-cow syndrome with attendant metabolic, digestive, and reproductive disorders near parturition. The resolution seems obvious. The cause in a number of cases is the availability of high-energy forage, especially corn silage, and too much grain. I do not advocate a rapid increase in grain feeding two weeks prepartum, but instead a constant amount of 4-6 lbs/day, assuming that the cow has already replenished the reserves needed for the coming lactation. Rapid increases in energy immediately prepartum provide energy that can only be deposited as fat. The fat requires that the metabolism of the cow reverse direction at parturition--from fat deposition to fat mobilization. This probably creates some additional metabolic stress. In addition, the cow is usually off-feed the day prior to calving and must begin coming back onto feed immediately postpartum which is probably easier if the preceding diet has been high in forage. A recent study from Minnesota (Johnson and Otterby, 1981) (table 1) illustrates some of the effects of excess energy (and maybe protein) prepartum. The increased retained placentas by the low-grain group are diffi-

TABLE 1. INFLUENCE OF PREPARTUM DIET ON REPRODUCTIVE EFFICIENCY OF HOLSTEIN COWS[a]

| | Diets | | |
	Alfalfa-grass hay	Alfalfa-corn silage + 12%C	Alfalfa-corn silage - 47%C
Older cows	25	25	25
Heifers	11	11	11
Diet crude protein	18.3	12.2	11.3
Diet crude fiber	31.7	22.5	14.7
Days from parturition to conception	128.3	107.4	121.6
Services/conception	2.06	1.48	2.30
Mammary edema	1.44	2.13	2.27
Retained placenta	6	16	3
Metritis	6	9	6
Displaced abomasum	0	0	0

[a]Johnson and Otterby (1981).

cult to explain, but the high incidence suggests that low selenium may have been involved. The objective is to move the cow from the high-forage prepartum diet to the high-concentrate postpartum diet as quickly and smoothly as possible.

Older Cows - Postpartum

Spalding et al. (1975) found that cows whose production averaged more than 2,000 lb of milk per lactation greater than herdmates had a first service conception rate that was 20.5% lower than that of a base group that had an average of 0 to 2000 lbs below herdmates. The high producers also had 13 days longer to first service than did herdmates. This supports the opinion that there is some inherent physiological conflict between high milk production and reproductive performance. An Israeli team (Francosi and Rattner, 1975) working with high producing cows noted that, "Generally it may be expected that cows producing more than the herd mean will conceive later than their herdmates that produce less." A number of reports show that cows losing weight have a lower conception rate than do cows gaining weight. McClure (1970) in Australia has emphasized that the lower fertility of lactating cows with low blood glucose reflects low energy consumption relative to energy demands for milk secretion.

In effect, the physiological drive of the high-producing dairy cow to secrete energy into milk has outdistanced her ability to consume energy from feed despite the highest-energy, free-choice diets. Therefore, most healthy dairy cows with high genetic merit have a period of negative energy balance during much of the first trimester (figure 1). By drawing on energy from fat reserves to supplement their insufficient dietary energy, most high producers achieve great milk output in early lactation. One hundred pounds of fat provides the energy for about 1000 lb of milk. We feed such cows in an attempt to stimulate even greater production or to minimize negative energy balance that appears to be detrimental to ovarian function and conception rate.

A recent study by Butler et al. (1981) relates to this point. Daily calculated energy balance and serum progesterone (every other day) were monitored for 84 days postpartum in 13 Holstein cows to determine the relationship between negative energy balance, milk production, and ovulation. "Average energy balance (cumulative daily calculated energy balance divided by days in milk) during the first 20 days of lactation was inversely related to days to normal ovulation $(r = -.60)$ and to milk production $(r = -.80)$. Milk yield during this period was not closely related to days of ovulation $(r = -.30)$. On the average, ovulation and the initiation of the first normal luteal phase occurred approximately 10 days after energy balance began returning

GENERALIZED MILK ENERGY - FEED ENERGY RELATIONSHIP

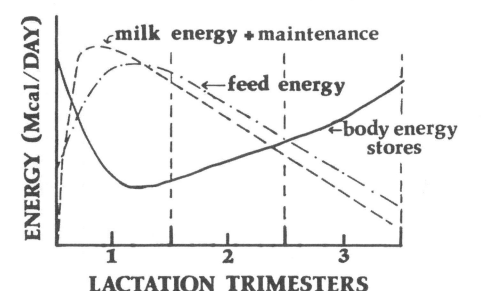

LACTATION TRIMESTERS

Figure 1. Relationship between feed energy, milk energy, and body energy stores by high producing cows

toward zero. This study suggests that energy balance during the first 20 days of lactation is important with respect to initiating ovarian activity postpartum." Consequently, we need all nutritional strategems available to maximize energy intake immediately postpartum. In many areas this suggests use of whole cottonseed at 5 to 8 lbs/cow/day at a forage-to-concentrate ratio of about 35:65, dry basis. In other areas, added fat of up to 5% should be used with the highest percentage of concentrate (65% to 70%) consistent with main-tenance of fat test. Sodium bicarbonate at about 1.0% to 1.5% of the total diet has helped (Erdman et al., 1980). Ideally, the concentrate should be blended with the forage and offered ad libitum throughout the day, which will result in numerous small meals and a relatively uniform rumen fer-mentation. Where a complete ration system is not feasible, there is great merit in devising a feeding plan that results in concentrate being fed from 4 to 6 times per day. This smooths out the rumen fermentation pattern and probably increases the amount of concentrate a cow can eat without digestive upset. Maintenance of milk fat test may be more consistent with multiple concentrate feedings.

In addition to its main function as an energy supplement, the concentrate mixture usually serves as a carrier of supplemental protein, minerals, and vitamins. Because the cow does not have the ability to store and mobilize most nutrients to the degree that she can store and mobilize fat, the diet for the cow in negative energy balance should be enriched in other nutrients about 20% above that needed by the cow in mid-lactation. With cows of high genetic ability for milk production, one cannot overfeed energy during the first trimester.

Protein

Although the 1978 NRC publication for dairy cattle suggests that excess protein can be fed without harm, several recent studies indicate otherwise. From Oregon State, Jordan and Swanson (1979) fed diets of 12.7%, 16.3%, or 19.3% crude protein (CP) and observed an increase in days open, in services per conception, and a decrease in days to first observed estrus with increasing dietary protein (table 2). A later study from Israel (Folman et al., 1981) compared diets with 16% CP from soybean meal protected to minimize rumen degradation (CPSP) to 16% CP and 20% CP unprotected soybean meal. Table 3 shows that the results were a higher conception rate, fewer inseminations per conception, and fewer days open for the cows fed the 16% CPSP diet.

TABLE 2. EFFECT OF DIETARY CRUDE PROTEIN ON REPRODUCTIVE EFFICIENCY[a]

Protein group (%)	Reproductive measures			
	Days open	Days to first observed estrus	Services per conception	Days to ovulation
19.3	106	27	2.47	16
16.3	96	45	1.87	28
12.7	69	36	1.47	18

[a]Jordan and Swanson (1979)

The latter group gave slightly more milk (nearly 90 lb/day) than did the other two groups, so the lower reproductive function was not an effect of higher milk production by the 16% and 20% CP groups. Chalupa (1982) suggests the adverse effects of protein in reproduction can be explained by differences in the dietary protein susceptibility to rumen degradation to ammonia. He suggests 2 dependent postulates: (a) protein rapidly degraded in the rumen releases ammonia that, if excessive, is absorbed and may cause metabolic and cellular disruptions; (b) additional absorbed amino acids change the balance of amino acids and net energy to result in a relative energy deficiency. The lack of an effect on reproductive performance in long-term and extensive studies (Ryder et al., 1972) of urea feeding does not support the idea of excess ammonia causing impaired fertility. The Israeli study (Folman et al., 1981) supports the first suggestion that gives greater emphasis to research on rumen-protected protein, appropriate levels, etc. One point is clear: feeding excessive levels of protein is neither justified nor economical. However, excess protein can occur if inaccurate estimates are made of forage protein levels such as in highly fertilized grasses (e.g., oats) pastured at an immature stage.

TABLE 3. EFFECT OF PROTECTED SOYBEAN MEAL AND PROTEIN PERCENTAGE ON REPRODUCTIVE PERFORMANCE[a]

	16% CPSP	16% CP	20%CP
No. of cows	20	19	20
Days to 1st estrus	36.8	40.4	37.5
Conception rate (%)	69	56	44
Service/conception	1.45	1.79	2.25
Days open	83.7	97.5	102.1
Milk production; lb/day	88.9	85.6	84.5

[a]Folman, Neumark, Kaim, and Kaufmann (1981).

MINERALS

Macro

I am using phosphorus as an example of the macroelements not because it is necessarily more important than the other six, but because its cost is greater than the other six and it is probably the first to be limited in pasture-fed heifers. Morrow (1969) found that services per conception were 3.7 before phosphorus supplementation and 1.3 after supplementation in dairy heifers grazing low phosphorus pasture. Clinical signs included rough hair coat, depraved appetite, and infertility. For heifers fed grain, a solution is to include a phosphorus supplement because free-choice consumption is much more erratic. Excess phosphorous from over-formulation is unlikely because of its cost unless using the thumb rule of adding 1% of a phosphorus supplement to a high phosphorous grain mix.

A recent study from Michigan State (Carstairs et al., 1981) reported that first lactation cows fed 138% of phosphorus requirements from week 5 through week 12 gained 4 lb less milk/day than did cows fed at NRC (1978) recommendations. This work supports the maxim that deficiencies as well as excesses should be avoided.

Computer-generated formulations can be expected to prevent most deficiencies of mineral elements that are part of the matrix. Of the 7 required macro elements, seldom are

308

more than 3 or 4 considered, but all are essential. A recent NRC (1980) publication on mineral tolerances emphasized
that not only is there a deficiency level but a toxic level
as well (figure 2). Even fewer computer programs set upper
bounds on mineral elements. With greater use of by-products
and materials not formerly used as feedstuffs, the importance of extensive nutrient profiles can hardly be over-
emphasized. The new NRC bulletin (1982) on feed composition
lists 14 minerals elements--7 macro and 7 trace. Although
there are many blanks, the publication is the most complete
now available.

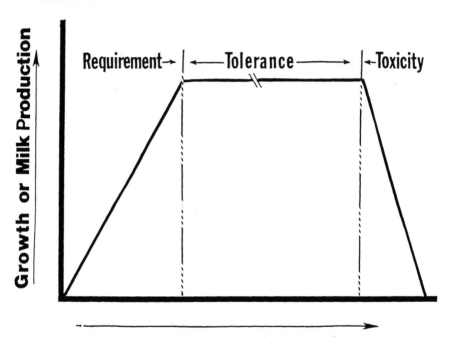

Nutrient Concentration in the Diet

Figure 2. For some mineral elements, the tolerance zone is
wide; for others, it is narrow.

Recently Mongin (1980) has suggested that the sum of
potassium and sodium minus chloride (K + Na - Cl) in mq/100
g of feed, of 25 results in maximum growth in poultry. If
some relationship among these 3 elements is beneficial to
dairy cows, formulation bounds on these 3 elements would be

necessary, none of which has bounds today in most formulations for dairy cattle.

Micro or Trace Elements

Most trace elements function as cofactors or activators of enzymes (Hideroglu, 1979). Dietary deficiencies in these elements probably depress fertility by causing disruptions in these enzymatic reactions. Forage tests routinely available give values for iron, manganese, zinc and copper. Laboratory tests for selenium, iodine, and cobalt are not routinely available because of their cost. McCullough of Georgia (1979) has made a reasonable suggestion regarding trace element supplementation when no forage test analyses are available: add 50% of the requirement if the basal diet has no toxic levels. Recently Jimenez (1982) has recommended adding 100% of the requirement, using the amounts in the ingredients as a safety margin. The cost of this fortification was about $.10 cow/month.

The traditional panacea of 1% trace mineral salt in the concentrate mixture rarely should be relied upon (McCullough, 1979). Many trace mineral salt formulations provide little trace element supplementation.

Vitamins

Although vitamins, particularly vitamin A, have long been recognized for their roles in fertility and reproductive performance, the routine low-cost addition of supplemental vitamin A and D to commercial feed mixtures has virtually eliminated the possibility of deficiencies.

Plants, however, contain no vitamin A but only its precursor, β-carotene, which the animal converts to vitamin A. Recent German studies suggest that the β-carotene may play a role in fertility that cannot be assumed by vitamin A. The corpus luteum of cattle contains a high concentration of β-carotene but no vitamin A, which suggests that β-carotene may play a specific role in ovarian function. Most forages that contain green-colored pigments contain high amounts of β-carotene and, moreover, its fat soluble nature allows extensive body stores, which can be used during periods of dietary shortage. Other work from Israel (Folman et al., 1979) does not support the idea of superior fertility with a supplement of β-carotene, but all studies so far have involved only a few cows. At this time, the practical significance of a specific β-carotene requirement for dairy cattle remains to be clarified.

Formulation Precision

One point seems clear from the previous discussion. We need to be much more precise in diet formulation to avoid

TABLE 4: EXAMPLES OF NUTRIENT VARIATION IN FORAGES[1].

Forage	Crude protein	Acid detergent fiber	Estimated TDN	Calcium	Phosphorus	Magnesium	Zn
	-------------------------------- (%) --------------------------------						(ppm)
Mixed, mainly grass hay							
Mean (1,187)[2]	10.5	40.4	59.7	.65	.22	.15	19
Normal range[3]	7.2-13.7	37.2-43.5	56.7-62.6	.39-.90	.17-.26	.08-.21	13-34
Corn silage							
Mean (2,434)[2]	8.5	27.5	65.6	.21	.20	.15	23
Normal range[3]	7.4-9.5	22.6-32.3	62.9-68.2	.13-.30	.15-.24	.11-.18	11-36
Mixed, mainly legume hay							
Mean (1,095)[2]	15.1	38.1	61.4	.99	.26	.20	20
Normal range[3]	11.4-18.7	33.7-42.4	58.5-64.2	.64-1.34	.21-.30	.12-.27	13-26
Mixed, mainly legume silage							
Mean (1,401)[2]	15.8	40.0	60.0	1.01	.28	.20	25
Normal range[3]	11.9-19.6	34.6-45.3	56.3-63.6	.69-1.34	.22-.33	.12-.27	2-47

[1]Composition on a dry mater basis. From NYDHIC Forage Testing Laboratory, 1980.
[2]Number of samples in the mean.
[3]Normal range is defied as ±1 standard deviation of the mean.

either deficiencies or gross excesses. This precision can only be achieved with extensive forage and concentrate ingredient testing. Table 4 shows an example of the variation in forage nutrient content from one forage testing laboratory. By-products and feedstuffs not commonly used for feed should be tested for their nutrient composition. The option of extensive mineral element analyses should be taken at every opportunity. The computer must be used to obtain a least-cost ration and to consider the many nutrients that ought to be included in the formulation for high-producing dairy cows. An ideal feeding system is one that allows precise control over the composition of the consumed diet; when this diet is offered ad libitum, many small meals will be consumed throughout the day.

CONCLUSIONS

Deficiencies and excesses of many nutrients have been shown to influence fertility and reproductive performance in dairy cattle. The major culprit appears to be energy: in growing heifers, either too little or too much is detrimental, and both conditions are practical field problems. Unfortunately, there is no simple measure of body fatness or condition in live cattle. In adult cows, energy is also the nutrient of greatest imbalance. In dry cows, excess energy is more likely to be the problem; whereas in postpartum, high-producing cows, energy deficit is common and has no simple resolution. Negative energy balance appears to be detrimental to ovulation, which helps to explain why very high producers may have longer calving intervals. New strategems are needed to induce greater energy consumption in high producing postpartum dairy cows. Mineral and vitamin needs can be met by detailed feed analysis and rational supplementation.

312

REFERENCES

Anonymous. 1978. You can breed your heifers to calve at a younger age. Hoard's Dairyman 123:778.

Butler, W.R., R.W. Everett, and C.E. Coppock. 1981. The relationships between energy balance, milk production, and ovulation in postpartum Holstein cows. J. Animal Sci. 53:742.

Carstairs, J.A., R.R. Neitzel, and R.S. Emery. 1981. Energy and phosphorus status as factors affecting postpartum performance and health of dairy cows. J. Dairy Sci. 64:34.

Chalupa, W. 1982. Energy and protein effects upon reproduction in dairy cattle. Feed Management 33: No. 5, 80.

Chase, L.E., R.D. Smith, and C.J. Sniffen. 1979. The impactof nutrition on reproduction. Proc. 11th Biennial Ruminant Heath-Nutr. Conf., Syracuse, NY. p 1.

Erdman, R.A., R.L. Botts, R.W. Hemken, and L.S. Bull. 1980. Effect of dietary sodium bicarbonate and magnesium oxide on production and physiology in early lactation. J. Dairy Sci. 63:923.

Folman, Y., H. Neumark, M. Kaim, and W. Kaufmann. 1981. Performance, rumen, and blood metabolites in high-yielding cows fed varying protein percents and protected soybean. J. Dairy Sci. 64:759.

Folman, Y., I. Ascarelli, Z. Herz, M. Rosenberg, M. Davidson, and A. Halevi. 1979. Fertility of dairy heifers given a commerical diet free of β-carotene. Brit. J. Nutr. 41:353.

Francos, G. and D. Rattner. 1975. On the relation between milk production and fertility in Kibbutz dairy cattle herds. J. Agric. Sci. Camb. 85:527.

Hidiroglou, M. 1979. Trace element deficiencies and fertility in ruminants: A review. J. Dairy Sci. 62:1195.

Jimenez, A.A. 1982. Don't skimp on trace minerals for dairy cows. Feedstuffs 54:26, p 11.

Johnson, D.G. and D.E. Otterby. 1981. Influence of dry period diet on early postpartum health, feed intake, milk production, and reproductive efficiency of Holstein cows. J. Dairy Sci. 64:290.

Jordan, E.R. and L.V. Swanson. 1979. Effect of crude protein on reproductive efficiency, serum total protein, and albumin in the high-producing dairy cow. J. Dairy Sci. 62:58

Maynard, L.A., J.K. Loosli, H.F. Hintz, and R.G. Warner. 1979. Animal Nutrition. 7th Ed., McGraw-Hill Book Co., NY.

McClure, T.J. 1970. A review of developments in nutrition as it is related to fertility in cattle. N.Z. Vet. J. 18:61.

McCullough, M.E. 1979. Don't overlook minerals and vitamins. Hoard's Dairyman 124:1126.

Mongin, P. 1980. Electrolytes in nutrition. A review of basic principles and practial application in poultry and swine. Proc. Third Ann. Int. Minerals. Conf., Orlando, FL p 1.

Morrow, D.A. 1969. Phosphorus deficiency and infertility in dairy heifers. J.A.V.M.A. 154:761.

National Research Council. 1978. Nutrient requirements of dairy cattle. 5th Rev. Ed., Nat. Acad. Sci., Washington, DC.

National Research Council. 1980. Mineral tolerance of domestic animals. Nat. acad. Sci., Washington, D.C.

National Research Council. 1982. United States-Canadian Tables of Feed Composition. 3rd Rev. Ed., Nat. Acad. Sci. Washington, D.C.

Otterby, D.E. and J.G. Linn. 1981. Nutritional effects on reproduction in dairy cattle. Proc. 42nd Minnesota Nutr. Conf. p 9.

Ryder, W.L., D. Hillman, and J.T. Huber. 1972. Effect of feeding urea on reproductive efficiency in Michigan Dairy Herd Improvement Association herds. J. Dairy Sci. 55:1290.

Schultz, L.H. 1969. Relationship of rearing rate of dairy heifers to mature performance. J.Dairy Sci. 52:1321.

Spalding, R.W., R.W. Everett, and R.H. Foote. 1975. Fertility in New York artificially inseminated Holstein herds in dairy herd improvement. J.Dairy Sci. 58:718.

31
REDUCE HEAT STRESS IN HIGH-PRODUCING DAIRY COWS WITH FEEDING AND NUTRITION MANAGEMENT

Carl E. Coppock

INTRODUCTION

Hot weather has a debilitating effect on high-producing dairy cows through its effect on intake and subsequently, on milk production and composition. And as production levels increase, the problem is becoming more severe. The higher the cow's milk production, the greater the energy consumption--and the greater the heat production associated with the digestion and metabolism of feed. The mammalian system places a high priority on maintaining body temperature within narrow bounds (homeothermy)--a priority that outranks productive functions such as lactation. Under conditions of heat stress, productive functions suffer as the cow attempts to balance heat production with heat loss (Collier, 1979). The high-producing dairy cow produces from three to four times as much heat does as the dry cow, and an increase of 1° C or less in body temperature has been reported (McDowell, 1972) to reduce performance in most species of livestock.

Alleviation of heat stress in dairy cows can be considered in three categories: (1) reduction of heat transfer to the body, (2) devices to increase heat loss from the body, and (3) lower heat production by the cow while maintaining milk production.

It has been noted (Collier, 1979) that an unshaded cow may receive a radiant heat load from the sun that exceeds the cow's metabolic heat production. Florida researchers (Collier et al., 1981) have noted that a simple shade structure can reduce the radiant heat load by 30% or more, through blocking direct solar radiation. They found rectal temperatures of 38.8°C and 38.5°C in shaded, lactating Holstein and Jersey cows compared to 39.7°C and 39.2°C in similar cows without shade. Respiration rates averaged 114.8/min for unshaded cows compared to 78.5/min for those with shade. These measures of heat stress were manifested in lowered milk production, although milk composition was not affected significantly. These researchers emphasized that the primary advantage of shade was that it reduced the total heat load and preserved the sensible modes of heat loss.

In addition to shade, Arizona researchers (Wiersma and Stott, 1973) used evaporative cooling that was effective in a relatively low humidity region and lowered temperatures by 10°C to 12°C in the cooled area. Milk production was reduced less during the summer months and there was a carryover benefit that extended into the winter months of December and January. Increased conceptions per service also were noted as compared to those of shade-only cows. Fans alone also were helpful, particularly in high humidity regions that do not obtain benefit from evaporative cooling.

The genetic trend for increased milk yield continues upward, implying greater heat production with the associated metabolism of more feed nutrients. If we assume that total environmental control is economically infeasible, that shade is feasible everywhere, and that evaporative cooling is feasible in regions of low humidity and high ambient temperature, are there other feasible strategems to reduce heat stress in high-producing dairy cows? In addressing the question, this paper describes some feeding schemes to facilitate heat loss in the cow and some diet formulation protocols to reduce the heat increment of feeding (HIF).

FEEDING MANAGEMENT STRATEGEMS

1) Feed fresh feed after body temperatures begin returning toward normal. Cows protected by shades were shown by Wiersma and Stott (1974) to reach maximum body temperatures of 3°F to 4°F above normal at about 1400 to 1600 hours under Arizona conditions on days when maximum ambient temperatures consistently peaked at 105°F at about the same time of day. Body temperatures declined very gradually but more sharply about 0300 to 0600 hours. Minimum ambient temperatures had a marked influence on the time of sharp declines in body temperatures.

Ideally, the cow should eat many (10 to 12) small meals throughout the day. The rumen is a continuous fermentation system and should be fed frequently to sustain a nearly uniform fermentation medium. Kaufman (1976) has demonstrated that a higher feeding frequency will allow a higher intake of concentrates without a decrease in pH and that the ratio of acetic to propionic acid tends toward 3:1 instead of narrower ratios, which prevent a decrease in milk fat content. The point here is that two large feedings of concentrate twice daily (as in parlor grain feeding) are contraindicated because they disrupt uniform rumen fermentation and result in sharp peaks in heat production. Based on the decline in body temperatures, a large fresh supply of feed should be available at 0200 to 0400 hours. This statement is based on the assumption that heat stress reduces feed intake and heat loss will occur most readily during the hours of minimum ambient temperature. Observations of Holstein cows indicate that they voluntarily graze 75% or more at night. Many feed bunks are unshaded and, consequently, cows either do not spend as much time at the bunks during the day as they would at shaded bunks, or they encounter heavy solar heat loads that increase their body temperatures, which must be resolved before the next day. Low ambient temperatures during the night greatly facilitate the discharge of body heat and the return of body temperatures to near normal (Wiersma and Stott, 1974). In every situation, it is well to keep in mind that it is always the highest-producing cows that produce the greatest amount of heat, suffer the greatest heat stress, and deserve first consideration.

2) Instead of feeding concentrate in the parlor, provision of fresh water (chilled water could be justified for the highest producers in the hottest weather) should have several advantages. The laggards requiring persuasion under almost all conditions will be encouraged to enter as they become thirstier with time. This system works well at Braums, a very large dairy near Oklahoma City.

3) Other feeding-management stratagems to reduce heat stress have been well described by Loper (1980) and Jimenez (1982) who suggest:

- Keep waterers clean and sanitized weekly and check daily for working order.
- Provide shade under many summer conditions. Do not require cows to expose themselves to the sun to eat during the day.
- Increase frequency of feeding to provide a more sustained rumen fermentation for feeds fed in limited amounts, and for those fed ad libitum to provide fresh feed to encourage greater intake.
- Add water to the complete ration or the concentrates (fed inside or outside) to reduce dust and increase succulence. Add chopped dry hay to the mixer wagon.
- Feed more succulent feeds (kept fresh) to encourage intake at the expense of dry hay.
- Keep mangers clean so that succulent feeds do not deteriorate rapidly in hot weather.

4) Social stress associated with regrouping cows can be assumed to increase heat production; therefore, transfer of cows to new groups should be minimized in hot weather.

5) It has been suggested (Duby, 1977) that it is possible to increase cow density in free stalls and reduce bunk space to about one-half the amount needed if all cows were to lie down at once or eat at once. The bunk space reduction seemed to work well as long as complete rations were offered free choice. The reason seems to be that cows soon realize there is little reason to rush to the bunk if the same feed is always available. Greater cow density means greater heat generation in that area, and much greater ventilation would be necessary to make this suggestion work well under conditions of heat stress.

DIET FORMULATIONS TO MINIMIZE HEAT PRODUCTION

Heat production increases following a meal. Heat increment (Maynard and Loosli, 1956) is associated with eating a meal. Webster (1979) partitions this heat increment of feeding (HIF) into: (1) the heat produced associated with eating and rumination; (2) heat produced by microbial fermentation in the gut; (3) heat from metabolic activities of gut tissues; and (4) heat from all other body tissues. In a previous paper, Webster (1976) noted that, "The initial invariable response to heat in a growing or lactating animal is to reduce food intake. This can be overcome somewhat by offering diets with a low heat increment." The same conclusion was reached by Fuller and Dale (1979) in the case of broiler diets. Lofgreen (1974) showed that diets formulated to reduce heat increment increased intake and gain in steers in the Imperial Valley. The issue is: What diet adjustments can be made to reduce the HIF in lactating dairy cows that will also maintain, or even increase, net energy intake?

Dietary Deficiencies

A cow's tendency to reduce feed intake under heat stress is a protective mechanism, but cows in early lactation that can continue negative energy balance (NEB) may sustain their milk production--because heat production associated with metabolism of body energy reserves results in one-half the heat produced by direct dietary energy conversion to milk. That is, an efficiency of 82% vs 64% for the two conversions is the same as a comparative heat production of 18% vs 36% of the metabolizable energy used. Nevertheless, it seems to be good strategy to feed as though we wished to eliminate NEB.

Reid (1972) noted that deficiencies of certain nutrients cause an inefficient use of absorbed energy or an increased heat production. These include protein, phosphorus, and possibly sodium, magnesium, cobalt, and copper. High-producing cows cannot long sustain their production when diets are deficient in these nutrients, but reduced intake of total feed (NEB) suggests an increased concentration of most nutrients in the feed dry matter is appropriate.

Diet Composition

Fiber. Feeding trials in the summer at Texas (Rupel et al.), Arizona (Stott and Moody, 1960), Louisiana (Breidenstein et al., 1960), and in the Philippines (Tsai et al., 1967) showed diets with a high percentage of concentrates and minimum fiber reduced heat stress and served to maintain milk production better than did conventional diets. Later, respiration chamber studies at USDA - Beltsville (Coppock et al., 1964) showed that less heat production occurred in the

utilization of metabolizable energy from high concentrate
diets as compared to that from high forage diets. Seventy
percent concentrate is as high as we can go. To some
degree, cows reduce their intake of fiber during heat stress
by reducing the amount of hay eaten--more than they reduce
silage or concentrate. This could be prevented by blending,
but it appears to be a part of the cow's protective mechanism. All strategems in this category, including use of the
best quality forage, minimum forage, etc., tend to depress
milk-fat test and involve a tradeoff. At least two features
of forage serve to maintain fat test: (1) Long fiber
results in greater rumination time with coincident saliva
production with its bicarbonate buffering action and (2) the
production of acetic acid. The logical addition of sodium
bicarbonate has helped maintain fat test and production in
hot weather under some conditions (Escobosa et al, 1983).
Fuguay (1981) reviewed the problem of cows pastured where
the fiber content increased as the season advanced.
Although an abundance of forage allows selection for a lower
fiber diet, the quantity available also often diminishes as
the summer advances. We do not know the fundamental reasons
why the energy of high-fiber diets is used less efficiently,
although the acetic-propionic acid ratio theory no longer
appears tenable (Webster, 1979).

Added fat. In their classic text, Maynard and Loosli
(1956) note that, "The greater-than-expected value of fat as
a source of energy can be explained on the basis that, with
equicaloric diets, increasing the fat component decreases
the heat increment." This suggests that added fat should be
more valuable in hot weather than in cold, and that one
should expect the greatest value from the maximum amount
that can reasonably be fed. Whole cottonseed (WCS) is a
feed ingredient highly valued by dairymen in the southern
and western regions of the U.S. where this product is available. Research at Utah State University, the University of
Arizona, and the University of California has shown that the
high energy and relatively high fiber content of WCS may
stimulate both milk yields and increase fat test. Because
of its high fat content, this ingredient should be particularly valuable in hot weather, which often reduces feed
consumption, milk yield, and milk-fat test. Moody (1962)
reported less heat stress as indicated by lower body temperatures in cows fed whole cottonseed in Arizona.
We do not know the maximum level of WCS that can be
fed. Most dairymen feed about 5 to 8 lb/cow/day. In recent
research trials in California, diets with 30% WCS were fed
for 15 wks postpartum, which resulted in intakes of up to 14
lb/cow/day. No ill effects were apparent in this trial
(Smith and Collar, 1980), which suggests considerable lati-
tude is possible when WCS is included at usual levels. A
field study (Smith and Collar, 1980) in California,
summarized level of WCS feeding in 55 commercial dairies (39
were feeding WCS) and cow health data; there was no apparent
relationship of WCS feeding to any recorded health disorder.

Those not fortunate enough to have WCS available should examine the work of Palmquist and Jenkins (1980) who obtained the consistent positive response from blended hydrolyzed fat. Palmquist (1979) suggests 5% to 6% added fat and calcium level of 1% in the total diet dry matter. To the degree that added fats or high fat ingredients decrease heat increment, they are undervalued by current feed evaluation systems (TDN or NE_l) when fed in hot weather.

Protein. Protein deficiency was mentioned earlier as a cause of increased heat production, but here the issues are (1) an excess and (2) amino acid balance. The work of Moe and Tyrrell (1972) showed decreased metabolizable energy intake by 12 kcal for each gram of digested nitrogen consumed above requirements. Thus, excess nitrogen will require energy for excretion and should be avoided. Excess is usually avoided because of the cost of protein supplements; but some temporary pastures are very high in protein (e.g., oats) and dietary adjustments may not be made. In other cases, book values are used that are grossly inaccurate. More frequently, systems of feeding and formulations with high levels of rumen soluble nitrogen are used that result in ammonia peaks after feeding, which exceed efficient use with consequent excretion. An example would be concentrate mixtures containing added nonprotein nitrogen that are fed twice daily in a milking parlor. Webster (1979) feels that the heat produced in gut tissue may account for up to 25% of total heat production in the animal. He suggests this high rate of heat production may be related to the high energy cost of protein turnover in the tissues of the gut lining.

Miller (1980) has recently pointed out the importance of amino acid balance and particularly the value of avoiding excess amino acids in reducing the heat increment of poultry diets during hot weather. Any reduction in heat increment by dietary adjustments will make more energy available for productive purposes. In ruminant nutrition, although we are not yet able to define protein requirements in terms of amino acids delivered to the small intestine, it seems clear that if (when) this is possible, a considerable reduction in HIF will occur. Ames et al. (1980) showed that with both sheep and cattle, protein could be reduced in proportion to expected reduction in daily gain as a function of thermal stress, without affecting gain. Thus, efficiency of protein use was increased. These results were obtained because of the increase in maintenance that results from thermal stress. A recent NRC publication (1981) suggests that a lactating cow's maintenance requirements for energy increases by about 20% at 35°C. Consequently, it is unlikely that additional dietary protein also is needed, unless cows are in negative energy balance.

Mineral Elements-Electrolytes. Baldwin et al. (1980) described opportunities to improve animal efficiency through adjustments in metabolism. In addition to selection of the most efficient animals, they suggested that manipulations of ion transport and protein turnover could yield maximum benefits of 10% to 30% in terms of basal maintenance-energy costs. A third suggestion was that manipulation of patterns of nutrient utilization to an optimum pattern must decrease heat increment of production in growing animals by up to 50%. These decreases would have particular value under conditions of thermal stress. These authors (Baldwin et al., 1980) further summarized the contributions of futile cycles to maintenance requirements—two having large influence: (1) NA+/K+ transport—20% to 30% and (2) protein turnover—10% to 15%. Although it was considered unlikely that these cycles could be eliminated, a substantial reduction in their cost would be greatly beneficial in hot weather. At this time, little is known about reducing these cycles, but it seems logical to avoid large excesses of the macro elements. Specifically, potassium, sodium, and chloride are often fed in great excess in dairy cattle diets; however, with high concentrate diets, marginal potassium is more likely. These three electrolytes also are closely related to acid-base balance. Mongin (1980) states that "the animal tries to keep its acid-base status as close as possible to normal in order to make base excess nearly zero...If the diet does not respect homeostasis, it will reduce the performance of the animal."

CONCLUSIONS

The simple, easy, inexpensive ways of reducing heat stress in lactating cows have been exploited. As production continues upward, we will need all available techniques to reduce the heat load. Although using low fiber and high fat diets to reduce metabolic heat production is not new, their significance in reducing heat stress is only now becoming widely appreciated. Obviously, nutrient deficiencies should be avoided for several reasons, but dietary excesses also may cause additional metabolic heat production as the cow strives to maintain homeostasis. Much remains to be learned about other dietary adjustments that will reduce heat production, but sufficient information is available to suggest that a substantial reduction is possible.

REFERENCES

Ames, D. R., D. R. Brink, and C. L. Wilms. 1980. Adjusting protein in feedlot diets during thermal stress. J. Anim. Sci. 50:1.

Baldwin, R. L., N. E. Smith, J. Taylor, and M. Sharp. 1980. Manipulating metabolic parameters to improve growth rate and milk secretion. J. Anim. Sci. 51:1416.

Breidenstein, C. P., J. E. Johnston, G. A. Hindery, and L. L. Rusoff. 1960. Effect of fiber content of the ration on the performance of dairy cows under hot and cool conditions. J. Dairy Sci. 43:443 (Abstr.).

Collier, R. J. 1979. Endocrinological and physiological effects of heat stress on lactation. Proc. FL Nutr. Conf. p 17.

Collier, R. J., R. M. Eley, A. K. Sharma, R. M. Pereira, and D. E. Buffington. 1981. Shade management in sub-tropical environment for milk yield and composition in Holstein and Jersey cows. J. Dairy Sci. 64:844.

Coppock, C. E., W. P. Flatt, L. A. Moore, and W. E. Stewart. 1964. Effect of hay to grain ratio on the utilization of metabolizable energy for milk production by dairy cows. J. Dairy Sci. 47:1330.

Duby, R. T. 1977. Can you crowd cows in free stall barns? Hoard's Dairyman. 122:1096.

Escobosa, A., C. E. Coppock, L. D. Rowe, Jr., W. L. Jenkins, and C. E. Gates. 1983. Effects of differences in dietary sodium and chloride on acid-base balance, blood electrolytes, blood metabolites, respiratory rate, and body temperature in lactating cows in hot weather. J. Dairy Sci. (submitted).

Fuguay, J. W. 1981. Heat stress as it affects animal production. J. Anim. Sci. 52:164.

Fuller, H. L., and N. M. Dale. 1979. Effect of diet on heat stress in broilers. Proc. GA Nutr. Conf. p 56.

Jimenez, A. A. 1982. Environment alters the needs of dairy cows. Feedstuffs 54:21, p 14.

Johnston, J. E. 1958. The effects of high temperatures on milk production. J. Hered. 49:65.

Kaufman, W. 1976. Influence of the composition of the ration and the feeding frequency on pH regulation in the rumen and on feed intake in ruminants. Livestock Prod. Sci. 3:103.

Lofgreen, G. P. 1974. Ration formulation for relief from heat stress. CA Feeders' Day Rep. p 81.

Loper, D. C. 1980. Another hot one--How to feed lactating cows during heat stress. Dairy Herd Mgmt. 17:18.

Maynard, L. A. and J. K. Loosli. 1956. Animal nutrition. McGraw-Hill Book Co., Inc., NY.

McDowell, R. E. 1972. Improvement of livestock production in warm climates. W. H. Freeman & Co., San Francisco.

Miller, R. F. 1980. Amino acid balance and hot weather nutrition-energy-the limit on performance. MSD-AG-VET, Merck & Co., Rahway, NJ.

Moe, P. W. and H. F. Tyrrell. 1972. Metabolizable energy requirements of pregnant dairy cows. J. Dairy Sci. 55:480.

Mongin, P. 1980. Electrolytes in nutrition: A review of basic principles and practical application in poultry and swine. Proc. 3rd Int. Minerals Conf., Orlando, FL. p 1.

Moody, E. G. 1962. Whole cottonseed in dairy rations. 11th Ann. Dairymen Conf., AZ State Univ., Tempe.

National Research Council. 1981. Effect of environment on nutrient requirements of domestic animals. Nat. Acad. Sci., Washington, D.C.

Palmquist, D. L. 1979. Feeding fat to dairy cows. Use less concentrate and increase milk production. Anim. Nutr. & Health, Nov-Dec:14.

Palmquist, D. L. and T. C. Jenkins. 1980. Fat in lactation rations: Review. J. Dairy Sci. 63:1.

Reid, J. T. 1972. Some feeding fundamentals. Unpublished mimeo. Cornell University.

Rupel, I. W., R. E. Leighton, and R. R. Harris. Low-fibre, high-fat, and restricted-roughage diets in relation to hot weather performance of dairy cows. Proc. XV Int. Dairy Cong. p 319.

Smith, N. E. and L. S. Collar. 1980. Whole cottonseed and extruded soybeans for lactating cows. Proc. CA Dairy Cattle Day. p 33.

324

Stott, G. H. and E. G. Moody. 1960. Tolerance of dairy cows to high climatic temperatures on low roughage rations. J. Dairy Sci. 43:871 (Abstr.).

Tsai, Y. C., L. S. Castillo, W. A. Hardison, and W. J. A. Payne. 1967. Effect of dietary fiber level on lactating dairy cows in the humid tropics. J. Dairy Sci. 50:1126.

Webster, A. J. F. 1976. The influence of the climatic environment on metabolism in cattle. In: H. Swan and W. H. Broster (Ed.) Principles of Cattle Production. p 103. Butterworths, London.

Webster, A. J. F. 1979. Energy costs of digestion and metabolism in the gut. In: V. Rucklebusch and P. Trivend (Ed.) Digestive Physiology and Metabolism in Ruminants, p 469. MTP Press Ltd., Lancaster, England.

Wiersma, F. and G. H. Stott. 1973. Evaporative cooling for improved performance in dairy cattle. Proc. Nat. Dairy Housing Conf ASAE. p 162.

Wiersma, F. and G. H. Stott. 1974. Dairy cow body temperatures in a hot climate. Trans. Amer. Soc. Agr. Eng. 17:745.

32
SURVIVE THE COST-PRICE SQUEEZE
WITH A NUTRITION AND FEEDING GAME PLAN

Carl E. Coppock

INTRODUCTION

With genetic trend increasing sharply in the dairy cow population, along with better control of mammary infections and metabolic disorders, inadequate nutrition is rapidly becoming the primary production constraint. Because feed costs represent 50% to 60% of production costs, they are a logical point of departure for reducing costs; there may be some reduction in milk yield, but this need not cause a loss in income over feed cost.

Nutritional and feeding system adjustments also must be made with an appreciation for their direct influence on reproductive performance. Because of the effect of high-peak milk production on total lactation yield, high intakes of energy during the first trimester seem imperative--almost independent of concentrate costs, although highest yields may not mean highest profits.

There are many opportunities to reduce feed costs at purchase, delivery, storage, processing, formulation, and on-the-farm delivery. In this article, my purpose is to point out some ways to reduce costs at each of these stages. I am assuming there is a milk production testing program that provides a record of individual cow performance. Some suggestions are more applicable to large herds fed in confinement, yet basic principles apply to all herd sizes.

FEED PROCUREMENT

If a dairyman is willing to assume some of the services and risks now provided by feed dealers, he can reduce the costs of his concentrate. He should be aware of these important functions that include (a) purchasing agent, (b) quality control specialist, (c) facilities, including a railroad siding, plus processing and distribution equipment, (d) formulation knowledge and experience, and (e) risk asumption associated with ingredient quality and compositon of the final mix. To take advantage of feed processing and mixing on the farm, a dairyman should write specifications carefully, buy in quantity, buy in season, buy direct, and test regularly to confirm the fulfillment of the nutrient specifications and the accuracy of the formulation.

As noted previously, computer formulations by linear programming not only give a least-cost formulation but identify the cost ranges of the ingredients used and the opportunity prices of available ingredients not used. Thus, not only are best buys identified, the degree or extent price change is given that must occur to cause a change in formulation. Good financial advice is crucial concerning forward contracting, purchasing in quantity, inventory costs (as well as losses during storage). Smaller dairies may combine to purchase in carload lots, but one large bulk truck load is the minimum tonnage that one should be able to receive. It is essential to find the services of a feed-testing laboratory that is both accurate and fast. Not only will one need to test purchased ingredients, but also it is essential to confirm the formulations that are prepared on the scene.

The "commodity barns" of California are aptly named. More by-products are being fed and their quantity will increase; indeed, if it were not for the livestock industry, their disposal would be very expensive. There may be some increase in fermentation by-products from gasahol production plants, and fossil energy costs suggest these by-products will be fed wet. A small percentage of dairymen will enjoy an advantage of location that will enble them to obtain large quantities of wet by-products at lower cost than alternative ingredients in dry form. But advantages are offset by greater transportation, storage, preservation, and handling requirements. A recent NRC publication (NRC 1981) describes the feeding value of some of these ethanol production by-products.

To further reduce costs, dairymen should look at nontraditional feedstuffs. For example, whey has been a disposal problem for many cheese manufacturers, but recently some dairymen have begun feeding it in its dilute form, even though handling and waste disposal may present problems (Nilson and Welch 1979). Cows may drink from 100 to 200 lb per day and, although uncondensed whey has only about 6.5% dry matter, the solids have a nutritive value equal to corn. The National Research Council is preparing a bulletin on materials underused as animal feeds, including industrial food processing wastes, nonfood industrial wastes, animal wastes, aquatic plants, crop residues, and forest residues. Important questions to be raised about any nontraditional feedstuff include the possibility of residues, heavy metals, or other undesirable materials that may appear in milk or tissues, as well as the acceptability and actual nutritive value of the material to the high producing cow. Some may have special value, e.g., some fluid whey may bypass rumen fermentation providing lactose as a glucose precursor to the small intestine. Other materials are appropriate only as roughage sources for cows in mid-to-late lactation. In the future, many materials will be used as feedstuffs that are not used today, and dairymen should be alert to these possibilities as good buys.

FEED FORMULATION

In this context, the feed refers to both forage(s) and concentrates even though they may be fed separately in the conventional way. Computer formulations and their features were discussed previously (Coppock 1983); however, at least one point is worth repeating. With new evidence concerning the value and quantity of rumen bypass protein (amino acids), and the possibility of B-vitamin supplementation being necessary, formulation for dairy cows is now much more complex. With higher and higher production, formulations should be much more precise and more nutrients should be considered. Only computers can really do this on a least-cost basis. Assitance with formulation is available from the extension service as well as from private consulatants.

FEED PROCESSING AND BLENDING

Many by-products need no further processing; although feed grains need some form of fracturing, rolling is adequate. Pelleting has significant advantages for feeding in the milking parlor because pellets are eaten faster than a mash--otherwise, pelleting has few advantages. In 1980, the cost of pelleting was estimated at $4.80/ton when 30,000 tons were produced per year (McEllhiney 1981). On-farm feed processing has considerable merit for large dairies. Based on 1979 costs and 400 tons of feed processed per year, cost estimates were $5.12 per ton for a small stationary electric mill and $7.29 for a mobile mixer-grinder, with both comparisons including on-the-farm grain storage bins (Benson 1980). The blending process should be quantitative and its performance easily monitored. Equipment should be available to add fat if whole cottonseed is not available. Whole cottonseed is undervalued by all ingredient evaluation systems, in both TDN as well as net energy. It is probably worth at least 20% more because of its positive effect on milk-fat test and lower heat increment in hot weather.

FEED DELIVERY

For larger herds, the complete ration (CR) system has much to commend it. This system's pros and cons have been described in detail (Coppock 1977). However, it does not fit smaller herds and particularly those that use pasture extensively. Apart from the CR system (an in the context of saving dollars and doing a better job of allocating concentrate) there are at least 2 good reasons to eliminate parlor grain feeding: (a) one has little control over the actual distribution of the concenterate, i.e., it's usually free-choice, and (b) considerable amounts are wasted on low-producers that do not need it and may become fat because of it.

The only good reason to continue feeding in the parlor is to encourage cow entry. And as production increases, more energy must be fed outside the parlor, so there is less incentive for the cow to enter the parlor because she is hungry. Alternatives to feeding concentrate in the milking parlor to induce cow entry include offering fresh drinking water, plus numerous strategems to make the milking experience enjoyable for the cows so that a conditioned response is developed in which the milking parlor is perceived as a final step preceding access to a fresh bunk of feed.

If all concentrates are fed outside and separately, lock-in stanchions have much to commend them, allowing individual feeding without serious time constraints on the cow. An understanding of taste preferences in cows has encouraged innovations and improvisations to take advantage of their preferred feeds where feeds are offered individually.

Savings also can be made in transportation and feed handling on the farm. In the summer of 1980, a survey was made of corral feeding and energy used for feed distribution by 175 herds in Tulare County, California (Shultz 1981). The herds were large and all were fed some hay, with the larger herds being fed more on-the-farm grain mixtures (table 1). The average daily fuel and labor cost per 100 milking cows shows a fuel cost advantage for separate feeding in 3 smaller herds, but a lower fuel cost for complete rations in larger herds. Dairies of similar size often differed greatly in the amount of fuel and labor needed to feed their herds. Some were adept at locating feed ingredients at efficient points, in matching equipment size to specific needs, and in conserving energy through innovation and attention to details.

PRODUCE OR PURCHASE A HIGH-QUALITY FORAGE

Super-quality forage has an important advantage apart from its immediate nutritive value and its greater energy consumption. It allows one the flexibility of a broad range in the forage-to-concentrate ratio--especially for cows, in the last 2 trimesters of lactation. Precise cost analysis suggests an emphasis on either maximum forage or minimum forage. Recently, dairymen who emphasized forage quality have demonstrated that over 17,500 lb of milk per cow per year can be produced while feeding less than one-half the amount of concentrates fed by an average DHI dairyman in

TABLE 1. TULARE COUNTY DAIRY CORRAL FEEDING PRACTICES

Herd size	Herds	Average cows milking	Production strings	Corral grain % Sepa-rate	Corral grain % Com-plete	Grain mix % Commercial	Grain mix % On-farm	Roughage used % Hay	Roughage used % H+silage	Roughage used % H+green	Cost/100 milking cows Fuel $ Sepa-rate	Cost/100 milking cows Fuel $ Com-plete	Cost/100 milking cows Labor Minute Sepa-rate	Cost/100 milking cows Labor Minute Com-plete
0–200	35	125	1 – 2	38	0	99	1	80	20	0	.80	—	55	—
200–400	79	253	2 – 3	57	3	46	54	42	55	3	.85	1.15	60	90
400–600	65	426	3 – 4	59	19	32	68	20	68	12	1.10	1.65	50	60
600–800	21	575	3 – 4	80	10	11	89	0	100	0	1.30	0.80	40	40
800–1000	15	705	4 – 5	70	30	30	70	10	60	30	1.10	0.95	40	45
1000–2000	18	1130	4 – 6	67	33	8	92	34	58	8	2.50	1.50	50	50

Source: T. Shultz. Corral feeding and energy costs. Proc. 12th Annual California Dairy Cattle Day (1981).

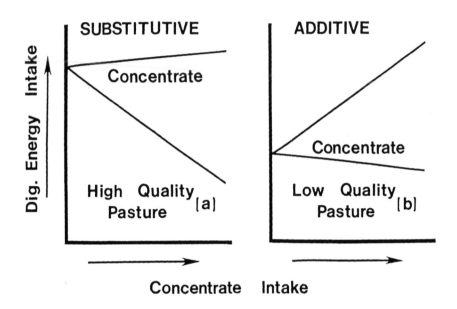

Figure 1. Effect of pasture quality and concentrate addition
on digestible energy consumed

their area (Coppock and Gallamore 1976). Mott and Moore
(1977) suggest that with low-quality pasture, concentrate
feeding is primarily additive for digestible energy intake,
but with high-quality pasture, concentrate feeding is largely substitutive (figure 1). Data from Lamb et al. (1974)
(table 2) show that concentrate feeding at relatively low
levels is primarily additive with respect to dry matter, but
changes to substitutive with forage at higher intakes, even
with high-quality alfalfa hay. The grass legumes have a
potential advantage over corn silage because immature plants
have a higher digestibility, which can approach that needed
in the total diet by most cows, especially in the last 2
trimesters (figure 2).

Concentrate as a small percentage of the diet is worth
considerably more than concentrate as a high percentage of
the diet. This is because (1) of the additive instead of

TABLE 2. EFFECT OF INCREASING CONCENTRATE INTAKE ON HAY CONSUMPTION AND MILK
PRODUCTION

Grain to Milk Ratio[a]	1:4.0		1:2.7		1:2.0		1:1.6
Grain recieved for:							
20# milk/day	0.0		0.0		0.0		0.0
40# milk/day	5.0		7.5		10.0		12.5
60# milk/day	10.0		15.0		20.0		25.0
80# milk/day	15.0		22.5		30.0		37.5
Cows per group	32		32		32		32
Hay consumed[b] lbs	11,715	(-457)	11,258	(-1359)	9,899	(-960)	8,939
Grain consumed[b] (lbs)	2,081	(+1252)	3,333	(+1609)	4,942	(+941)	5,883
Feed conumsed[b] (lbs)	13,796		14,591		14,841		14,822
Forage: concentrate	85:15		77:23		67:33		60:40
Milk/305 days (lbs)	13,878	(+783)	14,661	(+1287)	15,948	(+214)	16,162

Source: R.C. Lamb, G.E. Stoddard, C.H. Mickleson, M.J. Anderson and D.R. Waldo.
Response to concentrates containing two percent of protein fed at four rates for
complete lactation. J. Dairy Sci. 57:811 (1974).
[a]Applied only above 20 lb/day.
[b]Air dry feed consumed during the 305 day lactation.

substitutive nature of the small percentage and (2) the
depression in digestibility which occurs when mixed diets
are fed to high-producing cows. Under some conditions, this
depression appears to be directly related to the percentage
of concentrate in the diet (Wagner and Loosli 1967). The
dilemma is that to achieve high energy intakes with most
forages, it seems necessary to feed a high percentage of
concentrate. Wheeler and Noller (1976) showed that under
some conditions, up to one-half the depression in
digestibility may be due to starch passage, which can be
sharply reduced by use of effective limestone buffers.
These buffers appear to increase the pH of the small
intestine digesta to a level more advantageous for
pancreatic alpha amylase. An increase in starch digestion
by this or any other strategem that does not depress fat
test has value because it increases availability of glucose,
which may limit high-producing cows (Clark 1975), and it
increases efficiency of concentrate use.
     When economic conditions mandate high-forage diets
concern about the magnitude and duration of negative energy
balance is warranted. The physiological drive of the high
producing dairy cow to secrete energy into milk has
outdistanced her ability to consume energy from feed even

though offered a free choice of the highest energy diets
that can be formulated. Therefore, most healthy dairy cows
with high genetic merit assume a period of negative energy
balance during much of the first trimester. It is only by
drawing on energy from fat reserves to supplement their
insufficient dietary energy that most high producers achieve
their great milk output in early lactation. One hundered
pounds of body fat provides energy for about 1000 lb of
milk. We feed such cows in an attempt to stimulate even
greater production or to minimize negative energy balance,
which appears to be detrimental to ovarian function and
conception rate, although it is not clear why this is true.

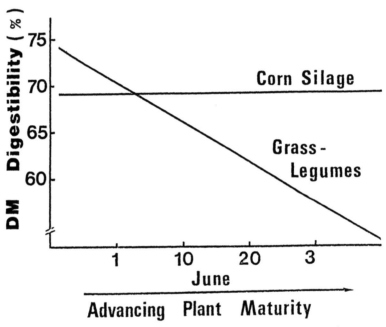

**Figure 2. Effect of cutting date and plant maturity on dry
matter digestibility of grass-legumes and corn
silage in the northeast.**

Consequently, we need all nutritional strategems
available to maximize energy intake immediately postpartum
and for the first trimester or until pregnancy is
confirmed. Then gradual transfer to a high forage diet may
cause little reduction in energy intake or milk production
(figure 3). If a quota system is used in your area, this
is one way to reduce production slightly, while maintaining
income over feed cost. Selling 10% of the lowest producing
cows will decrease milk production very little, especially
if their share of concentrate is fed to the remaining cows.
Another potential problem that may occur with
high-forage diets is an error in concentrate formulation,

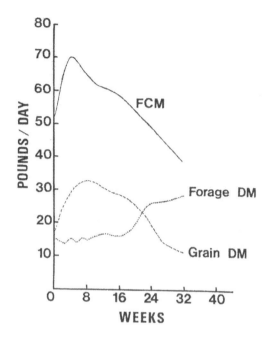

Figure 3. Effect of grain feeding regimen on ad libitum forage consumption during mid-lactation.

which becomes more critical as less is fed. In addition to providing energy (its primary function), a concentrate mixture usually acts as a carrier for protein, minerals, and vitamins. Also, the ability of the cow to store and mobilize several of these nutrients is limited. Therefore, the success of high-forage diets depends to a large degree upon the extent that needed fortification is carried by the concentrate mixture. For example, if 16% crude protein is desired in a total ration consisting of 20 lb of forage DM and 20 lb of concentrate DM and the mixed forage base contains 12% crude protein, then the concentrate fed should contain 20% crude protein. However, if concentrate is fed at the rate of 10 lb, even though voluntary forage consumption increases to 28 lb DM, the concentrate DM should contain 30.4% protein to meet requirements (table 3). At lower production levels and lower grain intakes, the effect becomes progressively greater.

Conversely, concentrates or some specific ingredient may become less expensive than forage and maximum concentrate-minimum forage use will be the objective.

Maintenance of milk fat test will usually constrain
concentrate used to less than about 70% of the diet, which
is less than the amount that often induces digestive upsets,
liver abscesses, and other disorders.

TABLE 3. EFFECT OF THE AMOUNT OF CONCENTRATE FED ON ITS
NEEDED NUTRIENT COMPOSITION[a]

	Feed DM	Crude pro- tein	Calcium	Phosphorus	Salt	Zn
	kg	--------------------%--------------				ppm
Nutrient level desired in diet DM		16	.80	0.45	.5	50
1. Alfalfa	5	16	1.25	.35	--	25
Corn silage	5	8	.25	.25	--	35
Concentrate	10	20	.85	.60	1.0	70
2. Alfalfa	7	16	1.25	.35	--	25
Corn silage	7	8	.25	.25	--	35
Concentrate	5	30.4	1.10	.96	2.0	116
3. Alfalfa	8	16	1.25	.35	--	25
Corn silage	8	8	.25	.25	--	35
Concentrate	3	42.7	1.33	1.40	2.0	116

[a]All values expressed on a dry matter basis.

CONCLUSIONS

Most dairymen with larger than average herds can cut
feed costs by purchasing feed ingredients in bulk, by
considering by-products and less commonly used feedstuffs,
and by processing and formulating diets on the farm. In so
doing, a dairyman should be conscious of the fact that he is
assuming an important role with many functions usually
performed by a feed dealer. There are also savings to be
made in transportation and feed handling on the farm. In
addition to its superior nutritive value, superior forage
allows wide substitution in the total diet. Dairymen who
pursue these strategems can reduce their feed costs
greatly.

335

REFERENCES

Benson, G.A. 1980. On-farm feed processing - will it pay? N. Carolina State Dairy Handbook B5.

Clark, J.H. 1975. Feeding top cows for even greater milk and milk protein production. Hoard's Dairyman 120:375.

Coppock, C.E. 1977. Feeding methods and grouping systems. J. Dairy Sci. 60:1327.

Coppock, C.E. 1983. More economical milk production with computer use in nutrition and feeding management. Dairy Science Handbook 16.

Coppock, C.E. and W.A. Gallamore. 1976. They get top production with limited grain feeding. Hoard's Dairyman 121:80.

Lamb, R.C., G.E. Stoddard, C.H. Micklesen, M.J. Anderson, and D.R. Waldo. 1974. Response to concentrates containing two percent of protein fed at four rates for complete lactations. J. Dairy Sci. 57:811.

McEllhiney, R.R. 1981. Feed pelleting - benefits and costs. Southwestern Poultry & Egg Assoc. Feed Natl. Mgmt. Sem.

Mott, G.O. and J.E. Moore. 1977. Existing and potential systems of finishing cattle on forages or limited grain rations in the tropical region of the south. S. Coop. Series Bull. 220.

National Research Council. 1981. Feeding value of ethanol production by-products. Nat. Acad. Press., Washington, D.C.

Nilson, K.M. and J.G. Welch. 1979. Feeding liquid whey to dairy animals. VT Agr. Expt. Sta. MP 103.

Shultz, T. 1981. Corral feeding and energy costs. Proc. 12th Ann. CA Dairy Cattle Day p 1.

Wagner, D.G. and J.K. Loosli, 1967. Studies of the energy requirements of high producing dairy cows. Cornell Univ., Agr. Exp. Sta. Mem. 400.

Wheeler, W.E. and C.H. Noller. 1976. Limestone buffers in complete mixed rations for dairy catttle. J. Dairy Sci. 59:1788.

33
SUPPLEMENTING DAIRY COWS
IN THE TROPICS

Manuel E. Ruiz

Milk production in any environment requires an ample
and constant supply of feed throughout the year. Any given
production system that has evolved in a particular location
is dependent upon the interaction of climate, the dynamics
of the soil-plant-animal complex and the socioeconomic
framework of the farmer (Swain, 1971).
Isolated reports from Australia (Cowan et al, 1974)
have indicated that it is possible to obtain high lactation yields with Friesian cows grazing tropical grass/legume
pastures. However, most observations are that milk yields
are low when cows graze solely on tropical pastures (Stobbs,
1976). This is a reflection of the low nutritional quality
of most tropical pastures. A relatively recent review compares research results from both temperate and tropical
regions (Stobbs, 1976, table 1).

TABLE 1. A SUMMARY OF RESEARCH FINDINGS ON TEMPERATE AND
TROPICAL PASTURES SHOWING DIFFERENCES IN DIGESTI-
BILITY AND MILK PRODUCTION

Diet	DM digestibility %	Maximum milk production[a] kg/cow/lactation
Tropical pasture		
1. Immature	60-65	1800-2200
2. Semimature	50-55	1000-1400
Temperate pasture	70-80	3300-3800
Concentrate ration	80-85	4400-4900

Source: T. H. Stobbs (1976).
[a]Jersey cows

ALTERNATIVES FOR INCREASING MILK PRODUCTION UNDER GRAZING CONDITIONS

If the data in table 1 are translated to daily production per cow, it becomes apparent that cows grazing unfertilized tropical grasses rarely reach levels higher than 6

to 7 kg milk/cow/day. Attempts to improve this situation have included the use of grass/legume associations and the application of N fertilizer to improved grasses. Comparison of the results of these alternatives with those obtained from tropical native grasses reveals a modest increase in milk production (table 2).

TABLE 2. MILK PRODUCING UNDER TROPICAL GRAZING CONDITIONS AND LOW STOCKING RATES, KG/COW/DAY

Sward	Breed		
	Jersey	Holstein	Crossbreds
Native	6.8	–	6.6
Grass/legume	10.9	13.3	8.0
Fertilized grasses	7.2	10.9	8.7

Source: T. H. Stobbs (1976).

A third alternative takes advantage of the extraordinary rate of growth of tropical grasses that results in production of biomass up to 6 times as much as is possible in temperate climates (Snaydon, 1981). The alternative consists of stocking up the sward to nearly its full capacity even if this means that the individual cow's milk production potential is not reached. The aim, therefore, is to maximize milk production per unit of area. To illustrate these points, table 3 shows milk production data as a function of the stocking rate.

TABLE 3. MILK PRODUCTION PER COW AND PER HECTARE BASED ON FERTILIZED TROPICAL GRASSES WITHOUT SUPPLEMENTS

Country	Sward	Stocking rate cows/ha	Milk production	
			kg/cow/day	kg/ha/year
Australia	Kikuyu	2.5	7.8	5,351
	(Pennisetum	3.3	7.1	6,227
	clandestinum)	4.7	6.9	9,000
Peru	Jaragua	1.9	10.7	4,266
	(Hyparrhenia rufa	2.1	10.2	5,002
	and Brachiaria	3.3	8.6	10,430
	(Brachiaria			
	decumbens)			

Source: T. H. Stobbs (1976).

In 1976, a system prototype for milk production in the humid tropics was developed in Turrialba, Costa Rica (Ruiz et al., 1980). The system was based on high stocking rates (5.5 cows/ha), daily pasture rotation, minimum supplementation, and genetic potential of the cow in balance with the environment and the quality of the pasture used (African Stargrass, cynodon nlemfuensis). Milk production varied between 15,000 and 16,000 kg/ha/year. The model has been used as the basis for technical assistance to farmers in Costa Rica and Honduras (Ruiz et al., 1980).

SUPPLEMENTATION: AN ALTERNATIVE FOR INCREASING MILK PRODUCTION

The fourth alternative for increasing milk production under grazing conditions is the use of supplements. This alternative has been one of dubious value in the past because of the variable responses to supplementation. For example, under temperate conditions, grazing cows have shown small responses to supplementation, averaging 0.30 kg of milk/kg of concentrate. In tropical countries, European breeds of dairy cattle, introduced to improve animal production, have usually received large amounts of supplementary concentrates but have shown only small responses (Herrera, 1973; McDowell et al., 1975; Phipps and Holmes, 1975; Salkeld, 1971). On the other hand, there have been reports of substantial increases (up to 24.5%) in milk production by feeding grain supplements (Hancock, 1958; Royal and Jeffery, 1972; Stobbs, 1971) to grazing cows.

The reasons for such contradictory results have not been studied in detail. However, various factors conditioning animal response to supplementation have been identified. First, as was noted in table 1, tropical grasses are usually of relatively low digestibility, which explains the limited levels of milk production (Menson and McLeod, 1970). Therefore, a large proportion of the ingested nutrients are used to meet the maintenance requirements and a very small proportion are available for milk production. This is in contrast to the situation with beef cattle because their requirements for weight gain are much lower than the requirements for milk production; as a result, it is not uncommon to see high rates of weight gain in beef cattle grazing tropical pastures. Schematically, these concepts are illustrated in figure 1.

Thus, energy is the major limiting nutrient for milk production in tropical and subtropical environment. Protein deficiencies will occur as the dry season settles and the supply of this nutrient must be ensured at this time.

Level of Supplementation

The amount of supplement given to milking cows is highly variable. In Latin America, in general, there is a tendency to use high levels of supplements for cows of European blood and small amounts or none for the native or crossbred cows. Very few trials have been conducted to define the proper level of supplementation in tropical conditions. In

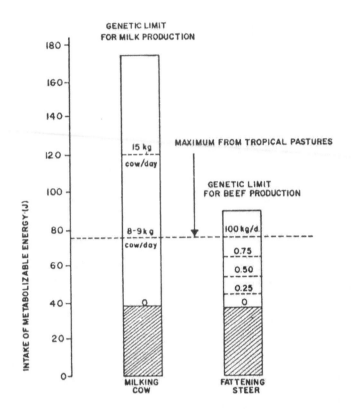

Figure 1. A simple representation of the levels of milk or
beef production obtainable from tropical grasslands at low stocking rates. The shaded areas
represent the proportion of the energy consumed
destined for maintenance; the unshaded areas indicate the energy available for production (based on
ARC [1965]).

table 4, a summary of various studies is presented indicating the stage of lactation at which the treatments were applied. An examination of the data in table 4 leads to the
conclusion that, under grazing conditions and low stocking
rate, supplementing dairy cows may cause an increase in milk
production. However, any response, will be modest,and the
level of supplement necessary to elicit such a response is
low. It seems that levels of supplement higher than 3
kg/cow/day (or 0.7 kg/100 kg LW/day) would not cause further
increases in milk output. As the responses are low, it
would be very difficult to justify the practice of supplementation of dairy cows from an economic point of view.
However, attention is called to the three reports that show

significant effects (table 4). These experiments were conducted at the initial stages of lactation while the others
used cows in the middle of their lactation. This observation seems to hold the key for the appropriate supplementation of dairy cows.

The Influence of the Lactation Stage on the Degree of Response to Supplementation

Perhaps the reason for the rather poor response to supplementation of grazing cows is the stage of lactation at
which comparisons were made. For statistical convenience,
most researchers have applied the supplementation treatments
to cows after they have passed the lactation peak, which is
usually attained 35 to 50 days after parturition. In fact,
switch-back designs require the use of cows that are already
in their downward phase of milk production. However, as the
cow nears the end of her lactation curve, she will be more
prone to utilize feed nutrients for body reserve replenish-
ment (i.e., weight gain) than for milk production.

In a recent paper (Combellas et al., 1979), the results
of twelve supplementation experiments with cows in midlactation were tabulated (reproduced here in table 5). As the
overall average shows, response to supplementation is only
0.34 kg milk per kg supplement. This closely agrees with
other results under both tropical conditions (Combellas et
al., 1979) and temperate conditions (Leaver et al., 1968).

On the other hand, as mentioned earlier, a few experiments have shown significantly improved milk production.
Work in Great Britain (Broster and Clough, 1974) and the
U.S. (McCullough, 1973) have led to the concept that the key
for significant increases in milk production consists of
ensuring proper nutrition during the first two months of
lactation when the cow has the greatest need for nutrients.
Broster and Clough (1974) state a rule of thumb: for every
kg of milk that the cow fails to produce at her lactation
peak, 150 kg of milk will be lost in the whole lactation.
Also, McCullough (1973) states that if a cow is well-fed in
the beginning, any decrease of milk production (due to poor
feed) at later stages can be quickly remedied by proper supplementation. This would not be true for a cow poorly fed
at the initial stage of lactation.

Raising the nutritional level at the beginning of the
lactation implies that the animal will consume whatever supplement she is given. However, food intake at this stage is
not enough to allow for adequate intake of nutrients (Bines,
1976); supplementation may need to start before parturition
to build up body reserves, especially if the cows are not in
good body condition (Greenhalgh and Gardner, 1958). These
considerations led to an experiment in Costa Rica (Cerdas,
1981) to determine whether or not the stage of lactation had
any influence on the response to variable levels of an
energy-rich supplement.

TABLE 4. MILK PRODUCTION BASED ON TROPICAL PASTURES AND SUPPLEMENTATION

Treatment	Sward	Production kg milk/cow	Notes and ref.
From day-30 to day-120 of lactation			
A. Concentrate, 1 kg/4 kg milk	Pangola,	8.5/day	A < B and
B. A + 0.5 kg molasses	Guinea and	9.2/day	A < C,
C. A + 1.0 kg molasses	molasses grass	9.1/day	(P ≤ 0.01) (9)
From day-84 to day-168 of lactation			
A. Grazing only	Pangola	9952/lact.	A < B < C
B. A + 1.9 kg conc./cow/day	grass	10648/lact.	(P ≤ 0.05)
C. A + 3.8 kg conc./cow/day		11557/lact.	(2)
Stage of lactation unknown			
A. Concentrate, 1 kg/cow/day	Native	7.0/day	N.S.
B. Concentrate, 3 kg/cow/day	grasses	6.0/day	Diff.
C. Concentrate, 0.5 kg/ liter above 4 liters		5.7/day	(8)
From day-90 to day-174 of lactation			
Five levels of cassave root meal (0-2 kg/cow/day)	African Stargrass	Maximum increase (14%) obtained at 0.7 kg cassava/ cow/day	N.S. Diff. (13)
From day-130 to day-230 of lactation			
Constant level of molasses (1.5 kg/cow) plus variable levels of green bananas (0-1.2 kg DM/100 kg LW/day)	African Stargrass	Maximum increase (12%) was obtained with 0.2 kg DM	N.S. Diff. (26)
For 90 days at various stages of lactation			
Constant level of molasses, urea (0.5 kg/cow) plus variable levels of green bananas (0-1.2 kg DM/100 kg LW/day)	African Stargrass	Maximum increase (20%) was obtained with 0.3 kg DM	Sign. Diff. (P ≤ 0.01) (5)

343

TABLE 5. RESULTS OF EXPERIMENTS ON TROPICAL PASTURES SHOWING THE
MILK-YIELD RESPONSE TO SUPPLEMENTATION IN MIDLACTATION

Country	Mean supplementation level (kg/day) High	Low	Period of lactation (weeks)	Milk yield of low level of supplementation (kg/day)	Response (kg) milk per kg supplement)
Trinidad	2.4	0	27	6.9	0.33
	2.5	0.5	27	6.7	0.20
Brazil	3.8	0	11-23	10.0	0.42
	1.9	0	11-23	10.0	0.37
Cuba	3.6	0	10-30	9.5	0.21
	2.7	0	10-30	9.5	0.28
	11.8	0	10-30	9.5	0.31
Australia	3.8	0	17-30	8.8	0.48
	2.7	0	17-30	8.8	0.41
	1.1	0	17-30	8.8	0.64
Brazil	4.1	0	10-22	10.8	0.35
	2.2	0	10-22	10.8	0.34
Cuba	6.0	0	8-23	7.9	0.07
	3.0	0	8-35	7.9	0.27
Venezuela	2.0	1.0	>6	8.7	0.29
	3.0	1.0	>6	8.7	0.34
Brazil	3.3	0	-	8.6	0.33
Cuba	3.7	0	9-24	14.1	-0.16
Uganda	2.7	0.7	9-26	8.6	0.25
Cuba	6.9	0	14-28	7.6	0.40
	3.7	0	14-28	7.6	0.40
	3.1	0	14-28	7.6	0.66
Australia	3.0	0	-	13.6	0.37
	3.0	0	-	11.4	0.50
Average					0.34 ± 0.17

Source: J. Combellas et al. (1979).

Sixty cows were grouped by the stage of lactation they were: at two months prior to calving, at calving, and at one month, three months, and five months after calving. Within each group, four levels of supplements were tested: 0, 0.3, 0.7, and 1.2 kg of bananas (dry matter basis)/100 kg LW/day. Supplementation was carried out for three months. The results are presented in tables 6 and 7. Table 6 shows that the lowest level of supplement was sufficient to cause the greatest response (across lactation stages).

TABLE 6. MILK PRODUCTION OF COWS GRAZING AFRICAN STARGRASS, SUPPLEMENTED WITH A CONSTANT LEVEL OF MOLASSES (1 KG/HEAD/DAY) AND VARIABLE LEVELS OF GREEN BANANA FRUITS

Level of bananas kg DM/100 kg LW/day	Average milk production kg/cow/day
0.0	7.7
0.3	9.2
0.7	7.9
1.2	8.0

Source: R. Cerdas (1981).

TABLE 7. MILK PRODUCTION OF COWS GRAZING AFRICAN STARGRASS AND RECEIVING 1 KG MOLASSES/HEAD/DAY AND 0.3 KG BANANAS (DM)[a]/100 KG LW/DAY

Point at which the 3-month supplementation is initiated	Milk production, kg/day	
	No supplement	With supplement
1 month prior to calving	7.9	10.6
At calving	8.9	10.3
1 month after calving	8.4	10.8
3 months after calving	6.4	8.0
5 months after calving	6.8	6.2
Averages	7.7	9.2

Source: R. Cerdas (1981).
[a]Equivalent to 1.5 kg green bananas (as-fed basis)/100 kg LW/day.

Looking at the response at the various stages of lactation, when 0.3 kg bananas (DM)/1090 kg LW/day is used, results were obtained as shown in table 7.

Clearly, if supplementation is to be effective, it must begin one month before parturition (especially if the cow is in need of improvement of her body condition), at calving, or at one month after calving. As the cow enters her fifth or sixth month of lactation, further supplementation would be inefficient. Obviously, since bananas (or any other tropical source of starch) are poor in protein, it is important to add a protein supplement. For Costa Rica, the recommendation is 40 g meat and bone meal and 2 g urea per 100 kg LW per day, mixed with a small amount of molasses to prevent rejections.

The above conclusions are supported by Broster and Clough (1974) who recommend that supplementation should start before parturition, increasing the level as the cow reaches her lactation peak and reducing it, afterwards, until it is eliminated from the feeding program. This plan depends, of course, upon the quality of the sward (figure 2).

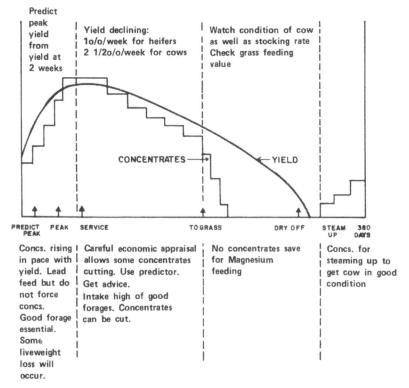

Source: W. H. Broster and P. A. Clough (1974).

Figure 2.   Main features of a feeding program for lactating cows under grazing conditions

REFERENCES

Agricultural Research Council. 1965. The nutrient require-
ments of farm livestock. No. 2 Ruminants: Technical
Reviews and Summaries. Agricultural Research Council,
London.

Aronovich, S., A. N. S. Correa, E. V. Faria, G. A. Dusi and
P. R. O. Nunes. 1966. Uso de concentrados na alimen-
tacao de vacas leiteiras en boas pastagens de capim
Pangola. I. Resultados de verao. Boletim do Instituto
de Pesquisas e Experimentacao Agropecuaria do Centro-
Sul (Brasil) 5:919.

Bines, J. A. 1976. Regulation of food intake in dairy cows
in relation to milk production. Livestock Pro. Sci.
3:115.

Broster, W. H. and P. A. Clough. 1974. Feeding dairy cows.
The Esso Farmer 26:1.

Cerdas, R. 1981. Banano de desecho (Musa acuminata) como
suplemento a vacas lecheras en pastoreo en diferentes
estados de lactancia. M. S. Thesis. Turrialba, Costa
Rica, UCR/CATIE.

Combellas, J., R. D. Baker, and J. Hodgson. 1979. Concentrate supplementation, and the herbage intake and milk
production of heifers grazing Cechrus ciliaris. Grass
and Forage Sci. 34:303.

Cowan, R. T., P. O'Grady, R. J. Moss and I. J. R. Byford.
1974. Milk and fat yields of Jersey and Friesian cows
grazing tropical grass-legume pastures. Tropical
Grasslands 8:177.

Esperance, M. and C. Esquivel. 1974. Influencia de diferentes niveles de concentrado a vacas en pasto natural
en la produccion de leche. Boletin P-3 de la Estacion
Experimental de Pastos y Forrajes. Serie Tecnico Cien-
tifica 1974:3.

Felix, A. 1968. Efectos de la melaza de la racion sobre la
produccion de las vacas lecheras en el tropico. M.S.
Thesis. IICA, Turrialba, Costa Rica.

Greenhalgh, J. F. D. and K. E. Garnder. 1958. Effect of
heavy concentrate feeding before calving upon lactation
and mammary gland edema. J. Dairy Sci. 41:822.

Hancock, J. 1958. The conversion of pasture to milk. The
effect of stocking rate and concentrate feeding. J.
Agr. Sci. 50:284.

Herrera, H. 1973. Costo de produccion de leche en la region Central. Seminario Produccion de Leche en Venezuela. Consejo Nacional de Investigaciones Agricolas. pp 435-451.

Lazarte, M. A. 1978. Efecto de la suplementacion con yuca (Manihot esculenta Crantz) como fuente de almidon sobre la produccion de leche en vacas en pastoreo. M.S. Thesis. UCR/CATIE, Turrialba, Costa Rica.

Leaver, J. D., R. C. Campling and W. Holmes. 1968. Use of supplementary feeds for grazing dairy cows. Dairy Sci. (Abst.) 30:355.

McCullough, M. E. 1973. Optimum feeding of dairy animals for meat and milk. University of Georgia, Athens.

McDowell, R. C., H. Cestero, J. D. Rivera-Anaya, F. Roman-Garcia, J. A. Arroyo-Aguilu, C. M. Berrocal, M. Soldevila, J. C. Lopez-Alberty and S. W. Metz. 1975. Tropical grass pastures with and without supplement for lactating cows in Puerto Rico. Bulletin No. 238, Agricultural Experiment Station, University of Puerto Rico.

Minson, D. J. and M. N. McLeod. 1970. The digestibility of temperate and tropical grasses. In Proceedings 11th. International Grassland Congress, Queensland, Australia.

Phipps, R. J. and W. Holmes. 1975. Supplementary feeding of grazing dairy cows in Uganda. Tropical Agriculture 52:59.

Royal, A. J. E. and H. Jeffery. 1972. Energy and protein supplements for dairy cows grazing tropical pasture. Proceedings of the Australian Society of Animal Production 9:292.

Ruiz, M. E., G. Cubillos, O. Deaton and H. Munoz. 1980. A system of milk production for small farmers. In Animal Production Systems for the Tropics. International Foundation for Science, Provisional Report No. 8, IFS, Stockholm, Sweden. p 246.

Salkeld, J. N. 1971. Supplementary feeding for dairy production in the tropical region of Australia. 2. Review of farming practice. Tropical Grasslands 5:221.

Snaydon, R. W. 1981. The ecology of grazed pastures. In F. H. W. Morley (Ed.) Grazing Animals. pp 13-32. Elsevier, Amsterdam, Holland.

Stobbs, T. H. 1971. Production and composition of milk from cows grazing siratro (Phaseolus atropurpureus) and greenleaf desmodium (Desmodium intortum). Australian J. of Experimental Agr. and Animal Husbandry 11:268.

Stobbs, T. H. 1976. Milk production per cow and per hectare from tropical pastures. In Produccion de Forrajes. Seminario Internacional de Ganderia Tropical, Acapulco, Mexico. Memoria. p 129.

Swain, F. G. 1971. Dairy production systems relevant to the tropical regions of Australia. 1. Review of the literature. Tropical Grasslands 5:269.

Villegas, L. A. 1979. Suplementacion con banano verde a vacas lecheras en pastoreo. M.S. Thesis. UCR/CATIE, Turrialba, Costa Rica.

Part 7

HEALTH, DISEASE, AND PARASITES

34
INFECTIOUS LIVESTOCK DISEASES: THEIR WORLDWIDE TOLL

Harry C. Mussman

By good fortune, plus nearly a century's worth of cooperation between stockmen and government, the United States has managed to wipe out or keep out the most serious of the world's livestock diseases.

Contagious bovine pleuropneumonia, introduced in the mid-1800s, was the first major disease we battled together and eradicated. Hog cholera was the last to be wiped out-- just four years ago. In between these two efforts were nine outbreaks of foot-and-mouth disease, which has not been back since the last remains were buried in 1929.

These diseases and others still bring major losses and reduce livestock productivity in much of the world. This is particularly true in the developing nations of Africa, Asia, and Latin America. The popularly held notion that Third World livestock are infected with all of the most dreaded diseases is not true, but many of these countries do live with at least a few of them, often superimposed one on the other. Losses due to animal disease in some developing countries are estimated at 30% to 40% annually--twice as great as losses recorded by most industrialized nations.

Looking at meat and dairy productivity, developing nations produce one-fifth the beef and veal per animal that developed countries do, one-half the pork, one-half the eggs, and one-eighth the milk. North America and Europe are roughly four times more efficient in mutton, lamb, and goat meat production than is South America, and 25% more efficient than Africa.

The so-called Third World is hungry already, and is projected to host 90% of the global growth in human population expected by the year 2000. Obviously, greater and more reliable sources of protein must be found and developed, or many of these people will not survive. A diseased animal that may not survive can hardly be counted a productive resource. A virus or other disease agent can wipe out entire herds in a very short time--or, in a chronic state, can leave them debilitated, more expensive to feed, and capable of producing far less milk or meat.

Of all the factors involved with production, animal diseases have by far the most dramatic impact. Unfortunately, disease control is not improving rapidly enough in the Third World. And many of the most feared diseases--because of the tremendous economic toll they can take--are appearing in locations where they never existed before or reappearing in places where they had been eradicated years ago.

Some of these diseases are listed next.

Foot-and-Mouth Disease. Foot-and-mouth disease (FMD) is the most feared of all the animal diseases worldwide because of its ease and speed of spread and the susceptibility of all clovenhooved animals. It has long been established in livestock populations of the Mideast causing abortions, deaths, weight loss, mastitis, and lowered milk production. Europe has managed to keep these Asian strains at bay in Turkey, however.

Prior to World War II, FMD swept through Europe every 6 to 8 years, severely reducing livestock productivity. Finally in the 1950s, better quality vaccine was introduced and stemmed the spread of the epizootics. Outbreaks that occurred in the early 1960s, and again in 1973, were halted by teams from the United Nations working with the affected countries and regions. Then, in March 1981, FMD was confirmed in France, in Brittany, and near Cherbourg. Within a few days it broke out on the Isle of Jersey and the Isle of Wight in Great Britain. Just prior to that positive diagnosis came in from the south of France, from Portugal, and from Spain--outbreaks that most likely originated in Spain. Austria, free of FMD for six years, was infected during the winter of 1981 from imported Asian buffalo meat.

After 10 years of freedom from the disease and despite a vigilant quarantine program, Denmark discovered an outbreak of FMD on the Isle of Fyn just last March. The origin is unknown, but suspicions are that it came from one of the Eastern bloc countries where severe FMD outbreaks were reportedly occurring just prior to Denmark's infection. The loss Denmark has suffered from those outbreaks is estimated at almost a quarter of a billion dollars. Most of the loss has come from the cessation of fresh pork sales to the U.S., Canada, and Japan, although sixteen other countries were also compelled to embargo Danish meats and animals. The cost of eradication alone was also significant.

U.S. policy prohibits imports of fresh, frozen, or chilled meat products, or swine, or ruminants for a full year from any country that experiences an FMD outbreak. We had to refuse United Kingdom exports for a year following their 1981 outbreaks. We also refused farm products from across the Canadian border in 1952 following an outbreak there--which also curtailed big game hunting for U.S. hunters up north. (Certified "clean" animals, however, may be brought in from FMD-infected countries through the Harry S. Truman Animal Import Center on Fleming Key, Florida, after

three months of high-security quarantine and testing at the
importer's expense; or they may be shipped directly to an
approved zoo.)
Some call this "politics," aimed at protecting the home
meat industry. U.S. meat producers do gain a competitive
edge from such an embargo, certainly, but that is not the
aim. The productivity of more than 200 million fully susceptible cattle, swine, sheep, and goats is at stake. The
collective worth of that stock is estimated at $24 billion
--if they stay healthy. Healthy animals with their high-
quality protein will keep America healthy, help alleviate
worldwide hunger, and help keep the U.S. trade balance
healthy.
If FMD were to invade the U.S. and enter a major market, it could spread to over a dozen states within 24
hours. By the end of the first year, direct losses would be
an estimated $3.6 billion--indirect losses to allied indus-
tries and curtailed exports could climb to $10 billion.

Rinderpest and Peste de Petits Ruminants. Another
serious, almost 100% fatal among cattle and resurging as a
serious problem in Africa, is rinderpest. As with FMD, the
United States (by law) cannot import cattle or other ruminants from rinderpest-infected countries (unless, as with
FMD, they are brought in through the Truman Center on Fleming Key or are destined for an approved zoo.)
Rinderpest was essentially eliminated from Africa in
1975, following a decade of intensive and widespread vaccination. A team of international cooperators trudged from
West to East Africa, vaccinating some 80 million head of
cattle at a cost of $30 million. Unfortunately, though, the
people became complacent once the disease had been sup-
pressed. As the livestock populations were built up again,
the young were not all vaccinated and were allowed contact
with remaining carriers and wild animals.
So, the incidence is climbing again, especially in West
Africa, and the virus is spreading east. Egypt is affected
now, and the Arab Gulf States are reporting such frequent
outbreaks that it appears the disease has become enzootic.
Peste de petits ruminants is a virus quite similar to
rinderpest and readily infects goats and sheep in West
Africa. It is actually classified as a strain of rinderpest
that has lost its ability to infect cattle under natural
conditions.

Rift Valley Fever. Another disease that has moved up
from Kenya through the Sudan into the Sinai is Rift Valley
fever. RVF primarily affects sheep but may also affect
goats. It is also a serious human health problem. An epi-
demic in Egypt a few years ago killed nearly 600 people,
along with a large number of livestock. The disease can
spread rather easily. A human body incubating the virus
could carry it anywhere in the world or a mosquito could
carry it while hitchhiking on an airplane.

Israel has been vaccinating its animals against RVF for self-protection as well as to help curb further spread of the disease around the Eastern Mediterranean Basin. Because of grave concern over Rift Valley fever, the U.S. Department of Agriculture has developed diagnostic capability of its own and is making preliminary vaccine studies. Work on possible domestic insect vectors is also planned.

African Swine Fever. Over the past 20 years, swine-producing nations have faced a rising risk of African swine fever infection. ASF has been established for many years in wart hog, bush hog, and giant forest pig populations in tropical Africa, south of the equator. Other animals may not have the same tolerance to the virus: in 1909 Europeans tried bringing domestic hogs into Africa, and the hogs were dead from the virus in a matter of a few days.

The first appearance of ASF in Europe was in Portugal in 1957 where the disease was at first mistaken for classic hog cholera and finally recognized as ASf. More than 16,000 pigs became infected before the virus was recognized as ASF and finally stamped out. In 1959, the disease was found in Spain, and in 1960 reappeared in Portugal. France was invaded by ASF three times in the 1960s and 1970s, though the French managed to eradicate it each time at an early stage--a tribute to their veterinary surveillance system.

Italy became infected in 1967 through the Rome airport. More than 100,000 swine were slaughtered before the disease was wiped out. Finally, about four years ago, Malta and Sardinia became infected. In Malta, the entire swine population was slaughtered to achieve eradication--at a cost of some $50 million (U.S.). Italy is still working on eradication of the disease on Sardinia.

The Western Hemisphere managed to avert an invasion of the dread ASF until 1971 when the virus hit Cuba. Approximately half a million pigs died or were destroyed before the disease was eradicated. Then, another Cuban outbreak occurred in 1980 and was successfully eradicated in about six weeks. ASF appeared in Brazil 5 years ago and is still there today. Positive diagnosis came from the Dominican Republic in 1978 and in Haiti early in 1979.

The U.S., Canada, and Mexico are particularly and acutely concerned about African swine fever in the Western Hemisphere. All three countries are free of the disease but are working together in the Dominican Republic toward eradication. The program used a hard strategy: they completely depopulated the country's swine farms after which they restocked with healthy, imported pigs. Not a trace of the virus has reappeared since the restocking was completed. The Haitian effort repeated the model program (and first of its kind in the West) completed in the adjacent Dominican Republic last year. Equal success is anticipated in Haiti.

Success is crucial to the three big cooperators. Mexico's Yucatan Peninsula is just a short hop across the

Caribbean from the Dominican Republic; Canada has trade
interests all over Latin America and the Caribbean; and the
threat of ASF lies right at the doorstep to the United
States. The U.S. strictly prohibits swine imports from ASF-
infected countries, but the virus could enter through contaminated food waste from ships or planes, through farm soil
still clinging to a returning traveler's shoe, or as a
result of illegal immigration. The investment in such an
eradication program is sizable. Direct costs for the Dominican Republic--shared by the UN's Food and Agriculture
Organization, the Inter-American Development Bank, and the
United States--amounted to approximately $20 million. The
bill for Haiti should run about the same.

The cost of chronically battling a disease like ASF can
be far higher, however. Spain, for example, has spent
roughly $315 million fighting it over the past 20 years. In
addition, exports from Spain (along with Portugal and
Brazil) have been restricted because of ASF-infected status.

Heartwater Disease. Latin America was jolted last year
by the diagnosis of heartwater disease in a goat on Guada-
loupe, an island in the eastern Caribbean. This was its
first appearance in the Western Hemisphere and the implications are serious. We are cooperating with other countries
in the region to make certain that it does not spread.

Heartwater is native to southern Africa and Madagascar
and may be a major cause of animal deaths reported in West
Africa. We have no simple and reliable means of diagnosing
the disease; a blood test, especially, is badly needed.

Contagious Bovine Pleuropneumonia. Contagious bovine
pleuropneumonia (CBPP) recently broke out in France for the
first time in years. The disease may have been smouldering
in the Basque region in Spain for some time. CBPP was the
first major exotic disease to invade U.S. shores back in the
1800s. Cattle ranching had become so productive after the
Civil War that a surplus of more than a quarter million head
was available for export to England. When pleuropneumonia
showed up in the animals, however, Britain closed off the
market. That was the spark that set off the first animal
disease eradication program in the United States.

Goat Diseases. In the United States evidence is accum-
ulating of two possibly serious diseases of goats: contagious caprine pleuropneumonia and caprine arthritis encephalitis. CCPP is well established in much of the developing
world--the regions that depend heavily on goats. The true
impact of the disease is clouded as yet, largely because of
common and widespread disagreement among diagnosticians.
More research is needed to develop definitive diagnostic
tests and effective vaccines.

Caprine arthritis encephalitis was only recently recognized in the United States and may exist yet undiscovered in
other countries also. Research may give us some answers to
this relatively new entity.

Trypanosomiasis. A full third of Africa--taking in most of the wide band of savannah and forest across the middle of the continent--is plagued by infestations of the tsetse fly. A blood parasite of animals as well as people, the tsetse fly transmits trypanosomiasis, or animal sleeping sickness, from animal to animal. If the tsetse-infested areas could be reclaimed for cattle production, it is estimated that annual meat production in Africa could double.

Bacterial Disease, Intestinal Parasites, and Others. The more dramatic virus diseases, such as FMD, ASF, rinderpest, bovine pleuropneumonia, and external parasites that transmit East Coast fever and trypanosomiasis, are the most highly visible of the livestock plagues worldwide.

Much more research remains to be done on bacterial diseases and intestinal parasites, particularly with small ruminants. Also more research is needed to understand the incidence and significance of the clostridial diseases--tetanus, anthrax, blackleg, and malignant edema--and paratuberculosis, and brucellosis. In the U.S., eleven states are free of cattle brucellosis now and another 25 have an extremely low incidence. Eradication of the disease requires only time. The damage done by different strains of the disease worldwide, however, is not entirely known. Since all strains can infect people as well as animals, the public health problem is always important. Anaplasmosis and babesiosis also are economically significant blood-parasite diseases in many tropical areas of the world, but the actual extent of the damage they do is not fully appreciated, much less known.

Another virus disease--one less dramatic in its toll, but gaining more and more visibility--is bluetongue. Bluetongue affects cattle, sheep, goats, and wild ruminants. It is particularly damaging to sheep, with a mortality rate running as high as 50% among affected animals; in cattle and goats, bluetongue lowers reproductive ability. Currently, U.S. cattle cannot be exported directly to Europe because of bluetongue in the South and southwestern states. Canada has a competitive edge over the United States in the European livestock trade--with the exception of the Holstein dairy breed--because the small gnat that carries the disease does not appear that far north.

Like caprine arthritis encephalitis in the United States, the possibility of diseases yet undiscovered still remains--diseases that may have been smouldering since the first goat herds on record were tended some 10,000 years B.C.

THE NEED FOR STRONGER CONTROLS

Proper vaccines, a solid border inspection and quarantine program, and accurate diagnosis could take care of many of the most severe disease problems in the world.

Vaccine

The alarming resurgence of rinderpest in Africa shows what can happen when a successful vaccination program breaks down. On the other hand, a laboratory opened in Mali just 5 years ago has been researching, tracing, and producing vaccine for several diseases rampant in West African livestock --rinderpest and bovine pleuropneumonia. The Central Veterinary Laboratory (CVL), as it is called, is also coordinating regional vaccination programs, and CVL veterinarians and technicians are being welcomed now by Malian herdsmen as they drive their cattle back and forth for the grazing season. The incidence of disease appears to be declining, at least locally.

Quarantine

Contributing to the rinderpest problem in Africa was the failure to keep sick animals from the healthy. Quite simply, quarantine can prevent the spread of many infectious diseases. For example, double fencing, as practiced in some parts of Africa, effectively protects domestic swine from African swine fever.

Border Inspection

Strict import controls over meat, feed, and animal products and a tight border guard are essential to keeping disease out. Years ago when most travel was by boat, a virus disease could probably not survive a transocean trip. Today a few hours on a jet plane are little threat to virus survival.

Customs and Agriculture inspectors at all 82 U.S. ports-of-entry inspect as much luggage and as many carry-on bags and travelers as possible. The goal of 100% inspection is a difficult one to achieve, however. International travel rose 8% in the United States last year, and 6% worldwide. Also more and more cargo is containerized every year.

Diagnosis

Research must proceed and develop simpler and more effective means of diagnosing heartwater disease, contagious caprine pleuropneumonia, and other diseases. Clinical signs of certain diseases are changing--and we must keep up with them. Genetic research must continue so that animal science can continue to breed for resistance to disease. Biological, chemical, ecological, and other means of controlling insect vectors are unfolding through field and laboratory trials.

CONCLUSION

Each country or region of the world will have its own formula for prioritizing the danger that different animal diseases present. Generally speaking, the most dangerous diseases are those that are spread not only by animal-to-animal contact (contagious diseases) but whose virus, mycoplasma, bacteria, or rickettsia may be transmitted by live vectors, inanimate fomites, or meat scraps. The most dangerous diseases do significant economic damage to producers and exporters and carry a human health hazard, such as brucellosis or Rift Valley fever.

Most important before the animal disease situation in the Third World can improve, veterinary services will have to be strengthened. Developing countries have nearly 50% of the world's total livestock population but less than 20% of the world's veterinary forces. A number of developing nations, in fact, have little or no organized veterinary structure at all. Certain nations in Africa, in fact, have only one or two veterinarians working the entire country; and in others they are often young, with little or no field experience. The expertise of technicians, if it can be found, is usually limited--gained through in-service training. Veterinary technicians have no reliable means of communication or transportation--yet national animal health policy decisions are based on their input. Thus, the global data on disease prevalence, productivity losses, and effectiveness of control measures on which many veterinary directorates rely is woefully deficient.

It becomes clear why diseases such as contagious bovine pleuropneumonia, foot-and-mouth, and rinderpest still exist and spread. With existing reliable diagnostic tests for both, as well as effective vaccines, eradication of CBPP and rinderpest would require little more than organization and finance, backed up by the government's strong commitment. With heavier investment in livestock health programs--and greater cooperation among nations--the great toll that diseases take can be curbed. The people can be fed.

35
IMPACT OF ANIMAL DISEASES IN WORLD TRADE

Harry C. Mussman

IMPACT OF ANIMAL DISEASES ON WORLD TRADE

Animal diseases are an important factor inhibiting world trade and hampering the free movement of both live animals and animal-derived products. The foot-and-mouth disease outbreak in Denmark in early 1982 provides a recent example of the impact of animal disease on trade. At one point, Denmark's export trade in meat and dairy exports was suffering badly--$7 million was lost each week because of the outbreak. The U.S.--as well as several other countries --placed Denmark on the list of countries from which animals and animal products cannot be imported until free status is regained. Overall costs of the outbreak will run in excess of one billion dollars.

Another example closer to home is our bluetongue situation. In 1980, the European Economic Community banned animal imports from the U.S. because this cattle and sheep disease was found in a portion of our country.

U.S. COMPETITIVE EXPORT POSITION

Despite the bluetongue situation and other domestic diseases such as brucellosis, tuberculosis, and leukosis, the U.S. export position remains competitive. In 1981, our animal and animal-product exports had a market value of $3.24 billion; and, in the same year, we had a $307 million trade surplus in these commodities, the first since 1977.

A U.S. animal export health certificate enjoys high credibility. One reason is that we have eradicated 12 major animal diseases that still plague many other countries of the world--diseases such as foot-and-mouth, rinderpest, hog cholera, and contagious bovine pleuropneumonia. Only a half-dozen or so other countries can claim freedom from these diseases. Largely because of our strict animal import health requirements and procedures, we are successful in keeping animal diseases out of our country.

IMPORT PROCEDURES

Livestock destined for this country must first be examined, tested, and certified by government veterinary officials as being healthy and meeting U.S. requirements in the
country of origin. The foreign exporter must obtain health
certification papers from his government and, in a majority
of cases, an import permit in advance from us. At the U.S.
port of entry the livestock must be examined again, this
time by our veterinarians. This examination sometimes
includes further testing and port-of-entry isolation,
depending upon the kind of animal and the country it came
from.
Although we are strict in our import controls, we try
to meet the needs of American importers, particularly U.S.
livestock breeders needing new bloodlines and exotic breeds
of cattle. We recently opened a specialized import-
quarantine facility--the Harry S. Truman Animal Import Center, at Fleming Key, Florida. Imported cattle are carefully
tested and held there in quarantine for 3 months. They are
mingled with a select "sentinel" group of susceptible U.S.
cattle and swine to make sure they will present no disease
threat to U.S. livestock.
Because some swine diseases, such as African swine
fever, could devastate our swine herds, we do not accept
swine imports from most of the world. Imports of sheep and
goats are also severely limited because of the threat of
scrapie, which has an incubation period of up to four years
or more.
Restrictions on horse imports are generally less strin-
gent than those required for swine, cattle, sheep, and
goat. Still, incoming horses are tested for such diseases
as dourine, glanders, equine infectious anemia (EIA), equine
piroplasmosis, and contagious equine metritis (CEM).
Because of Venezuelan equine encephalomyelitis, horses
from all Western Hemisphere countries--except Canada and
Mexico--are quarantined for at least a week. Horses from
countries known to have African horse sickness are quarantined for two months.
Contagious equine metritis is a recently identified
disease of breeding horses. It has been found in parts of
Europe, Japan, and Australia. As a result, horses cannot
be freely imported from these countries.
Stallions and mares can be imported only after extensive treatment and negative culturing of the genitalia, both
in the country of origin and again in the U.S. while under
quarantine.

PROCEDURES FOR IMPORTS FROM CANADA AND MEXICO

The entry procedures for animals from Mexico and Canada
are generally less strict than those for animals from overseas nations because Mexico and Canada have animal disease

situations relatively similar to ours. The exception is hog cholera in Mexico, and for that reason we do not import swine from that country. In the case of other animals, however, entry quarantine and advance import permits are not required for either Mexico or Canada.

The U.S. has 16 crossing points on the Mexican border and 43 on the Canadian border where APHIS veterinarians examine and process animal imports. Entry procedures vary. For example, cattle from Mexico must be dipped in a pesticide solution as a precaution against cattle fever ticks and scabies.

ANIMAL SEMEN, EMBRYOS

Since animal semen is as much a potential disease threat as live animals, it must be subjected to strict standards for collection, handling, and shipping. These standards are spelled out in agreements between USDA and the foreign countries involved. While the technology for testing animal semen is well researched and established, the same cannot be said for embryos, the newest practical method for exporting animals. More research is needed to determine the diseases to which embryos are immune.

Our restrictions on animal imports may seem extreme to some. But we have a major responsibility for maintaining the health of our livestock--for example, a $9 billion swine industry, a $35 billion cattle industry, and a $10 billion poultry industry. Considering what is at stake, our restrictions are reasonable.

EXPORT PROCEDURES

Just as we insist that only healthy livestock be imported into this country, we have an obligation to see that only healthy animals are exported to other countries.

We ensure the health status of our exported animals in two ways. First, we establish our own health rules for exports; second, we cooperate fully in meeting the import rules of receiving nations.

All animals we export are subjected to special testing and certification requirements to indicate freedom from certain diseases found in the United States: bluetongue for cattle, sheep, and goats; brucellosis for cattle and goats; anaplasmosis for cattle; equine infectious anemia for horses; and pseudorabies for swine.

Our own export rules are necessary because we do have some health problems, i.e., EEC bluetongue of concern to foreign importers. We want to avoid damaging our position in the world market by making sure the animals and animal products we export are free of disease.

The animal health requirements of foreign countries vary, reflecting the particular animal disease problems and

362

danger they face in their part of the world. Testing and
certification are performed by private veterinarians accredited by us (usually at the shipper's expense) who normally
conduct their tests on the farm or cattle ranch. Adult
dairy and breeding cattle and goats must be tested and found
free of brucellosis and tuberculosis within specified time
limits before shipment. An animal health certificate is
endorsed by the APHIS area veterinarian after the private
accredited veterinarian completes his testing. The endorse-
ment certifies that the private veterinarian is qualified to
conduct the examination and tests.

Once the certificate is endorsed, livestock can move to
a port of embarkation. There they must rest for 5 hours at
a USDA-approved facility while the animals and paperwork are
given a final check by an APHIS veterinarian. If the animals are healthy, if they are properly identified, and if
the health certificate is in order, they are loaded on a
ship or aircraft under APHIS supervision for export.

COMPLEXITY OF FOREIGN REQUIREMENTS

The health requirements imposed by foreign governments
can be quite complex, so whenever possible we work out
agreements with these countries. We do all we can to nego-
tiate standard requirements, but we have over 150 agreements
with some 70 foreign governments, and they can be changed on
short notice. It is nearly impossible for an exporter or an
examining veterinarian to keep track of the different rules
for every overseas livestock shipment. To avoid delay and
frustration, the exporter and the accredited veterinarian
are urged to contact the APHIS Veterinary Services office in
their state before attempting the process livestock or animal products.

Each APHIS Veterinary Services office keeps a current
file of foreign animal import health requirements. Each
area office of APHIS Veterinary Services also has a veterinarian assigned to work with exporters and their veterinarians. His job is to check the export health tests and
certifications and place the final endorsements on the
health papers before the livestock can leave this country.

CANADIAN, MEXICAN EXPORTS

As with our imports, our exports to Canada and Mexico
are handled more simply and quickly than those to overseas
nations. Exported livestock do not need an APHIS veterinary
examination at the port of export. Once the health tests,
certification, and APHIS endorsements are completed, the
animals move directly to the border where they are examined
by the Canadian or Mexican officials. Canada and Mexico are
by far our biggest customers for exported livestock (and
poultry as well), so it is important to devote some special
attention to health matters on shipments across our borders.

Exports to Mexico move with few special problems. However, Canada has some requirements that exceed our own export rules. The Canadians are particularly concerned about chemical residues in cattle shipped to slaughter, so feed additives and antibiotics should be withdrawn from livestock within the recommended time limits. This is the responsibility of the exporter.

CLOSE COORDINATION REQUIRED

Successful U.S. exports, particularly those to overseas nations, require close coordination between the exporter, the private veterinarian, APHIS officials, the forwarder, the broker, the insurance underwriter, and the carrier. Among the most common causes of costly delay is the failure to conduct all required tests and failure to allow enough time for completion of the tests at the diagnostic laboratory.

Even if a plane or ship is waiting, animals cannot move to the port of embarkation unless APHIS endorses the health papers. Therefore, exporters should be aware of all the requirements and plan to allow sufficient time for testing when making plans to ship animals to another country.

INTERNATIONAL INVOLVEMENT

The U.S., along with its major partners, is actively involved in international organizations such as the Food and Agriculture Organization (FAO), the Office of International Epizootics (OIE), and the General Agreement on Tariff and Trade (GATT) in dealing with animal diseases worldwide and taking steps to assure the expeditious movement of healthy livestock and livestock products. We are actively involved in international health programs because the more we can do to reduce diseases worldwide, the more freely our own animals will move in international markets--an advantage of the U.S. exporter.

CONCLUSION

APHIS does all it can to help the stockman with his exports. We safeguard his markets by making sure no diseased animal gets out; and we make sure no diseased animal gets in to infect his livestock. We are against the unduly restrictive animal health import requirement that functions as a nontariff barrier and, more often than not, serves to protect a country's livestock industry--more from foreign competition than from foreign animal diseases.

The U.S. strongly supports "free trade" and endeavors, whenever possible, to make it a dominant principle in world

36
BIOLOGY AND CONTROL
OF INSECT PESTS OF DAIRY CATTLE

R. O. Drummond

Dairymen have to continually battle a variety of flies, grubs, lice, ticks, and mites that bite, sting, annoy, injure, lower milk production of, transmit diseases to, and suck blood from dairy cattle. Annual losses incurred by United States dairymen reach the hundreds of millions of dollars despite the use of insecticides to control these pests on dairy cattle and in livestock buildings (including milk rooms). Awareness of the biology, life history, and local abundance of external parasites and of the proper use of chemicals and other methods of control is the most effective way dairymen can prevent losses, reduce costs of treatment, and control ectoparasites without hazard or danger to their cattle, the insecticide applicator, the consumer, or the environment.

Because many dairies are located in or near populated areas, it is important that dairymen adequately control flies and other highly mobile insects that can annoy and irritate people who live near dairies. Dairymen have 2 types of control problems: 1) those associated with the milking herd -- lactating cows in limited areas on highly nutritious rations; 2) those associated with the heifer herd and dry cows -- nonlactating animals on larger pastures that are handled more like a beef herd. Each situation calls for a special type of ectoparasite control.

Dairymen must exercise extreme caution when selecting and applying insecticides to dairy cattle. Certain insecticides should be applied to nonlactating animals only, because their application to lactating cattle or soon-to-freshen cattle would create illegal residues of insecticides in milk and milk products. Other insecticides, when applied as recommended, can be used directly on lactating dairy cattle without creation of residues.

This article contains information about the biology of a number of pests of dairy cattle, describes accepted techniques for their control, and lists precautions for the safe use of insecticides on dairy cattle.

EXTERNAL PARASITES

Flies

The most common insect pests of dairy cattle are blood-sucking flies (commonly called "biting flies") such as stable flies, horn flies, horse flies, deer flies, and mosquitoes; nonbiting flies such as house flies, face flies, blow flies, and screwworm flies; and nonfeeding flies such as heel flies or gadflies.

Biting flies. The stable fly, Stomoxys calcitrans (L.), the most important bloodsucking fly that attacks dairy cattle, is found around dairies, dry lots, and other areas where dairy cattle are confined. The adults, about the size of a house fly, usually suck blood from cattle once or twice a day and spend the rest of the time resting on fences, and buildings. Female stable flies lay eggs on and maggots develop in moist straw, hay, spilled feed, and other decaying organic matter that is mixed with manure.

The first and most important step in control of the stable fly is to remove all the breeding material, usually manure-contaminated wastes, from the area. This material should be spread and dried so it will no longer be a breeding site. Breeding material can also be treated with insecticides. Surfaces of barns, stalls, fences, and other structures can be treated with insecticides that have a long period of effectiveness to kill flies that contact treated surfaces. Finally, dairy cattle may be treated with insecticides to repel or control stable flies. However, these treatments tend to be relatively ineffective or give protection for only a short time. Insecticide-impregnated ear tags may be used as an aid to control stable flies. An insecticide given daily to dairy cattle as a feed additive or supplement controls stable fly larvae in manure of treated cattle. Insecticidal baits of the type used to attract and kill house flies around pens, lots, etc., are not effective against stable flies.

The horn fly, Haematobia irritans (L.), is found on pastured dairy cattle throughout the U. S. Thousands of these small dark flies may be present on the heads, shoulders, and backs of cattle. Adult flies may suck blood from cattle as often as 20 times/day. The constant irritation produced by these flies causes infested cattle to shake and toss their heads and to switch their tails--as a result, they do not graze normally. Female horn flies lay eggs on freshly dropped manure, and the larvae develop in undisturbed manure pats. The life cycle takes about 2 weeks. The species has the potential to build up very large populations on cattle in a short time. Horn flies, however, are not a problem in confined areas such as dairy lots because the larvae cannot complete development in manure pats that are disturbed, broken, or mixed with urine.

In general, horn flies are controlled by treating dairy cattle with dips, sprays, dusts, or "pour-ons" of insecticides. Often treatments have to be repeated every 2 or 3 weeks during the horn fly season, which lasts from last frost in the spring to frost in the fall. Daily mist sprays can also control horn flies. Insecticide-impregnated ear tags provide control of horn flies for several months. Some dairymen place backrubbers or dust bags in their pastures or at openings to water and mineral feeders so that the cattle treat themselves with insecticides in oils or in dusts. Also available are feed additives and supplements that contain insecticides. Cattle consume enough treatment so the manure is toxic to horn fly larvae.

Other biting (bloodsucking) flies, mosquitoes, horse flies, deer flies, gnats, and black flies are difficult to control. For example, mosquitoes are best controlled by eliminating or treating the water in which the larvae are found. Repellents applied to dairy cattle usually are short-lived and do not provide satisfactory protection against feeding of these other biting flies although daily application to lactating dairy cattle may afford some protection. Insecticides that control horn flies and stable flies on dairy cattle may kill other biting flies found on dairy cattle.

Nonbiting flies. Although the house fly, Musca domestica (L.), does not suck blood from cattle, it can be a considerable nuisance to dairymen and their neighbors. House flies breed in decaying organic matter, and large numbers are often found around dairy facilities. Also, flies may be a nuisance to people living near dairies. As with stable flies, the most effective method of control of house flies is removal of breeding material -- manure, contaminated feed and hay, and rotting organic matter. Also, breeding material may be treated with insecticides. The surfaces of livestock-holding structures may be sprayed with insecticides, or insecticide-treated baits may be used to attract and kill the flies. Dairy cattle can be treated with sprays or mists of insecticides to control house flies. Also a feed additive/mineral supplement that contains an insecticide, if consumed daily, will control house fly larvae in manure of treated cattle.

The face fly, Musca autumnalis De Geer, a recent introduction into the U. S., is found in all states except Texas and New Mexico. This fly, about the size of the house fly, is an important pest of pastured dairy cattle because it feeds on liquids around the eyes and nostrils of cattle. This feeding activity irritates cattle and interferes with normal grazing, thus lowering milk production; it also increases the incidence of eye problems such as pinkeye. Female face flies lay eggs on freshly dropped manure and larvae develop. Treatment of the head and neck of cattle with insecticides in dusts, smears, sprays, and ointments, may give some short-term control of face flies. Daily

application of insecticides to lactating dairy cattle may
afford adequate control. Certain insecticide-impregnated
ear tags will control face flies, while others can be used
as an aid to control face flies. There is an insecticide
that, when added to the diet of cattle by feed or by
mineral-salt block can control face fly larvae in the
animal's manure. This treatment kills local flies, but it
will not prevent adults from migrating from nearby untreated
cattle unless treatment is on an area-wide basis.

The screwworm fly, Cochliomyia hominivorax (Coquerel),
has been eradicated from the U. S. and is currently the subject of a highly successful eradication campaign in Mexico.
Although screwworms have been eradicated from the U. S.,
dairy cattle producers along the Mexico border should routinely examine their animals for wounds, and if the wounds
contain larvae, some should be collected and sent to the
Screwworm Eradication Program, P. O. Box 969, Mission, Texas
78572, for identification. Wounds should be treated
thoroughly with a spray, dust, aerosol, or smear of insecticide in order to kill maggots and protect the wound from
reinfestation.

Other flesh flies and blow flies may be found in wounds
on cattle but generally these flies lay eggs on dead or
decaying flesh and do not create wounds as does the primary
screwworm. Larvae should be controlled by treating wounds
with insecticidal dusts, aerosols, smears, and sprays.

Nonfeeding flies. Dairy cattle, especially young
stock, can be infested with larvae of 2 species of cattle
grubs: the common cattle grub, Hypoderma lineatum (de Vil-
lers), is found throughout the U. S., and the northern cattle grub, H. bovis (L.), is found in the northern two-thirds
of the U. S. Cattle grubs are the larval stages of adult
flies called heel flies or gadflies because they "chase"
cattle to lay their eggs on the hairs of cattle. After 3
days, larvae hatch from eggs, penetrate into the animal's
body, and then migrate during a period of 3 to 6 months to
the esophagus (common cattle grub) or the spinal column
(northern cattle grub). After several weeks, the larvae
migrate to the animal's back where they secrete an enzyme
that dissolves the hide from the inside, thereby creating a
hole; then the grubs are found in warbles or wolves, which
are characteristic swellings, usually along the middle of
the back. Several weeks later, cattle grubs leave the
animal's back and drop to the ground, where they pupate.
The pupal period may last 2 to 6 weeks or more, depending on
temperature. Then the nonfeeding adults emerge from pupae
and mate, and females chase cattle that usually respond with
typical gadding. The life cycle takes about a year, so in
any given location the same stage -- adult activity (called
heel fly season) or warbles in backs -- is found about the
same time each year.

The annual multi-million-dollar losses due to gadding
(decreased grazing and milk flow) and to the presence of

grubs (meat trim and hide damage) can be prevented by treating nonlactating dairy cattle with animal systemic insecticides during the period after the end of heel fly season and
before warbles appear in the animals' backs. These insecticides, applied in the feed, in a salt-mineral supplement, or
as a spray, dip, "pour-on," or "spot-on" travel throughout
the system of the cattle and kill the migrating larvae.
Since the larvae do not finish their migration, losses in
hide value and from meat trim are eliminated. However, the
time of treatment of cattle with animal systemic insecticides is very critical, and it is important for each rancher
to know when the heel fly season is over in his region.
Usually this information is available from the local agricultural advisor, county agent, or extension official.
Since these insecticides travel through the blood system of
cattle, they cannot be applied to dairy cattle being milked
because the insecticides would be found in the milk.

Lice

Dairy cattle can be infested with 2 types of lice, biting lice and sucking lice. All lice have a similar life
cycle in that all stages are found on the host, females
attach eggs (called nits) to hair, and immature forms
(called nymphs) hatch from eggs and molt one or more times
before they become adults.

Biting lice. Dairy cattle are infested with one species of biting louse, the cattle biting louse, Bovicola
bovis (L.). Biting lice live off skin scales, debris, hair,
and other matter on the surface of the animal's body.
Usually biting lice are found in long hair on the shoulders
and neck of cattle. Infestations are largest in cooler
months when cattle have their heaviest hair growth.

Sucking lice. Dairy cattle are infested with 4 species
of sucking lice: the longnosed cattle louse, Linognathus
vituli (L.); the so-called little blue louse, Solenopotes
capillatus Enderlein; the shortnosed cattle louse, Haematopinus eurysternus (Nitzsch); and (rarely) the cattle tail
louse, Haematopinus quadripertusus Fahrenholz. All these
species suck blood from cattle, and massive infestations can
cause anemia, inefficient feed utilization, loss of condition, a scruffy appearance, and loss of hair. They may lead
to the death of the animal. As with biting lice, infestations of sucking lice are usually greatest in the winter.
Certain animals, called louse carriers, normally have greater infestations than others and often serve as sources of
reinfestation for the remainder of the herd.
In general, all species of lice on dairy cattle can be
controlled by the application of effective insecticides to
the body of the animal. Cattle may be dipped, sprayed, mist
sprayed, or dusted thoroughly for louse control. Self-
application by backrubbers and dust bags may aid in louse

control, but these treatments are not as thorough as sprays, dips, or dusts. "Pour-on" or "spot-on" treatments may also control lice.

Ticks

Dairy cattle that graze in pastures are parasitized by a variety of ticks. These 8-legged relatives of spiders may cause heavy losses because ticks suck blood, transmit diseases, cause paralysis, and create unthrifty animals as a result of massive infestations and constant irritation ("tick worry").

Soft ticks. Dairy cattle are infested with 2 species of "soft tick," a title given to a large group of ticks because of the wrinkled, leathery texture of their "skin." The most common species is the spinose ear tick, Otobius megnini (Duges), which lives deep in the ears of dairy cattle. Adults of this species are free living (they do not feed) and are found in protected places such as cracks and crevices of buildings, fences, under salt troughs, etc. Females are bred and lay eggs in these protected places. Small 6-legged larvae hatch from eggs, seek cattle, attach deep in the animals' ears, feed for a short period, and molt to the spiny-appearing nymphs, which may feed for several months. The species is found on cattle in most States but is common in the southwestern states.

Spinose ear ticks are usually controlled by the thorough treatment of ears of dairy cattle with insecticides such as dusts, low-pressure sprays, aerosols, and smears. Effective treatments will provide adequate control for a month or longer. Insecticide-impregnated ear tags will control spinose ear ticks for several months. Attempts to control the adults in the environment are generally unsuccessful.

A second soft tick that is a pest of dairy cattle is the pajaroello tick, Ornithodoros coriaceus Koch, a vector of epizootic bovine abortion (EBA) in California. This tick feeds for a very short time on pastured cattle and thus is very difficult to control. Ranchers may reduce losses to EBA by not exposing first-calf heifers to ticks during early pregnancy.

Hard ticks. Most ticks in the U. S. have a hard covering on all or part of the back and thus are called "hard" ticks. These ticks have 2 types of life cycle. One is the "1-host" cycle in which larvae (or seed ticks) attach to a host, engorge on blood, and molt to the next stage (the nymph), which engorges and then molts to the adult male or female. These adults mate on the host; the females fill with blood, detach, drop to the ground, find a secluded spot, and lay eggs from which larvae will hatch. All the molting and engorging take place on a single host. The other cycle is the "3-host" cycle. Larvae of a 3-host tick

feed on a host and, when fully fed, drop off the host and
molt on the ground to nymphs. The nymphs find another host
on which they engorge. Fully engorged nymphs drop off the
host and molt on the ground to the adult stage. The adults
find a third host, and mate; then the females engorge, drop
off, and lay eggs.

Each region of the U. S. has its own group of hard
ticks that attacks pastured dairy cattle. The Pacific Coast
tick, Dermacentor occidentalis Marx, is found on the Pacific
Coast west of the coastal mountains. The winter tick, D.
albipictus (Packard), is generally found throughout the northern tier of states and as far south as Texas. The Rocky
Mountain wood tick, D. andersoni Stiles, is usually distributed in the northern Rocky Mountain states. The American
dog tick, D. variabilis (Say), is found generally distributed over the eastern half of the U. S. The blacklegged
tick, Ixodes scapularis Say, is found on cattle in southcen-
tral U. S. The lone star tick, Amblyomma americanum (L.),
is found throughout the southeastern one-third of the U. S.
The Gulf Coast tick, Amblyomma maculatumKoch, is limited to
south Atlantic and Gulf Coast States, although large populations are found in eastern Oklahoma and surrounding
states.

Because of the variety of tick species, their variable
life cycles, and their seasonal abundance, the dairyman must
know which species is attacking his animals so that he can
apply the most effective treatments at the proper time of
the year. Generally tick control consists of treating pastured dairy cattle thoroughly with insecticides applied as
dips or sprays. "Pour-ons," dusts, or self-treatment
devices usually do not provide adequate control. Insecticide-impregnated ear tags effectively control Gulf Coast ear
ticks because they attach to ears of cattle.

In some situations, ticks may be controlled by the
application of approved insecticides to the ground to kill
larvae, newly molted forms (of 3-host species), and engorged
females.

Mites

Dairy cattle are infested with several species of itch,
mange, or scab mites. These very tiny species live on or in
the skin and can cause intense irritation, itching, loss of
hair, thickening of skin, and considerable discomfort to
infested cattle. One species, the common scab mite, Psorop-
tes ovis (Hering), is the subject of a national eradication
campaign, and cattle infested with this species are subject
to quarantine and compulsory treatment in order to eliminate
infestations. Other species of mange, itch, or scab mites
on confined dairy cattle can be controlled by the application of insecticides by whole-body spray. Less thorough,
but more frequent, low-volume mist treatments are only partially effective.

USE OF INSECTICIDES ON DAIRY CATTLE

Treatments

Insecticides remain our first line of defense against insect pests of dairy cattle. Properly used, approved insecticides can kill the pests and thus prevent or reduce losses due to the infestations of flies, lice, ticks, mites, etc. It is necessary to seek the advice and recommendations of local officials such as county agents and agricultural advisors who have current information on kinds of pests in the area, correct techniques of application, recommended times of treatment, and a variety of other facts about local pests, their biology, and their control.

To avoid residues of insecticides in milk, restrictions have been placed on their use for the control of insect pests of milking cattle. For example, no insecticides are recommended for the treatment of cattle grubs in milking cattle. Also, insecticides should usually be applied to milking cattle when the animals leave the milking parlor. It is important that the dairyman apply only those insecticides to milking cattle that are specifically recommended for that use. Milking cattle are usually not dipped.

The nonlactating dairy herd may be treated with a larger variety of insecticides. Nevertheless, it is important for the dairyman to observe the stated period after treatment of nonlactating cows before they freshen to avoid insecticide residues in milk.

Dairymen should be aware that certain insecticides can be applied to milking parlors and milk rooms for the control of flies. However, only specific insecticides applied in limited amounts can be used to prevent contamination of milk.

Precautions

The insecticides that can be used to kill insect pests of dairy cattle also can be toxic to the cattle and the humans who apply them. In addition, these insecticides can create illegal residues in milk and tissues of treated cattle and can be destructive to the environment if not used and handled in a safe manner. The following are a few precautions to follow when choosing and applying insecticides for the control of insect pests of lactating and nonlactating dairy cattle:

- Use only those insecticides recommended and approved for use on dairy cattle by a recognized authority, usually a government official, such as an agricultural agent or advisor.
- Use a formulation of the insecticide that is approved and designed for use on cattle. In dipping vats use only those formulations designed specifically for dipping vats.

- Follow the label directions <u>exactly</u>. The label contains all the information on dilution, time of retreatment of animals, antidotes for poisoning, methods of disposing of unused insecticide, intervals between treatment and freshening, and other important facts.
- Avoid treating cattle in cold, stormy weather, and avoid treating stressed, overheated, or sick animals.
- Be sure that spraying equipment is clean, working properly, and provides sufficient agitation to allow for thorough mixing of insecticides.
- Be aware of safe practices when mixing and applying. Wear protective clothing; do not smoke, drink, or eat while applying insecticides. Do not contaminate feed or feed and water troughs.
- Learn to recognize signs of insecticide poisoning in livestock (and humans) to avoid delaying antidotal measures.
- Store all insecticides in original containers. Do not store insecticides with food or where they can be reached by children, animals, or unauthorized persons.

37
POTENTIAL AND ACTUAL ECONOMICS OF A HERD HEALTH PROGRAM

Jenks Swann Britt

Almost any magazine, field day, or educational program dealing with animal agriculture will feature one or more topics on Herd Health. During the past 15 years, herd health, and preventive medicine have come to the forefront in food-animal production.

Our four-man veterinary practice in South Central Kentucky handles all species of animals but about 75% of our work deals with food animals; namely, dairy cattle, beef cattle, and swine. We have developed herd health programs that involve 40 dairy herds, eight farrow-to-finish swine units, and several small- to medium-size beef herds.

In developing programs for dairy farms, the dairyman almost always begins some phase of herd health medicine because of a specific need in his herd such as mastitis, poor breeding, or calf death loss.

In 1979, we decided to evaluate the effectiveness of our dairy herd health service in giving the dairyman a good return on his investment.

MODEL HERD

Our approach was to set up a model herd with certain production and health problems. This herd had the following health problems:
- Rolling herd average of 11,000 lb for 100 cows. (At that time the average cow in Kentucky produced a little less than 9,000 lb of milk.)
- Average somatic cell count of 700,000.
- Projected calving interval of 13.5 months.
- 2.25 services per conception.
- 20% calf mortality.
- First-calf heifers calving at 32 months.
- No balanced nutrition program.
- No regular parasite control program.
The problem areas were defined as mastitis, reproduction, calf raising, nutrition, and parasite control, each described below.

HERD FACT SHEET

Name_____Date_____Doctor_____

Address_____

Number employees_____Number Cows_____Owner milks yes

no

D.H.I.A.yes or noPounds Day Tank_____Pounds Per Cow_____

MACHINE FUNCTION

PUMP MAKE_____PUMP CONDITION_____

CFM @ PUMP_____CFM NEEDED @ PUMP_____

TYPE VACUUM CONTROLLER_____NUMBER CONTROLLERS_____

CONTROLLER FUNCTION_____

RESERVE WITH MACHINE READY TO MILK_____RESERVED NEEDED___

TYPE PULSATORS_____PULSATOR SPEED_____

NUMBER UNITS_____CLAW SIZE_____

CLAW VENT_____TEAT END VENT_____

INFLATION BORE_____INFLATION CONDITION_____

LINE SIZE = MILK_____VACUUM LINE SIZE_____

DISTRIBUTION TANK SIZE_____LINE CLASS OR STAINLESS_____

DOUBLE OR SINGLE SLOPE_____WEIGH JARS_____

LOW OR HIGH LINE_____WHERE MILK FILTER_____

FLUCTUATION_____RECOVERY TIME_____M/R RATIO_____

LINE VACUUM_____TEAT END VACUUM_____

OTHER_____STRAY VOLTAGE_____

MILKING TECHNIQUE

TYPE WASHING_____TYPE DRYING_____

AVE. TIME BEFORE UNIT ON_____AVE. TIME ON_____

STRIPPING TIME_____HOW REMOVED_____

UNITS PER MAN_____DETACHERS_____

Other_____

SANITATION

UDDER WASH_____MACHINE WASHING_____

BARN LOTS_____TYPE HOUSING_____

TYPE BEDDING_____FLY CONTROL_____

OTHER_____

PREVENTION & CONTROL

AVERAGE LEUKOCYTE COUNT_____CMT_____

AVERAGE BACTERIA COUNT_____CLINICAL TREAT._____

DRY TREATMENT_____TEAT DIP_____

Mastitis

A mastitis control program is implemented in the herd. This program involves dipping every teat on every cow after each milking. All cows to be turned dry are dry treated in all four quarters. Milking technique is evaluated and proper procedures are explained to the people involved with the herd. These procedures include cow preparation, drying the udder, milking machine application and removal, proper machine stripping and handling. The milking equipment is completely evaluated using a dual-channel recorder, air flow meter, mercury column, level, watch, and other necessary equipment. Evaluations are made of vacuum pump capacity, reserve vacuum, pulsator function, regulator function, claw size, hose size and length, pipeline size, and equipment design. The herd fact sheet (figure 1) is used to evaluate the milking-parlor situation. Minimum standards that we recommend for machine function are listed in (figure 2).

Individual CMT samples and composite cow milk samples for culture reveal what infections are present and which cows have these infections. Followup monthly somatic cell counts on each cow and the bulk-tank cultures and cell counts will enable us to monitor the herd.

*Pump capacity	10CMF/unit ASME
Vacuum reserve	5CMF/unit ASME
Recovery time	3 sec.
Teat end vacuum fluctuation	2 in.
Teat end vacuum	12" ± 1"
Stray voltage	0.5 volts
Liner slippage	None
Pulsator speed	45-60
Milk rest ratio	60/40 to 40/60
Pulsator line size	Looped 1 1/2 to 2"**
Milk line size	1 1/2" - 2 units
	2" - 4 units
	3" - 9 units

* Pump capacity for over 8 units may decrease below 10 CFM/unit.

**Pulsator line size 2 in. limit 16 units.

Figure 2. Standards for milking machine function

By following this program, our model herd should reduce its somatic cell count in a year from 700,000 to as low as 350,000. DHIA records show that this cell count decrease should result in an increase of 5 lb of milk per cow daily.

At the 1979 price of $12.00 cwt the increased income from mastitis control would be: 5 lb x 305 days x $.12/lb x 100 cows = $18,300.

REPRODUCTION

The reproductive health program involves monthly visits to the farm to do pregnancy exams, check fresh cows for reproductive problems, treat any problem cows, and assist in other aspects of the herd reproductive program. An individual record (figure 3) is kept on each cow.

Diseases that may cause breeding problems or abortions are controlled by herd vaccination. These diseases include brucellosis, leptospirosis, IBR-BVD, campylobacter, and listeria.

Heat detection is discussed and evaluated. If necessary, a teaser animal may be infected to aid in heat detection. Kamar heat or crayon may be used to mark the rumps of cows expected in heat.

AI technique and semen handling is discussed. If natural service is used, the bulls are evaluated using scrotal measurement and electroejaculation. The society for theriogenology score card for evaluating bulls is used (figure 4).

An evaluation of heat detection is made using the DHIA record (figure 5).

Our goal is to reduce the calving interval by 14 days and to reduce services per conception by .25. A shortened calving interval of 14 days would result in $1.50 per day in savings.

14 days x $1.50/day x 100 cows = $2100.00/year.

The savings of .25 services per conception at $6.00 per semen straw would be:

.25 x 100 cows x $6.00/straw = $150.00/year.

The total savings using our reproductive program is $2250.00.

Calf Raising

The third phase of our program with our model herd involves improving the calf raising. Our model herd had a mortality of 20%.

The calf health program starts with proper dry-cow nutrition and vaccination. Calves must be born in a clean environment; they are force-fed 6 to 8 lb colostrum during the first 12 hours after birth. Calves from first-calf heifers are force-fed frozen colostrum from older cows. Calves are moved to a clean hutch or clean calf-raising area. Calves are vaccinated for roto virus, corona virus, brucellosis, IBR-BVD, haemophilus, black leg, and malignant edema. The hope is to reduce calf mortality by at least 10%. The extra return would be:

Figure 3

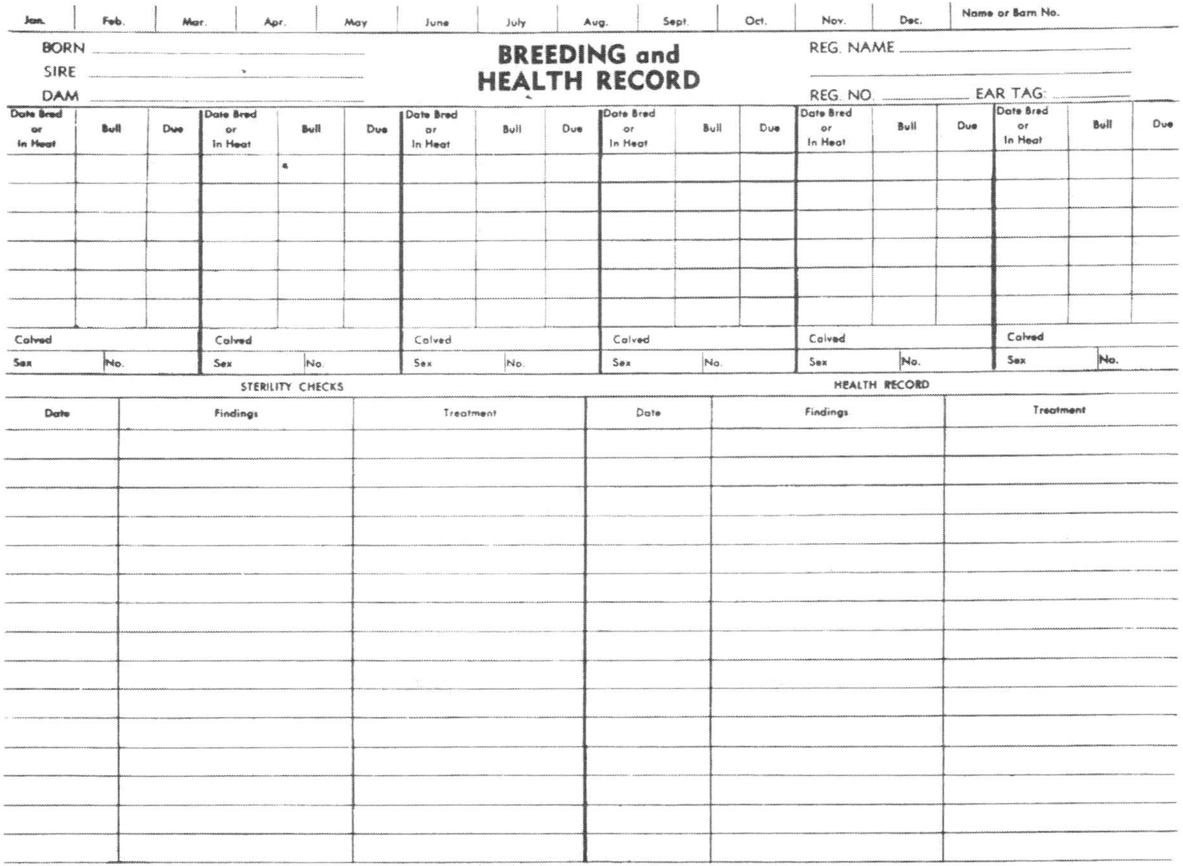

BREEDING and HEALTH RECORD

Jan. | Feb. | Mar. | Apr. | May | June | July | Aug. | Sept. | Oct. | Nov. | Dec. | Name or Barn No.

BORN
SIRE
DAM

REG. NAME
REG. NO.
EAR TAG

| Date Bred or In Heat | Bull | Due | Date Bred or In Heat | Bull | Due | Date Bred or In Heat | Bull | Due | Date Bred or In Heat | Bull | Due | Date Bred or In Heat | Bull | Due | Date Bred or In Heat | Bull | Due |

Calved | Sex | No. | Calved | Sex | No. | Calved | Sex | No. | Calved | Sex | No. | Calved | Sex | No. | Calved | Sex | No.

STERILITY CHECKS

| Date | Findings | Treatment | Date | Findings | Treatment |

HEALTH RECORD

| Date | Findings | Treatment |

Classification	Motility Score
Very Good	"20"
Good	"12"
Fair	"10"
Poor	"3"

	Morphology		
	Primary	Total	
Classification	Abnormalties	Abnormalties	Score
Very good	< 10	< 25	"40"
Good	10-19	26-39	"24"
Fair	20-29	40-59	"10"
Poor	> 29	> 60	"3"

Spheroids
Less than 5/HP field = Occasional = + 5% Primary Abnormality
5 to 15/HP field   = Few   = +15% Primary Abnormality
15 to 25/HP field   = Many   = +25% Primary Abnormality
More than 25/HP field= Multitudes = +35% Primary Abnormality

SCROTAL CIRCUMFERENCE			Age (Mos.)		
All breeds except Brahma	12-14	15-20	21-30	30+	Score
Very good	>35	>37	>39	>40	"40"
Good	30-35	31-37	32-39	33-40	"24"
Poor	<30	<31	<32	<33	"10"

Brahma	Age (Mos.)	Scrotal Circumference
	14	29.9 ± 2.7
	14-17	27.4 ± 3.0
	17-20	29.4 ± 2.2
	20-23	31.4 ± 2.6
	23-26	31.7 ± 1.9
	26-30	33.5 ± 1.4
	30-36	34.7 ± 3.1
	36	36.7 ± 2.7

KEY
Satisfactory Potential Breeder   60-100 total points
Questionable Potential Breeder   30-59 total points
Unsatisfactory Potential Breeeder 0-29 total points

Figure 4. Scorecard for evaluating bulls

	1. Early postpartum breeding							
	Heats observed (%)							
Services Per conception	100	90	80	70	60	50	40	30
				Days open				
1.3	67	70	74	79	86	95	107	129
1.4	69	72	76	82	89	99	112	136
1.5	72	75	79	85	93	113	118	148
1.6	74	77	82	88	96	107	123	136
1.7	76	79	84	91	100	111	128	152
1.8	78	81	87	94	103	126	133	152
1.9	80	84	89	97	107	120	138	160
2.0	82	86	92	100	110	124	143	177
2.1	84	88	95	103	114	128	148	184
2.2	86	91	97	106	117	132	153	191

	2. 60-Day postpartum breeding							
	Heats observed (%)							
Services per conception	100	90	80	70	60	50	40	30
				Days open				
1.3	81	84	88	93	100	109	121	143
1.4	83	86	90	96	103	113	126	150
1.5	86	89	93	99	107	117	132	157
1.6	88	91	96	102	110	121	137	164
1.7	90	93	98	105	114	125	142	171
1.8	92	95	101	108	117	130	147	177
1.9	94	98	103	111	121	134	152	184
2.0	96	100	106	114	124	138	157	191
2.1	98	102	109	117	128	142	162	198
2.2	100	105	111	120	131	146	167	205

Figure 5

```
    - 5 bull calves    @  $ 70.00  =  $ 350.00
    - 5 heifer calves @  $200.00   =  $1000.00
      Total calf return---------- =  $1350.00
```
    In addition to baby calves, efforts are made to keep heifers healthy, growing, and calving at 24-26 months rather than 32 months.

<u>Nutrition</u>

    Our nutrition program starts with an analysis of all feeds used on the farm (figures 6 and 7). These results are then fed into our TI-59 programmed minicomputer and a feeding program is generated. The cow's body weight, daily production, fat test, maintenance factor and lead factor are used to compute the cow's needs. The ration is balanced to a minimum of 17% crude fiber, 72 Mcal/100 lb of feed, .65% calcium, .5% phosphorus, .22% magnesium, and .2% sulfur. Protein is adjusted to the production level. A printout is transferred to a recommendation chart (figure 8), which is sent to the farmer.
    Our goal is to increase milk production 3 lb per cow daily or to produce the same amount of milk on less feed. Our increased income from projected milk increase from the nutrition program is:

AGRILAB SERVICES, INC.

P. O. Box 1085

Hopkinsville, KY 42240

(502) 365-2004

Owner_____   Submitted By_____

Address_____   Address_____

_____   _____

SAMPLE IDENTIFICATION

(HAY)		(HAYLAGE)	(FRESH FORAGE)

( ) Alfalfa   ( ) Timothy

( ) Clover   ( ) Orchardgrass

( ) Fescue   ( ) Brome Grass

( ) Alfalfa-Orchardgrass

( ) Alfalfa-Timothy   If legume grass mixture indicate:

( ) Clover-Fescue

( ) Clover-Orchardgrass   _____ % legume

( ) Other_____   _____ % Grass

Cutting:( )1st ( )2nd ( )3rd

( )Other_____

STAGE OF MATURITY

LEGUMES   GRASSES

( ) Bud   ( ) Pre-head

( ) 1st Bloom   ( ) Heads Emerging

( ) Mid Bloom   ( ) Headed

( ) Full Bloom   ( ) Seeds Formed

SILAGES

TYPE	STAGE OF MATURITY	SILAGE ADDITIVES

( ) Corn   ( ) Barley   ( ) Pre-tassel or head

( ) Sorghum   ( ) Wheat   ( ) Tassel or head

( ) Oats   ( ) Brome Grass   ( ) Milk

( ) Sorghum-Sudan   ( ) Soft dough or early dent

( ) Other_____   ( ) Hard dough or late dent

( )YES   ( )NO

KIND_____

GRAINS AND CONCENTRATES

( ) Dry Shelled Corn   ( ) Dry Ear Corn   ( ) High Moisture Shelled Corn

( ) High Moisture Ear Corn   ( ) Milo   ( ) Oats   ( ) Barley   ( ) Wheat   ( )Rye

( ) Other_____

CHECK DESIRED TEST

A.___Moisture, Protein...$ 5.50

B.___Moisture, Protein,Calcium,Phosphorus....................................$13.50

C.___Moisture,Protein,Acid Detergent Fiber (ADF).............................$14.60

D.___Moisture,Protein,Acid Detergent Fiber (ADF),Calcium,Phosphorus..........$18.75

E.___Moisture,Protein,Ash,Fat..$12.00

F.___Moisture,Protein,Ash,Fat,Calcium,Phosphorus............................$19.00

G.___Calcium,Phosphorus..$ 8.00

H.___Digestible Protein (for heat damaged samples)...........................$15.00

I.___Moisture,Protein,Acid Detergent Fiber (ADF),Calcium,Phosphorus,
     Potassium,Sulfur,Magnesium,Zinc,Copper,Nitrate-qualitative..............$32.50

Figure 6

AGRILAB SERVICES, INC.
P.O. Box 1085
HOPKINSVILLE, KY 42240

SUBMITTED BY _DR. J.S. Britt_ CASE NO. _82-0471_

OWNER _J & W DAiRY_ DATE _4/26/82_

SAMPLE _FResh Cut Rye_ _4/30/82_

		A-F	M-F	A-F	M-F	A-F	M-F
MOISTURE	%	70.33	0				
DRY MATTER	%	29.67	100				
CRUDE PROTEIN	%	3.01	10.14				
FIBER	%	10.99	37.06				
FAT	%						
ADF-NITROGEN	%						
ASH	%						
NFE	%						
TDN	%	19.21	64.73				
DIGESTIBLE PROTEIN	%						
DIGESTIBLE CAR-HYD	%						
N.E. GAIN	MCAL/LB						
N.E. MAINT	MCAL/LB						
N.E. MILK	MCAL/LB	0.18	0.59				
NITROGEN	%						
CALCIUM	%	0.17	0.56				
PHOSPHORUS	%	0.07	0.25				
MAGNESIUM	%	0.06	0.20				
POTASSIUM	%	0.59	1.98				
SULFUR	%	0.04	0.14				
IRON	PPM						
ZINC	PPM	22	75				
COPPER	PPM	6	20				
NITRATE							

Figure 7

A-F AS FED
M-F MOISTURE FREE

384

JENKS S. BRITT, D.V.M.

LOGAN COUNTY ANIMAL CLINIC DAIRY NUTRITION RECOMMENDATIONS

502-726-3511, RT. 1, RUSSELLVILLE, KY. 42276

OWNER _____ DATE _____

ADDRESS _____

On _____ your ration was formulated using our microcomputer.
The ration formulation was made from: (1) actual forage analysis, (2) NRC tables, (3) both.

Your ration was based on:
Cow's weight _____ Daily Milk Production _____

Fat Test _____ Lead Factor _____

Heifers _____ Dry Cows _____

YOUR GRAIN MIX		**FORAGES (Daily)**
Daily	**Ton**	

Corn _____   Hay _____

Barley _____   Corn Silage _____

Other _____   Haylege _____
                         _____ lb.
Other _____   Custom   Other _____
                         Mineral Mix
Soy Bean Meal _____   Other _____

Other _____   Other _____

Deflou. Rock. Phos._____ ____ _____

Dynamate _____ _____   **RECOMMENDED GRAIN FEEDING**

Dical. _____ _____   **lbs. Milk**   **lbs. Grain Mix**

Monoammonium   30
Phosphate _____ _____   40

Ground Limestone _____ _____   50

Magnesium Oxide _____ _____   60

Custom Mineral Mix _____ _____   70

Other _____ _____   80

Bicarb. _____ _____   90

T M Salt _____ _____   100

Vit. A & D _____ _____ Over 100

Cows over _____ lbs. should be on a Magnet Feeder. The Magnet Feeder ration
should contain _____ % protein.

☆ See back for other notes

Figure 8

3 lb milk/day x 305 days x $.12/lb x 100 cows = $10,980.00

Parasite Control

The final phase of the model herd health program is to control internal and external parasites. Internal parasites are controlled in the adult cows by deworming at calving. Heifers are dewormed at 4 months, 12 months, and at breeding.
External parasite control consists of lice control during the winter months using a dust or liquid mop. Pour-on is used on nonlactating cows during the summer and fall. Fly tags are used in the ears on selected heifer groups and cow groups. Mop applicators, manure clean up, and larvacides in the feed are preferred in summer months for fly control. The projected increase in milk production from parasite control is:
1 lb/cow/day x $.12/lb x 305 days x 100 cows = $3660.00
Our total projected milk and income increase for our model herd is as follows:

	lb/day/increase	$ Yearly
Mastitis control	5 lb	$18,300.00
Reproduction		$2,250.00
Calf raising		$1,350.00
Nutrition	3 lb	$10,980.00
Parasite control	1 lb	$3,660.00
TOTAL	9 lb/day	$36,540.00

ACTUAL HERD RECORDS

To see if our herd model was accurate in its prediction, we compared the ledger card records of 18 herds of similar size in our practice. These herds were almost equal in size as a group. Nine herds were on a regular monthly herd-health program in which they received a reminder card of the monthly activity that would be taking place (figure 9). Clients received these cards 5 to 10 days before the scheduled visit.
The other nine herds were herds that called us for emergency calls only and were not on a true herd-health program, although some of them follow a partial program.
The emergency-call-only herds ranged in size from 40 to 90 cows. Yearly veterinary cost ranged from $6.60 per milking cow to $54.43 per milking cow. This total was reached by dividing the total amount of dollars spent on veterinary service and drugs at our clinic by the number of lactating cows only, thus heifer and calf costs were included in the cow total. We figured the herd averages of these herds

using DHIA records or daily bulk-tank weights. The range in production was from 9,000 lb to 15,000 lb with the average being 11,770 lb. The dollars spent on veterinary service averaged $17.06 per cow.

The herd-health herds ranged in size from 28 to 140 cows, the production ranging from 11,000 to 18,000 lb and an average of 14,670 lb. The per cow veterinary cost ranged from a low of $17.01 per cow to a high of $82.64 per cow. The average was $33.74 per cow yearly.

Logan County Animal Clinic
HERD HEALTH AND VACCINATION REMINDER

cows
calves
On _____ your horses were vaccinated

for _____ , _____ , and _____

These vaccinations should be repeated in the month listed below.

Luckett
Hatchett
Britt
On _____ Dr. James will be to

your farm to do fertility work or _____ .

Please call the clinic to schedule this work.

Repeat _____

Figure 9

If the herd-health herds had not been on the program, then their veterinary cost for emergency calls would have been $17.06 per cow, the same as the emergency-call herds. The extra dollars spent for herd health amounted to $16.68 per cow ($33.74 minus $17.06) yearly. The herd health herds produced 2,893 lb more milk per cow than the emergency call herds. This milk was worth $347.16 per cow more than the emergency-call herds. If the herd had 100 cows as our model herd had then the increased income would be $34,716.00. This figure is very close to our predicted increase in income from the model herd, which was $36,540.00.

These figures indicate that the dairyman has a potential return on his herd-health dollar of about 20:1. Herd-health costs were $16.68 per cow and the return from increased production was $347.16 per cow.

Two other interesting findings revealed the reason for increased herd-health costs in some herds. Purebred breeders spent more on herd health per cow that did those with grade herds. As herd numbers increased in both the herd-health herds and emergency-call herds, average cost of veterinary service also increase (figure 10).

	Cost Per Cow	
Production Level	Herd Health Program	Call Basis Herd
16,000 - 18,000	$66.44	-
14,000 - 16,000	$34.23	$33.26
12,000 - 14,000	$22.28	$6.60*
Below 12,000	$18.05	$14.83

* Only 1 herd in this group.

Figure 10. Health program and call basis costs per cow under different production levels

38
A HEALTH PROGRAM FOR DAIRY CATTLE

Samuel B. Guss

In the past thirty years, tremendous changes in the size, productivity, housing, and nutrition of dairy cattle herds have caused equally dramatic changes in health care. In the first half of the twentieth century, veterinarians were concerned primarily with diagnosis and treatment of individual cows. As herds became larger and breeding improved, productivity of cows, health care, disease prevention and treatment, and reproductive management changed drastically. But dairymen still call for veterinary service on an individual cow basis so that the veterinarian has little opportunity or motivation to apply his knowledge to the herd situation.

As herds become larger and management becomes the responsibility of the dairyman himself, the need for the advice and services of a dairy herd health specialist becomes increasingly essential. These veterinarians closely supervise both preventive medical aspects and disease and reproductive management aspects of large dairy cattle herds. The veterinarian makes regular visits to survey the health and reproduction program. He often supplies the medication used and controls administration of medication by the selection of products on their merit and economy, and he trains the farmer in how to use them.

This is the modern, positive, and constructive way to provide health care and reproductive management of dairy cattle herds. Some essentials are necessary for success: (1) there must be continuous frank and open communication between the veterinarian and the herdsman. (2) There must be agreement on day to day management improvements that will minimize the medication used for the herd.

When a competent dairy veterinarian becomes involved in a herd health program for a large dairy herd, his services, advice, and supervision should be the most economical way to get the job done. Dairymen are not interested in subsidizing and supporting veterinarians. They are forced to invest in veterinary services that will minimize losses from disease and improve health and production of their herds.

The cooperation of the dairyman and his veterinarian can achieve satisfaction in both profits and good relations. From this author's viewpoint, there are three ways of managing large dairy herds: In the "do it yourself way," the dairyman uses huge amounts of medication, often wastefully, at excessive cost and with disappointing results. In a second management system, sick cows are treated on an individual call basis by the veterinarian. Little sound reproductive and disease prevention managment occurs because the veterinarian and the herdsman lack good communication on such matters and herd health management is a visit-to-visit enterprise.

In the third approach to herd health management, the herdsman and the veterinarian work together and enjoy mutual respect and more profit. The "traveling drugstore" may still stop regularly at this farm, but the herdsman and the veterinarian control the purchase and use of the medication. My discussion here is a fervent plea to dairymen and veterinarians alike to seek a mutually profitable relationship for the ultimate profit and success of the herd.

THE EFFECT OF HOUSING AND FACILITIES ON DAIRY HERD HEALTH

Loose Housing

Free-stall housing is proving to be the best housing for dairy cattle over a wide range of climates. In good free-stall housing, no attempts are made to close the barn, even in severe winter weather. The ideal free-stall barn is open under the eaves on both sides, with an open space at the ridge proportional to the size of the barn and the number of cows in it.

In Pennsylvania, free-stall barns holding 100 cows have a ridge space open at least 15 in. Thus, in winter, warm air rising off the cows is directed out of the ridge vent. At the same time, cold air entering at the eaves replaces the warm moist air and eliminates the condensation on animals, the building interior, and the bedding. Good free-stall barns have greatly reduced pneumonia in Pennsylvania dairy herds.

Stalls

In a good free-stall barn, stalls are 4 ft by 7 ft. They should have an 8 in. curb at the rear and sides that readily can be dismantled, if necessary. The base of a good free-stall may be made of tamped clay or fine limestone quarry dirt; concrete or blacktop flooring is also quite satisfactory.

Bedding

Two kinds of bedding are infinitely better than a single type. For example, shredded corn fodder and chopped straw, shavings and straw, chunk corn cobs and chopped straw, or peanut hulls and chopped straw are combined. Some Pennsylvania dairymen bed the stalls with sand in the summertime, if it is cheap enough.

Bedding in free-stalls should be higher in front than behind. Desirable floors prevent the formation of holes and afford good footing when cows lie down or rise.

The type of bedding used is important to dairy herd health. I have seen a fine dairy herd literally crippled with one or two severely injured hocks in every cow because kiln-dried hardwood shavings were used for bedding on hard smooth floors. The slippery shavings made it almost impossible for cows to rise or lie down without falling on the hard surface, thus injuring the lateral (outer side) surface of their hocks. Production and reproduction suffered as a result of the painful arthritic situation.

Two truckloads of beautiful, white, shiny, dry, no-coat shavings cost the owner of those 80 purebred cows a tremendous amount of money. Similarly, in Pennsylvania, rotten hardwood (oak) sawdust can be had for the hauling of it except that it seems to be an excellent medium for the propagation of the klebsiella bacteria. This can cause horrendous mastitis losses when the sawdust becomes contaminated with feces or urine.

Barn Floors and Alleyways

In free-stall barns the floors and alleyways are an important building consideration. Concrete finished too smoothly produces severe spraddle-leg injuries and udder injuries as a result of excessive foot growth. The use of carborendum chips on the surface of concrete floors or grooves cut into the surface of concrete floors are excellent for preventing fall injuries.

If ramps are used for moving cows from one level to another, they should be long, have gentle slopes, and be narrow (30 in. w.) so that cows move in single file and cannot turn around. Better than ramps are steps with less than 10 in. treads that are roughly surfaced and wide enough to allow a cow to turn around without falling.

The end-of-the-concrete area is the major health hazard of free-stall housing. It is usually a foul mess and attempts often are made to fill these holes with crushed stone. Cows passing through this area can injure the skin of their feet and at the same time be exposed to ideal conditions conducive to foot rot. A simple solution to this sanitation problem is to make a fan-shaped fill at the end of the concrete walk with fine, packed gravel or quarry dirt.

Fencing

Fencing is another dairy herd facility that directly affects herd health. Board fences are excellent but costly for installation and maintenance. Smooth, high-tensile, steel wire electric fence is the most practical and safe way to fence in dairy cows. Cows learn to respect the high voltage current, which even electrifies weeds growing up through the fence.
Barbed wire and woven wire fencing are sources of wounds and traumatic pericarditis (hardware disease) and have no place on a dairy farm.

Feeding and Watering Troughs

Feeding and watering troughs should provide ample space and be designed to prevent cows from defecating into them. Fence-line feeders and water troughs outside the loafing area result in better over-all sanitation and better consumption of feed and water.

General Sanitation

Clean cows are healthier and produce cleaner milk. This basic premise of dairy farming is often circumvented with attempts to wash filth off cows bodies either in the holding area outside the milking parlor or in the milking parlor itself. For the control of mastitis there are simply no teat dips, udder germicides, or mastitis treatments that supplant milking clean, dry udders and using clean, dry milking equipment.

ESSENTIALS FOR HERD HEALTH MANAGEMENT

Know the Situation

The environment, facilities, source of the animals, quality of food, water and forage provided should all be understood by both the owner and the veterinarian. Ideally, dairy cow herds should be the home-reared progeny of superior sires.
Dairy farms should have adequate sanitary housing and yarding facilities. A good soil and forage-production system is imperative. How many dairymen do you know who are good cowmen and lousy farmers or vice versa? We all know dairymen who like their cows, who have an avid interest in the breeding and conformation of their cattle, but can't buy or produce good forage for their cows. How about the farmers who are "on the ball" as crop farmers? They produce excellent cow forage but lack the "cowmanship" that makes a good farmer a successful dairyman. I have seen this many times during my career as a veterinary practitioner and extension veterinarian. With today's larger herds, a highly

successful combination is a good cowman with a good farmer son or vice versa. That combination is not uncommon.

Keeping Records

Dairy farming has never been profitable without good record keeping. In modern, large dairy herds, there is no way to eliminate the keeping of good records without "going broke."

Production Records

Owners of purebred herds who are interested in type and production improvement--as they relate to sales of stock-- cannot operate progressively without DHIA records. Commercial dairymen may find Owner Sampler or AM-PM Testing useful enough for measuring productivity of their cows.

All three types of records are used by the veterinarian for surveillance of metabolic disease, assessment of changes in feeding, medication, or other aspects of management. Without good records, both the dairyman and the veterinarian "operate in the dark."

Reproductive Management Records

These are completely separate sets of records with completely separate methods for keeping them. Both the dairyman and the veterinarian keep these records cooperatively and both use them continuously.

THE MILKING HERD

The approach to herd health management discussed here identifies components of the usual large dairy herd and makes health recommendations for use throughout the year.

All cows are brought into clean, disinfected maternity stalls to freshen. Cows in loose housing should be moved away from the herd to freshen in a clean, shaded pasture lot close enough so that they can be watched or brought into clean tie stalls or maternity pens large enough to provide for the cows' comfort and the safety of the newborn calves.

Feeding

Balanced grain mixtures are required to complement quantity and quality of forages fed. Forage test results should dictate the quality, composition, and amounts of concentrates fed. Cows should be fed as individuals or in groups consistent with the stage and level of milk production. Challenge feeding pays off in higher milk production and less metabolic disease.

The good dairyman feeds each group of cows--recent fresheners, peak lactation, middle lactation, late lactation, and dry cows--as if they were individual animals. At all stages of the cows' annual production cycle it is imperative that the cows remain in lean, active, alert condition without becoming over-conditioned (fat) or too thin.

Maximum forage intake should be the goal during the entire cycle. This is accomplished by feeding various forages many times during each 24-hour period. Feeding amounts of forage that can be eaten quickly as often as ten times daily, and varying the kinds of forage should be both profitable and healthful for the herd. Today, this is made possible by efficient, automatic feeding devices such as silo unloaders, automatic mixers, and conveyors.

It is virtually impossible to properly feed good dairy cows in large herds, unless the animals are divided into groups according to stage of lactation and level of production. This grouping enables the dairyman to feed the cows more closely according to their needs, with the greatest economy and efficiency.

In Pennsylvania, Penn State dairy extension specialists, who worked with many dairy herds, improved production by more than a ton of milk per cow per lactation, with tremendous savings in grain feeding just by feeding the cows in groups according to their production level and stage of lactation.

Previously, when loose-housed cows were fed by mechanical means, the inadequate feeding of concentrate feeds at the peak of lactation and excessive feeding of those feeds later often resulted in a "fat cow syndrome". Due to this syndrome, many dairymen were put out of business, many over-fat unproductive cows were slaughtered and many good cows died from late pregnancy ketosis, downer syndrome, displaced abomasum, and lactation ketosis.

The incidence, morbidity, and mortality of metabolic disease in dairy herds is directly related to nutrition. Thus, cows should be fed according to their needs at all stages of annual lactation-pregnancy cycle. Modern feed and forage analysis pays good dividends in economical use of concentrates for maximum production. The whole basis for good dairy feeding is maximum utilization of farm-grown, good-quality forage supplemented by intelligent use of low-cost concentrate.

DRY COW MANAGEMENT

Milking cows should be dried off by the 220th day of pregnancy. Those producing less than 45 lb of milk per day can be dried off simply by withholding concentrate feed and discontinuing milking.

Heavier producers may be more easily dried-off by removing concentrate from the ration, feeding low-quality grass hay for a few days and getting them out of sight and sound of the milking routine. Cows that have tolerated at least 5 days without milking should be stripped-out and dry-treated against mastitis.

"THE EYE OF THE FEEDER" AND THE HEALTH OF CALVING COWS

Dry cows should be kept in a group and observed at least once each day. Ideally, they should be in lean, active condition. Judicious feeding of typical milking-cow concentrates (to supply adequate protein and energy) and abundant feeding of good quality, (but not necessarily the best) mixed-grass-legume hay will develop rumen capacity and motility to meet the needs of the postparturient (fresh) cows at the onset of the next lactation cycle.

The single most common cause of abomasal displacement is failure to feed enough long-fiber forage to develop greater rumen capacity and motility during the dry period. I am aware of Pennsylvania dairy herds where 25% of the early lactating nature cows required displaced abomasum surgery as a result of poor, dry-cow management.

VITAMIN A, D, E, SELENIUM INJECTIONS

In selenium deficient areas, administration of one of the commercially available solutions of sodium selenite and alphatocopherol (selenium and Vitamin E) may be helpful for preventing problems related to deficiency (retained placenta, weak calves, etc.).

In the eastern U.S., where grass hay may be of less than desirable quality for feeding cows in late pregnancy in late winter or spring, administration of 1 million units of Vitamin A with D and E has been found to be a practical, inexpensive aid to cows and their newborn calves. The ADE mixture can be given intramuscularly within 4 to 2 weeks before calving.

CALF MANAGEMENT

Calves should be born in clean, disinfected maternity stalls or large, wide (4 1/2 ft by 9 ft) tie stalls. As soon as they are born and cleaned off by the cow, they should be moved to clean, disinfected, off-the-floor tie stalls or off-the-floor group pens with slatted or expanded metal floors.

Baby calves should be reared in a draft-free, well-ventilated, dry environment with dry bedding. Disinfecting navels with 2% tincture of iodine is a good precaution for disease prevention. Feeding clean colostrum as soon as possible after birth (within two hours) gives calves passive antibody protection for at least the first six weeks of their lives. Fermented colostrum is a practical way of utilizing the best available disease prevention/nutrition source for young calves.

At birth, vaccination against neonatal (at birth) virus diseases may be indicated in some herds but should depend upon the diagnosis of suspected virus disease by a competent veterinary diagnostic laboratory.

These vaccines are sometimes over-promoted and sold by manufacturers and some veterinarians. Their use should be determined solely on laboratory-substantiated evidence.

Feeding Calves

Losses of calves can be kept as low as 3% or become as high as 60% to 70%, depending upon who looks after them.

In Pennsylvania, a survey made at least 10 years ago showed that average losses of calves cared for by the dairyman's wife were 5% and losses of calves fed and cared for by the hired men ranged as high as 60%. Empathy, understanding, and common sense are the key words for this dairy-farm enterprise.

HEIFERS

Heifers are best managed if they are kept in groups closely approximating the same age. Dehorning young calves demands good restraint, a cherry-red hot iron, and correct use of it.

Vaccination

Vaccination against enterotoxemia at 3 and 5 weeks of age may be necessary in some herds and administration of Bovine Mixed Pasteurella Multocida-hemophilus may be useful.

Strain 19 brucellosis vaccination should not be administered before 4 months of age and preferably not within 10 days of any other vaccination. All bacterins administered should be given twice at 2 to 3 week intervals.

Clostridial disease vaccination (blackleg, malignant edema, anthrax) can be acomplished at, or near, 4 months of age and certainly before heifers are put out to pasture (at about 6 to 7 months of age).

Vaccinate against bovine respiratory virus disease, IBR, and PI3 at or near 4 months of age.

In some large herds, calves are exposed to intranasal IBR-PI3 vaccine as often as three times before they reach 8 months of age. The repeated challenge of vaccine virus seems to promote excellent immunity. Intranasal vaccination seems most desirable. Because these viruses are airborn, vaccination of half of the calves with modified live-virus vaccine may be as effective for preventing these diseases as would be vaccinating all of them. A practical application of use of MLV IBR-PI$_3$ vaccine is often done in this manner. BVD vaccine is not considered necessary before calves are at least 6 months of age; it is thought that passive colostral antibody protection may last as long as 8 months against BVD.

In selenium deficient areas, administration of injectable selenium-Vitamin E is indicated before weaning and at 4 months of age. Heifer feeds containing 1% dicalcium phosphate and 0.5% FM salt will supply adequate mineral supplementation. All feeds fed to all classes of dairy cattle in selenium-deficient areas of the country should be "fortified" with selenium. This is available in premixes used in mixing feeds.

Multivalent leptospirosis bacterin injection is advisable for all heifers before they are put on pasture. In some areas, annual vaccination for the entire herd (over 6 months of age) is practiced.

Health Care

Inspection of heifers for genital abnormalities, extra or abnormal teats, should be done at weaning time. Trimming feet before 4 months of age may prevent serious foot problems later.

Herd-replacement heifers should be encouraged to consume palatable grain mixtures and good quality dry forage or silage as early as possible. Production of large, growthy, lean, active heifers should be the goal from the beginning. Fancy vitamin mineral mixtures are not necessary.

Not enough attention is given to the need for good pasture and adequate exercise for young stock on most dairy farms. Heifers over 6 months of age can be fed all the mixed-grass/legume hay they will consume while not on pasture. During the pasture season, clean pasture, shade, and clean potable water are adequate and more desirable than hay or silage feeding.

On some Pennsylvania dairy farms, subadult cattle are overfed on concentrates, good hay, and silage to the point where they are over-fat. At breeding time, these heifers may be subfertile as a result of excessive fat deposition around their ovaries and genital tracts. By the time they freshen, their udders may be surrounded and almost displaced by masses of fat. This seriously limits mammary tissue development and milk production.

BULLS ON THE DAIRY FARM

Ideally, the best bulls for use as herd sires are to be found in AI breeding services. Recently, the use of intact males in large herds has developed in an effort to provide heat detection. The risk of injured cows and injured or killed herdsmen has returned with the use of so-called "heat detector" intact males. The advantages, if there are any, certainly do not offset the risks. (See Reproduction Management)

MASTITIS PREVENTION AND CONTROL

There is now common agreement that mastitis is caused by specific pathogens (disease causing organisms) that can be spread from infected to susceptible cows by milking procedures and lack of specific sanitary precautions. Those who sell udder washes, teat dips, and antibiotics for treatment promote the notion that mastitis prevention and control requires only these things. In reality, basic prevention is to milk, wash, and dry udders with clean, dry equipment. Milking machines that do not injure mammary tissue, especially teat ends, minimize infection in susceptible cows.

A SIMPLE MILKING FORMULA

Multiply the number of units used in the milking system by the number of minutes required to milk each group of cows and you should come up with a milking time per cow of less than 6 min.

Regardless of whether you milk 20 cows with a single bucket milker or you use the newest milking system available, the milking time per cow (or machine on-time per cow) is the simple way to evaluate cow milking. You will find that some of the systems that make great claims for cow milking fail miserably when measured by this formula.

There have been many excellent innovations in milking systems and milking equipment, but no system has successfully eliminated the need for clean, careful, conscientious people operating the units.

In the past 15 years, many dairymen have found that women operators, especially older-women operators, can greatly improve cow milking. Cows move readily into the milking arena (parlor, carrousel, etc.), let their milk down better, and milk out cleaner and faster when women do the milking.

The milking-machine operator should be the highest paid worker on the dairy farm. Too often, the wrong workers are hired to milk cows.

Milking systems should have regular inspection and maintenance, preferably by a competent company representative. Be aware that stray voltage in the milking area has been a perplexing cause of udder health problems.

MASTITIS SURVEILLANCE

The strip cup and the California Mastitis Test are the two basic tools to use for mastitis surveillance. In many

herds, quarter sampling, laboratory culture, sensitivity, evaluation, and antibiotic treatment are used to set up a mastitis control program.
The regular use of the DMSCC (direct microscopic cell count) or electronic cell counts will identify problem cows. Sanitary milking, use of a nonirritant, germicidal teat-dip and dry treatment of all cows that have shown evidence of infection (high cell counts or clinical mastitis) will also identify problem cows. Although less expensive than laboratory diagnosis, it may take longer. In summary, for practical mastitis prevention and control: 1) practice sanitary, managed milking (clean, dry cows, adequate let-down stimulation), 2) use milking equipment that will remove milk rapidly without injury, 3) use the new tools (laboratory, CMT, DMSCC or other somatic cell determination, a strip cup, teat dipping and necessary dry-cow treatment) but remember that none of these, nor any item of milking equipment replaces the need for good milking-equipment operators.
Unusual types of mastitis require the services of a veterinarian and, possibly, a milking-system expert.

OTHER SPECIFIC DISEASE PROBLEMS .

Other disease problems may occur only in certain herds or in certain areas. They require the use of competent veterinary diagnostic assistance. Every animal that sickens and dies or is removed from the herd to slaughter because of sickness should be examined by somebody capable of recognizing abnormality and disease.
Dairy farmers and their veterinarians should learn to seek answers to problems that affect health, productivity, and longevity of cows. Losses of animals are accepted as inevitable by some dairymen, but the good manager always wants to know "why" and, "what could have been done to prevent it?"
The dairyman and his veterinarian can cooperatively provide the dairy herd with a health program that develops procedures to prevent problems.

REFERENCES

Alexander, G. I. 1978. An integrated approach to improvement in dairy cattle production. Bovine Practitioner, No. 13.

Avidar, Y. 1981. Possible relationship between fertility, feeding and blood parameters of milking cows. Bovine Practitioner, No. 16.

Baker L. 1968. Bovine Health Programming. United Publishing Corp.

Harrington, B. 1979. Preventive medicine in dairy practice. Journal AVMA 174:398.

Schnurrenburger, P.N. 1979. Defining preventive medicine in veterinary medicine. Journal AVMA 174:379.

Spice, M. F. 1981. Bovine immunization. Bovine Practitioner No. 16.

39

ACHIEVING A PROFITABLE CALVING INTERVAL THROUGH A REPRODUCTIVE HEALTH AND MANAGEMENT PROGRAM

Jenks Swann Britt

DHIA records show that maximum lifetime milk production occurs when dairy heifers calve at 20 to 24 months and calve every 385 days thereafter. For the commercial dairyman this calving interval would appear to be the most profitable. Dairymen engaged in the breeding and sale of purebred cattle may discover that a longer calving interval is just as profitable through the sale of breeding animals produced by embryo transfer.

Our reproductive health program is based on the following standards:
- Useful reproductive record systems.
- Pratical and aided heat detection.
- Prevention of reproductive disease by vaccination.
- Improvement of reproductive efficiency by balanced nutrition.
- Monthly treatment of problem cows.
- Early pregnancy diagnosis.
- Attempted diagnosis of abortions.
- Breeding evaluations of bulls used in natural service.
- Embryo transfer.

The record system that I find most useful (figure 1) gives 4 years of reproductive records for a cow. Some of our clients use other systems that work best for them. It is important that the dairyman have a system that he will use. Systems need to identify problem cows and give their reproductive history. The system should be easy to record and use and should identify groups of cows to be checked on the monthly visit. These groups include any cow that:
- Is fresh.
- Has been bred for 30 days.
- Has had 3 or more services.
- Is a problem (repeat breeder, cystic, anestrus, abortion, etc.).

We find that the dairymen with the best kept records usually have the shortest calving interval.

The DHIA herd summary gives useful herd information. This information includes average days since calving, average days open, average days to first breeding, services per conception, individual cow breeding intervals, percentage problem cows, projected minimum calving interval, and possible breedings that were serviced. Due to the way the computer is programed, the DHIA calving interval may be 0.5 months less than actual.

Figure 1

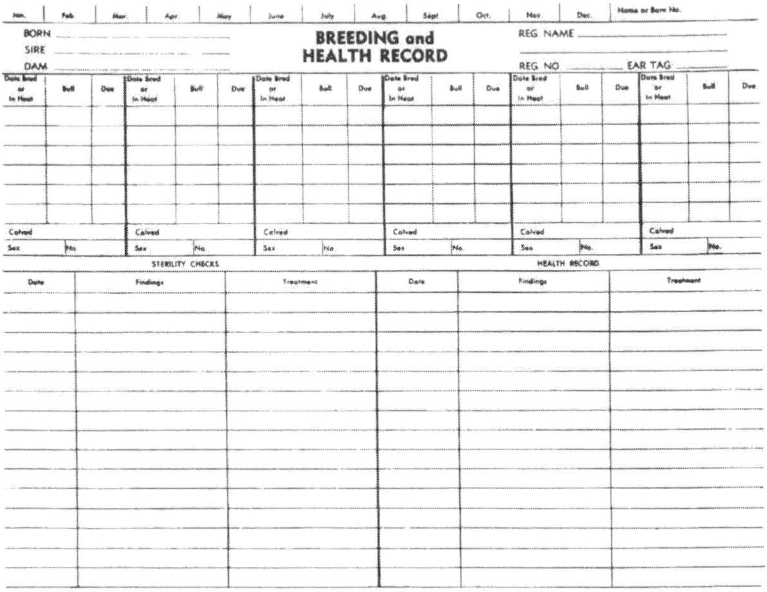

Jan. | Feb | Mar. | Apr. | May | June | July | Aug. | Sept | Oct. | Nov | Dec. | Name or Barn No.

BORN

SIRE

DAM

BREEDING and HEALTH RECORD

REG NAME

REG NO

EAR TAG

Date Bred or In Heat	Bull	Due	Date Bred or In Heat	Bull	Due	Date Bred or In Heat	Bull	Due	Date Bred or In Heat	Bull	Due	Date Bred or In Heat	Bull	Due	Date Bred or In Heat	Bull	Due
Calved			Calved			Calved			Calved			Calved			Calved		
Sex	No.		Sex	No.		Sex	No.		Sex	No.		Sex	No.		Sex	No.	

STERILITY CHECKS | HEALTH RECORD

Date	Findings	Treatment	Date	Findings	Treatment

The most common problem in all AI herds is heat detection. Four methods are used to aid visual heat detection. The best percentage of detected heats occurs when the dairyman can spend four, 15-minute periods daily in watching the activity of his herd. One of these periods needs to be after 8 p.m. or before 5 a.m. to achieve best results. Most of our clients are involved in raising tobacco so their time is often limited. The aids to visual observation include:

- Kamar heat detectors that are glued to the cow's tailhead. A capsule breaks and turns red when pressure is applied by a cow mounting the cow in heat. False positives may occur when cows are crowded into a holding pen and the cow being mounted is not in heat but cannot move away. Estrous interval should be checked on any cow wearing a Kamar, before she is bred.

- Tail marking with crayons, a method similar to the Kamar detector. A chalk mark is made on the cow's tailhead. Mounting by other cows rubs the paint off. It is necessary to mark cows every 1 to 3 days to keep the paint fresh.
- Heat prediction charts. These are calendars with 21-day columns that allow the dairyman to record information on a 21 day basis, thus following a cow's cycle across the chart (figure 2).

Figure 2

JULY 1980	JULY	AUGUST	SEPTEMBER	SEPTEMBER 1980
6	27	17	7	28
7	28	18	8	29
8	29	19	9	30
9	30	20	10	OCT. 1
10	31	21	11	2
11	AUG. 1	22	12	3
12	2	23	13	4
13	3	24	14	5
14	4	25	15	6
15	5	26	16	7
16	6	27	17	8

21-Day Reproduction Record
UNIVERSITY of KENTUCKY · COLLEGE of AGRICULTURE

- Teaser animals. Injecting a barren female or a steer with testosterone will cause that animal to mount cows in heat. A total of 1500 mg of repositol testosterone is given on the first day and smaller injections are given on 2-week intervals. Chin-ball markers are sometimes used on these animals.

A heat detection percentage chart (figure 3) will give the dairyman a guide as to his number of missed heats.

Figure 3

	1. Early postpartum breeding							
	Heats observed (%)							
Services Per conception	100	90	80	70	60	50	40	30
				Days open				
1.3	67	70	74	79	86	95	107	129
1.4	69	72	76	82	89	99	112	136
1.5	72	75	79	85	93	113	118	148
1.6	74	77	82	88	96	107	123	136
1.7	76	79	84	91	100	111	128	152
1.8	78	81	87	94	103	126	133	152
1.9	80	84	89	97	107	120	138	160
2.0	82	86	92	100	110	124	143	177
2.1	84	88	95	103	114	128	148	184
2.2	86	91	97	106	117	132	153	191

	2. 60-Day postpartum breeding							
	Heats observed (%)							
Services per conception	100	90	80	70	60	50	40	30
				Days open				
1.3	81	84	88	93	100	109	121	143
1.4	83	86	90	96	103	113	126	150
1.5	86	89	93	99	107	117	132	157
1.6	88	91	96	102	110	121	137	164
1.7	90	93	98	105	114	125	142	171
1.8	92	95	101	108	117	130	147	177
1.9	94	98	103	111	121	134	152	184
2.0	96	100	106	114	124	138	157	191
2.1	98	102	109	117	128	142	162	198
2.2	100	105	111	120	131	146	167	205

Vaccination can often prevent reproductive diseases that cause abortion or infertility. The vaccinations used in our practice are classified as (1) routine vaccinations recommended for all herds and (2) optional vaccinations used on selected herds.

Routine vacccinations include Brucella abortus vaccine, which is used in either a standard dose on calves 2 to 6 months or a reduced dose on calves 4 to 10 months of age. Leptospirosis vaccine is used on all herds. The 5-strain type vaccine is used. It is given to heifers twice before their first breeding and repeated at 6 to 12 month intervals for the rest of their breeding life.

Optional vaccines include Campylobacter fetus vaccine on herds with natural service. Infectious bovine rhinotracheaitis, bovine virus diarrhea (IBR-BVD) vaccine, is given at 8 and 12 months to open heifers and yearly to open cows in some herds. The newly available killed IBR-BVD vaccine is being used instead of MLV vaccine in herds where all cows would be vaccinated at one time. In herds where abortion due to listeria has been a problem, an autogenous vaccine is used on cows at freshening.

Figure 4

AGRILAB SERVICES, INC.
P.O. Box 1065
HOPKINSVILLE KY 42240

SUBMITTED BY Dr. J.S. Britt   CASE NO 82-0471
OWNER J & W Dairy   DATE 4/26/82
    4/30/82

SAMPLE   Fresh Cut Rye

MOISTURE	%	70.33	0	
DRY MATTER	%	29.67	100	
CRUDE PROTEIN	%	3.01	10.14	
FIBER	%	10.99	37.06	
FAT	%			
ADF-NITROGEN	%			
ASH	%			
NFE	%			
TDN	%	19.21	64.73	
DIGESTIBLE PROTEIN	%			
DIGESTIBLE CAR-HYD	%			
N.E. GAIN	MCAL/LB			
N.E. MAINT	MCAL/LB			
N.E. MILK	MCAL/LB	0.18	0.59	
NITROGEN	%			
CALCIUM	%	0.17	0.56	
PHOSPHORUS	%	0.07	0.25	
MAGNESIUM	%	0.06	0.20	
POTASSIUM	%	0.59	1.98	
SULFUR	%	0.04	0.14	
IRON	PPM			
ZINC	PPM	22	75	
COPPER	PPM	6	20	
NITRATE				
		A-F	M-F	A-F	M-F	A-F	M-F

A-F   AS FED
M-F   MOISTURE FREE

Figure 5

JENKS S. BRITT, D.V.M.
LOGAN COUNTY ANIMAL CLINIC DAIRY NUTRITION RECOMMENDATIONS
502-726-3511, RT. 1, RUSSELLVILLE, KY. 42276

OWNER _____ DATE _____
ADDRESS _____
    On _____
The ration formulation was made from: _____ your ration was formulated using our microcomputer.
(1) actual forage analysis, (2) NRC tables, (3) both.
Your ration was based on:
    Cow's weight _____   Daily Milk Production _____
    Fat Test _____   Lead Factor _____
    Heifers _____   Dry Cows _____

YOUR GRAIN MIX   **FORAGES (Daily)**
Daily   Ton

Corn _____   Hay _____
Barley _____   Corn Silage _____
Other _____   Haylege _____
Other _____   lb.   Other _____
Soy Bean Meal _____   Custom Mineral Mix   Other _____
Other _____   Other _____
Deflou. Rock. Phos. _____
Dynamate _____   **RECOMMENDED GRAIN FEEDING**
Dical. _____   lbs. Milk   lbs. Grain Mix
Monoammonium Phosphate _____   30
Ground Limestone _____   40
Magnesium Oxide _____   50
Custom Mineral Mix _____   60
Other _____   70
Bicarb. _____   80
T M Salt _____   90
Vit. A & D _____   100

Cows over _____ lbs. should be on a Magnet Feeder. The Magnet Feeder ration
should contain _____ % protein.
☆ See back for other notes

Cows will not cycle or breed efficiently if nutrition is not balanced for maximum performance. Forage analysis (figure 4) is made of all forages used on farms we serve. The results of these analyses are plugged into our minicomputer and a balanced ration is computed (figure 5). Special emphasis is placed on energy, crude protein, crude fiber, calcium, phosphorous, magnesium and sulfur. The total ration contains a minimum of:

Crude protein	16%	Phosphorus	.5%
Energy	72 Mcal/100	Magnesium	.22%
	lbDM	Sulfur	.22%
Crude fiber	17%	Calcium	.65%

Micronutrients including selenium are added to all rations.

The monthly or bimonthly herd visit is scheduled with a herd health reminder card (figure 6). Herds using AI are

Figure 6

Logan County Animal Clinic

HERD HEALTH AND VACCINATION REMINDER

cows
calves

On _____ your horses were vaccinated

for _____ , _____ , and _____

These vaccinations should be repeated in the month listed below.

Luckett
Hatchett
Britt

On _____ Dr.   James   will be to

your farm to do fertility work or _____ .

Please call the clinic to schedule this work.

Repeat _____

visited every 14 or 30 days and herds using natural service are visited every 60 days. Checking of the cows usually is done in the parlor, in lock up stanchions, or a breeding chute. The cows selected for checking on the monthly visit include any cow--

- Bred 30 days or more and not previously checked pregnant.
- Fresh since the last visit.
- Bred 3 or more times.
- With a retained placenta, cyst, anestrus, metritis, etc.
- Treated on a previous visit and not bred.
- That had been previously diagnosed pregnant but had been seen in heat.

Treatments used on retained placentas include a volume uterine infusion of 500 to 1000 ccs of a furacin-tetracycline mixture or tetracycline boluses. No attempt is made to remove the placenta prior to the time that the fetal cotyledon has released from the maternal caruncle.

Cows fresh more than 2 weeks are infused with a 30 to 120 cc antibiotic or iodine mixture if there is an increased volume of fluid in the uterus or a purulent discharge. Cows fresh over 30 days are infused only if the uterine horn lumen can be palpated or there is external discharge. Milk residue problems are monitored using a Delvo-P test kit.

Use of hormones is limited for cystic or anestrous problems.

Prostaglandin, GNRH, or chorionic gonadotrophin is used for anestrus or cystic conditions. Progesterone may be used for inactive ovaries.

Pregnancy diagnosis before 40 days postbreeding is important to determine if the cow is pregnant or will return to heat. Early palpation at 26 to 30 days will reveal a high probability of pregnancy by the experienced palpator.

All dairies will have a number of abortions during the year. Some of these are early embryonic deaths before 60 days of pregnancy, while others occur from 2 months until near term of pregnancy. Many abortions are unexplained and hard to diagnose. Our system of diagnosis is to attempt to recover the fetus as soon as possible and submit it to the laboratory for a workup. We also submit placenta and a maternal caruncle from the uterus if possible. Paired serum samples (taken 10 to 14 days apart) from the cow are submitted for laboratory serology. These are checked for brucellosis, leptospirosis, IBR-BVD, and listeria. Samples may be submitted for chlamydia. Feed samples are checked for mycotoxins and nitrates.

With all of this laboratory workup, we only can diagnose 25% to 30% of the abortions. Some herds will experience a few abortions and then not have any for several months or years.

If herds are using natural service, we like to do a prebreeding exam on the bull before he enters the herd for service. The society for theriogenology score card for grading bulls is used (figure 7). The bulls are tested for brucellosis and are vaccinated for leptospirosis and Campylobacter fetus. We encourage dairymen to start the bull slowly and to watch his activity to see if normal breeding is occurring. We recommend one mature bull per 40 cows with a rotation of the bulls if the herd is over 100 cows. If cows return to heat in 30 to 60 days after service to the bull, a trichomonas evaluation is made on the bull. This disease has not been a major problem in our practice area.

During the last 3 years, we have been using embryo transfer in our practice. Our success rate with nonsurgical transfer was about 15% during the first 6 months, but has been improving to date (July 1982). In the first quarter of

1982, the percentage of pregnancies from embryos transferred was 62%, with a range from 37% to 75%. During this same time we averaged 3.4 pregnancies per flush when checked at 40 days of pregnancy. We have found a normal abortion rate of between 40 days and term in the recipient animals.

We prefer to use recipient animals on the owner's farm. Our choice for recipient is the first-calf heifer, fresh 60 to 90 days; our second choice is the open heifer, and the second-calf cow is the third choice.

Classification	Motility Score
Very Good	"20"
Good	"12"
Fair	"10"
Poor	"3"

	Morphology		
Classification	Primary Abnormalities	Total Abnormalities	Score
Very good	< 10	< 25	"40"
Good	10-19	26-39	"24"
Fair	20-29	40-59	"10"
Poor	> 29	> 60	"3"

Spheroids
Less than 5/HP field	= Occasional	= + 5%	Primary Abnormality
5 to 15/HP field	= Few	= +15%	Primary Abnormality
15 to 25/HP field	= Many	= +25%	Primary Abnormality
More than 25/HP field	= Multitudes	= +35%	Primary Abnormality

SCROTAL CIRCUMFERENCE		Age (Mo)			
All breeds except Brahma	12-14	15-20	21-30	30+	Score
Very good	>35	> 37	> 39	>40	"40"
Good	30-35	31-37	32-39	33-40	"24"
Poor	< 30	< 31	< 32	< 33	"10"

Brahma	Age (Mo)	Scrotal Circumference
	14	29.9 ± 2.7
	14-17	27.4 ± 3.0
	17-20	29.4 ± 2.2
	20-23	31.4 ± 2.6
	23-26	31.7 ± 1.9
	26-30	33.5 ± 1.4
	30-36	34.7 ± 3.1
	36	36.7 ± 2.7

KEY
Satisfactory potential breeder 60-100 total points
Questionalbe potential breeder 30-59 total points
Unsatisfactory potential breeeder 0-29 total points

Figure 7

Donor cows are synchronized if we are going to do more than one the same day. Recipient animals are also given two injections of prostaglandin to assure that their estrous cycle will be near that of the donor. Recipient animals are used if they are in heat within 24 hours of the donor cow.

Donor cows are flushed about 60 to 90 days postpartum and may be flushed a second time in 40 days. Embryos are recovered from evaporation dishes; they are graded as A, B, or C embryos and washed with media containing antibiotics. Our current transfer media is 400 ccs TCM 199 to which we add 2% fetal serum and 5 ccs of antibiotic-antimycotic. For holding media, we add more fetal serum to make a 15% to 20% serum solution.

Embryos are loaded into 1/4 cc straws and the straw loaded into an AI gun. The embryo is placed in the horn of the recipient animal on the same side as the ovary with a corpus luteum.

Best results have been obtained by using sound breeding donor cows with no history of reproductive problems. Problem breeding cows have yielded poor results in most cases. Embryo transfer will aid in diagnosis of oviduct blockage or other uterine problems.

Donor cows that yield pregnancies may actually have a natural calving interval of 16 to 18 months--but will produce 2 to 8 embryo-transfer calves during the same period, which is a 2 to 3 month calving interval on actual calves produced.

Part 8

MASTITIS AND RELATED TOPICS

40

SOMATIC CELLS: THEIR ORIGIN AND
FUNCTION AND AN INTERPRETATION
OF THEIR NUMBERS

Don G. Rollins

Dairymen generally are confused by talk about relation of somatic cells, their origin, function, and the interpretation of their numbers. It is not too unfamiliar when a dairyman with a herd problem comes in to a veterinarian looking for that almighty shot. The dairyman has perhaps $250 million invested and 50% of the quarters in his herd are infected. He hasn't done many preventative acts to help himself, but he comes in and he wants a shot. It is very frustrating to talk to him an hour. Generally, he won't listen to what you have to say because he is seeking a "quick cure." This presentation will give you some rules of thumb that you can get people's attention with, a few analogies to show them the extreme economic importance of mastitis control rather than crises treatment.

Consider the situation in figure 1. When this cow comes into the barn, if everything sounds good, looks good, smells good, and tastes good, then the only way that we can foul up let-down is to not touch good. We see so many people who spend $35,000 or so to get a set of equipment in the milking parlor that has 6 or 8 on a side, and they go down through the line and wash six cows; then, if the dry udders, they dry six before they come back and start putting on units. As a result, there is no personal touching attention given to the udder so that the good message is sent up to the brain and oxytocin is sent back to wring out that "udder sponge." You have to impress upon the dairyman that you work with that, when he goes to those teats, he essentially has to make love to the bottom of that udder. He cannot go in there roughly because then he will get the adrenalin response and the cow will either fight or she will run (adrenalin response). If she cannot fight or run because she is tied up in the stall, she is going to stand there and the drug, adrenalin, will offset oxytocin so that you do not get that "wring out." You don't flush the udder.

Imagine getting hold of a teat and pulling out a branch. What I want to do is show you with a schematic drawing like figure 2 the origin of a somatic cell so that you can give the dairyman an idea of its function. The streak canal is 10 to 15 millimeters long or about one-

414

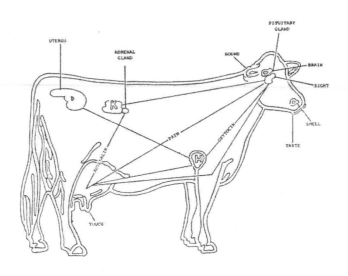

Figure 1. Let-down of milk

quarter of an inch. We are going to take a trip all the way
up to the alveolus right at the end of the ductwork and look
at where somatic cells come from, how we get them all the
way from the alveolus, and what their numbers mean after
they get out. In figure 3, there are four, five, six
lobules and we will cut one out and look at it. Dairymen,
in general, don't understand that once bacteria gets up into
that teat that they just float around in the milk. Each
bacterium carries a machine gun or a laser beam to point
wherever they want to go. Bacteria passing through the
walls of the duct, or via milk to the alveoli, irritate the
cells. The little cell hollers, "Help, help!" There are
normally somewhere between 5 and 13,000 leucocytes running
through the blood stream. When "help" is yelled, the somatic cells (leucocytes) squeeze into the alveoli lumen
between milk-producing cells. In addition to the leucocytes
squeezing between the milk cells, they allow the blood
plasma and salts to leak through.

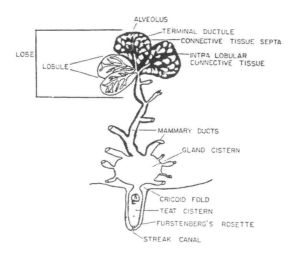

Figure 2.

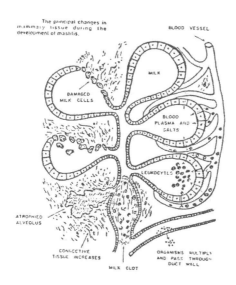

Figure 3.

Notice the "damaged milk cells." This is the other kind of somatic cell, the epithelial cell. There is a normal turnover, and a lot of these are regenerated during the dry period. Those that are damaged will float out in the milk and they will get down to the ductwork with a bunch of leucocytes. When we count somatic cells we are counting leucocytes and damaged milk cells. Some of them are going to get hung up in scar tissue and clots. They fail to produce milk because of the irritation. Little cells that are fighting a battle are sitting there not making milk. Some of them are scarred completely and scar tissue surrounds them, squeezes them off, and they won't make milk ever again. The net result is that we get milk clots made up of leucocytes, blood, and plasma, a little milk, damaged cells, and they get all hung up in the ductwork. This is why we need to capitalize on letdown and take advantage of Mother Nature squeezing the sponge. The guy doing the milking thinks the cow is failing to produce a bunch of milk--but she may be failing because the road is blocked. We cannot get milk from above the clot. If we capitalize on let-down by proper stimulation to the base of that udderly getting the unit on within 60 seconds of when we first touch her, we are going to be milking out all the pipelines at the same time that oxytocin is up there at the alveolus squeezing and filling the pipeline up again. We are apt to shoot that clot out this way. If we do not, bacteria above the clot will be sitting there where it is a warm 101.5° with all the food they would ever want to eat, and all they have to do is sit there and make babies.

Figure 4 is a 3D view of one alveolus with a section cut out of its side. The main thing to observe is that between milking times this alveolus fills up with milk that is extruded from each of the cells lining the lumen. Each cell is fed by the artery the amount and quality of food fed to the cows. Notice the arterial blood keeps getting smaller until it gets down to just where it is one cell thick. When the cell calls for so much protein and so much fat to make milk, it depends on what you are putting in the cow's mouth, which finally comes through this artery to be exchanged. If you are feeding the cow right, then this milk-secreting cell is going to have enough nutrients to be healthy and make good wholesome milk. When bacteria get up there and go to bugging her she doesn't just die, she has a little fight to her.

In figure 2, we saw the myoepithelial cells that looked like spiders lying on top of the alveoli. There are also some in the walls of the ducts down through the intralobular ducts where the milk is going. If the milker is washing or drying cow no. 6, and cow no. 1 is back there standing about ready to bust, then when we do get that unit on her the muscle cells are going to give out because the myoepithelials have had their oxytocin (half of it is gone in the first minute) and they have been up there just squeezing as hard as they can but all the ducts have been blocked with milk.

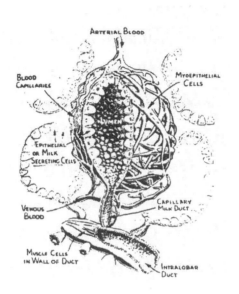

Figure 4.

In 3 min there is only 1 1/2% of the original oxytocin
left. If the milker delays 2 min or so getting the unit on,
it will dump out what is in the ducts and the myoepithelial
cells will be out of squeezing power. They just lay there
leaving milk in the lumen that will just dribble out. If
you have automatic takeoffs that catch the milk flow when it
gets to a certain level before it pulls the unit off, it is
going to get this dribble. But you are going to leave some
milk. Some herds will leave significant percentages of the
milk still in there. They will leave the bacteria in there
too! When they come back up to milk the next time, the cow
is going to have a lot more bacteria in there than she had
previously. The alveoli will have bacteria in them and they
are fighting the milk secreting cells. And the somatic cell
count is going to go up!
     Now, how do we count these cells? In a normal, unin-
fected cow, the somatic cell count will be 75% leucocytes.
With little irritation, some damaged cells, and someone hol-
lering for help, those leucocytes will go in there to clean
up the mess. Twenty-five percent of the numbers of a somatic cell count in a uninfected cow will be just old worn-out
cells. As we increase infections in the alveolus, the somatic cell count numbers become predominantly secreting cells
that have been killed.

The California Mastitis Text (CMT) may take a lot of time when done on a regular basis, but a dairyman can make more money with it and have a grade card of his management ability. His CMT records tell him where that quarter of that cow and his ability to manage her stand at any particular time. When you do a California Test on a herd, it is helpful to have someone writing and someone reading results. The person writing needs to know what you are reading and how to get it down quickly. To make the test put a squirt of milk in the cups, tip i almost vertical so there is very little left within the cups (just a half teaspoon of milk). Put in equal parts of blue reagent, swirl it, and call your decision in five seconds. Don't stand there and say you can't tell if that's a two or three. Call your diagnosis and record it because it is only a "ball park" figure to give you a general idea. Three, three, three, two. Whoever is writing should know that you mean left rear, left front, right front, and right rear. Put them down that way so that when you get to the house you know what you have. If you swirl it around you are going to make more clots or fewer clots, according to how you swirl it. The best way to explain this so that someone can understand it is: Get that half teaspoon of milk in there and add the blue stuff to it (diluted according to directions). That blue stuff is a soap and a blue dye. The soap in the reagent gets after the somatic cells, tears their little house down, and causes their guts to fall out. Then the blue stuff stains their guts. The more somatic cell guts you have, the more blue you are going to have. Everybody can understand that.

A dairyman can look at his milk to say, "That's perfectly good milk--that's not only selling milk, that's drinking milk." But when you do a CMT and have a blue reaction, it says, "Okay, the milk looks good but the test shows us that we still have a little irritation way up there because we see all these somatic cells rupturing and their insides locking together." This reaction shows that we have inflammation in the quarter. The CMT test allows us to give every quarter of every one of our cows a grade card and make a decision as to whether she should be next year's grand champion or tomorrow's McDonald burger.

The Wisconsin Mastitis Test takes the sample of milk like the California Mastitis Test and puts the milk in a tube. You don't use the blue dye on this one: you just put the soap in there and it busts the leucocytes and their guts fall out. However high the stack of somatic cell "insides" is, in millimeters, that's your WMT score. In other words, you get enough somatic cells in your sample of milk so that when the detergent busts the hide on all of them and their insides react together, they stand in the WMT tube in a column so many millimeters tall.

Comparative somatic cell counts, figure 5, compares the various methods. I would like to get away from the Wisconsin Mastitis Score because it is as broad-ranged as the

California Test. We have the trace on a bulk tank count, then it could be anywhere from a 3-12 on a WMT just by my opinion of what "blue" is or my opinion of what this "slime" sticking to the bottom is. You can see the wide variations and there will be further variations when we count them exactly.

Figure 5. Comparative somatic cell counts related to milk losses and cure rates

C.M.T.	Trace	1 Weak	Distinct 2 Positive	Distinct 3 Positive
W. M. T. score	3-12	12-21+	19	37
Cells/ML	150-500M	400-1500M	800-5000M	5,000,000+
Loss in production	6-9%	10-19.5%	16-31.8%	24.5-43.4%
Lb/qtr loss/day	0.1	1-2	2-3.2+	3.2-4.5
% of quarters infected	6.5%	27.3%	64.7%	71.3%
Treatment-cure	68%	67%	28%	13%

Figure 5 numbers are on bulk tank samples. If a sample had a CMT trace reaction, and if we did a WMT on it, it would be somewhere between a 3 and 12. We would take that sample and through a direct microscopic somatic cell count could have from 150,000 to 500,000 cells per c.c. or milliliter. We look at the whole herd that produced this sample of milk and those cows would be failing to produce between 6% and 9% of their regular production because of the irritation the cell count indicated in those quarters. The more irritation, the more cells. If we would go into that herd and look microbiologically at every quarter of that herd, we would find 6 1/2% of them to be infected with mastitis-causing organisms. Because of the environment where cows live, that is acceptable. We can see the range of spread as we get higher, as we get up towards regulatory limits, we have some broader ranges and it is reason for confusion with a lot of dairymen because they could call their count a 2 on a CMT and say "Boy, it don't look like it is all that bad." If we actually counted them, he could have from 800,000 to 5 million. His production could be down 30% because of irritation in those quarters. Every cow he is milking is giving 3 1/2 lb less of milk per day. Two-thirds of the quarters can be infected with mastitis organisms because of a little bit of irritation in a lot of the udders. So we are going to lose a few pounds of milk. Figure 5 assumes the infections are staph. If we identify those exact quarters that are infected and have the best available antibiotics today, with the treatment schedule, the rate, the dosage, and the duration correct, we could probably only cure (clinical cures during actual milking) 50% or less when the CMT score

exceeds a 1. On some farms, 70% of the quarters being
milked are infected mostly by staph, we would have a cure
rate of around 13%.
     As a cow gets older, the somatic cells increase. Let's
look at what some fellows found and relate it to infection
and the percentage of quarters infected. Figure 6 shows the
data obtained by two researchers from Pennsylvania who
looked at 3,130 cows from two years old to over seven years
old. All two-year-olds had an average somatic cell count of
232,000. All the three-year-olds had counts of 314,000.
Under general management conditions, every year a cows somatic cell count goes up 100,000. The cows that were not
infected just cruised along with Mother Nature. None of the
epithelial cells hollered for help. She wasn't infected; it
didn't matter if she was 14 years old. If she was not
infected, they found that all cows four years of age had
just about the same number of somatic cells as cows over
seven years of age. As a rule of thumb, all cows' average
somatic cell count will increase at about 100,000 per year.
As long as we don't have bugs in there tearing up everything, we don't have an increase in somatic cells.

Figure 6. **Mean somatic cell counts by cow age and infection**
status

Age (Year)	All cows	No infection	Minor infection	Major infection
		(1000's cells/ML)		
2	232	126	190	614
3	314	149	218	661
4	390	148	233	753
5	565	180	308	977
6	544	194	322	880
7	654	251	320	986
>7	868	113	519	1207

Source: Eberhart et al., Proc. NMC Annual Meeting, p 32,
     1979. Data from 3,130 cows.

     Let's look at a minor infection like some of the streps
that 'stay out there in the milk and bounce off the wall
every once in awhile. They don't cause all that much rise,
but they cause quite a bit more irritation. From two-year-
olds on, where you see minor infections, you will see
gradual increases in somatic cells. Let's now put in a
major infection, some staph or some strep ag--one that only
bounces off the wall but goes through it sometimes and tears
up a couple of cells on its way. They holler for help and
say, "We are being irritated, send us some help." Help is
sent and in high numbers. You can see from figure 6 that
there are 1 1/2 to 2 times as many somatic cells from major
infections over no infection for the average cow.

Now let's look at the money in figure 7. From this we can develop another kind of feel for the somatic cell relationship. Look at the relationship between somatic cell count, production loss, percent production loss in pounds, and dollar loss at $12/cwt. Let's take any cow that produces 50 lb of milk a day. This data comes from the 3,130 cows of figure 6. Cull out all the cows that were producing somatic cells at less than 300,000. They were uninfected cows so there was relatively no irritation up in the udders. They get a zero--no failure to produce because of mastitis. According to the data all those cows over 300,000 but not over 500,000 produced 5% less than the cows under 300,000--$0.36 worth of milk each day that was not milked. A count of 500,000 to 1,000,000 shows $0.60 per cow per day with a count of 1,500,000, the failure to produce is 25%, or $1.50 worth that this cow is not producing. If we grade her somatic cell count, she's got enough irritation to produce 1-3,000,000 somatic cells per c.c. of milk. Cows over 3,000,000 are not sick, not swelled up, not hurting, but have enough subclinical irritation up there to decrease production by 40%. It takes 10 low-count cows to dilute those counts out to keep you out of trouble. It takes 10 of these to offset the 3,000,000 count cows, so that the dairyman should develop patterns to help him identify the cow that is not hurting, not hot, and not swelled. She is failing to make you about $2.50 a day.

Figure 8 shows 13,500 cows that produced less than 200,000 somatic cells, and a lactation average of 15,170 lb. The researchers went to look for more cows and compared their production against these with low somatic cell counts. You can see that if the cell count is 200-400,000, the cow drops 900 lb of milk per year. If she goes 400-600,000, she produces 311 lb less than the base cows. Six to 800,000 cell count and almost 1,000 lb less milk is produced. When you get around regulatory limits, research shows us that subclinical mastitis in these high count areas costs us 1000, 1070, to 1200 lb of milk per cow per year. The 803 cows with counts between 1,500,000 and 5,000,000 produced 1,600 lb less per cow than did those cows that maintained less than 200,000.

Figure 7. Relationship between somatic cell count, production loss and economic loss in the dairy cow producing 50 pounds daily

Cell numbers (cows)	Production loss %	lb	Economic loss ($12.00/cwt)
< 300,000	0	0	0.00
3 - 500,000	5	-3	0.36
5 - 1,000,000	10	-5	0.60
1 - 3,000,000	25	-13	1.56
> 300,000	40	-20	2.40

When the cell count goes over 5 million, you lose 3,000 lb of milk per cow per lactation.
Figure 9 shows the losses from reduced milk production starting at 100,000 cell count.

Figure 8. Effect of milk cases of mastitis upon milk production

Cell count range	Average	milk	yield	(lb)
	Cows	Daily	Lactation	Reduction
0 - 200,000	13,587	39.7	15,170	
200,000 - 400,000	2,929	46.7	14,253	917
400,000 - 600,000	1,154	47.1	14,359	811
600,000 - 800,000	644	46.7	14,253	917
800,000 - 1,500,000	1,029	45.7	13,945	1,225
1,500,000 - 5,000,000	803	44.4	13,554	1,616
> 5,000,000	121	39.5	12,053	3,119

Source: The Virginia Dairyman, p 23, June 1979.

Figure 9. The effect of increasing somatic cell count upon milk production and economic loss from mastitis

Herd somatic cell count (1000's/ML)	Reduced milk production	
	Lb/cow	Percent
100	165	1.1
200	331	2.2
300	481	3.2
400	633	4.2
500	796	5.3
750	1,186	7.9
1,000	1,577	10.9
1,500	2,373	15.8

Source: Jones and Heald, VPI & Su, 1979; Moxley et al., J. Dairy Science 61:1637. 1978.

As we increase the somatic cell count by 100,000 in our bulk tank, we lose an additional percentage. You can tell a guy with a consistent fossomatic electronic count of 400,000 over the last six months that it is beautiful milk, but he missed getting 4%. Four percent of 10,000 lb of milk per day on some farms is place payment interest money. Figure 9 explains a dairyman's problem to him. A somatic cell count tells the dairyman just how much more milk is in those cows. If the cow maintains, for all practical purposes, a count under 500,000, then we generally don't figure a significant loss, even though we have already reviewed research numbers where they really got close and showed actual losses.

Figure 10 shows us that if you have one cow go over one million one time, one lactation, and if this is her first lactation, that she will produce 377 pounds less milk for that lactation. That's $50 worth of milk. Let's say we never quite knock out the subclinical infection--that we drive it back to where we're satisfied but subclinically it is still sitting there chewing and killing cells crying for help.

Figure 10. Pounds of milk lost per lactation due to level of SCCs

| | Lactation Number | | | |
	1	2	3	4
SCC less than 500,000	0	0	0	0
One test over 1 mil. in early lactation	377	393	421	349
Two tests over 1 mil. in months 2 and 4	881	918	983	815
Four tests over 1 mil. during lactation	1,132	1,130	1,264	1,048

If this cow bounces over a 1 million cell count 2 times, she will give less milk. It doesn't matter how old she is. If we don't knock out those subclinical infections, when she goes over a million somatic cells, we lose $100 worth of milk. If we let her bounce over four times during a lactation (and for a lot of them we can't help it,) the cure rate is 13% (figure 5) with a CMT of 3. Over 1 million, four times during a lactation, means $150 worth of milk. Our goals are to keep the level of irritation down and to use that somatic cell count as a grade card, but during lactation we have got a very poor chance because of the little milk clots in the ductwork up there on the way to the alveoli and we can't get our medicine through there. Even with sticking her in the butt and coming around from the back door, we have a poor chance of clinical cure during lactation. So we have to prevent new infections on a daily basis and knock out existing infections during the dry period when those glands are not working by laying in a big dose of long-acting antibiotic. Dry cow treatment gets most of the infection for the next lactation during those first two weeks when we shut them off and turn them dry.

Now if we look at the teat dipping issue, there is a researcher in England who followed these cows out to pasture. In herds that did not teat dip, she found staph organisms in the streak canal in 70% of the cow's teats. Then she went to a farm where they dipped the cow's teats

and found of those that were dipped with a safe and effective commercially available teat dip that 1% of these teat
ends were colonized with staphylococcus mastitis-causing
organisms. That research alone shows us that when we turn
those cows out dipped with safe and effective teat dip that
we will slow down, if not reduce, our new infection 50% in
the first year.

The teat canal is 10 to 15 ml long and contains ubiqui-
ton. A recent Canadian study showed that if we put a mastitis-causing organism 3 ml into the streak canal, regardless
of its name, we will infect 8% of the quarters. Remember
that when we get through milking there is a drop of milk
that forms there in the streak canal. If there is bacteria
swimming around in that drop, we are going to implant them
there as that drop of milk vaporizes, or drops off, or
absorbs, or is squeezed, or whatever. Let's go one more
millimeter up into this streak canal with a mastitis-causing
organism. With bacteria 4 millimeters up into the streak
canal, we will infect 31% of the quarters. If we go one
more millimeter, putting bacteria about one-third to one-
half the way up there, 91% of them will become infected. We
have got to dip those teats to form a barrier of sanitizer
solution to keep the bacteria from working up in there!

Figure 11 tells us why it is important that every quarter of every cow be dry-cow treated and every teat be dipped
between milkings. This is a study done in England. Mastitis is no different there than in southwestern Missouri or
northwestern Arkansas when it comes to going up into the
udders and trying to find the bugs and irritation. The
researchers found that of all the farms that did not dip
teats or that didn't dry treat, an average of 25% of the
quarters of cows turned dry were found to be infected with
mastitis-causing organisms. When those cows freshened, 30%
of them picked up new infections during the dry period.
During the next lactation, because they were flushing the
system twice a day, they flushed the bugs out, drained the
system, and kept ahead of new infection rates, actually
decreased to 25%.

During the next dry period when they were turned dry,
they bounced right back to 30%. The researchers went from
one farm to the other in a massive educational program tel-
ling dairymen the attributes of safe and effective teat dip
and dry cow treatment for all quarters, all cows. Afterwards they found that the percentage of quarters infected
dropped 50% in the first year and 50% more the second year.
They found that when they turned their cows dry after 3
years that every quarter, every cow dry treated, every teat
dipped with a safe and effective product every day when the
cows went dry, that only about 12% of the quarters were
infected. Because we laid a big dose of longacting stuff in
there, we not only killed that infection in some of the 12%
quarters infected, but we had a little insurance policy laying there so that before that little plug got in that streak
canal it killed those guys that crawled in there off the dry

pasture. When the cows freshened, 6% got infected quarters. During lactation, even with Mother Nature helping to
flush the system twice a day and dipping teats to prevent
infection on a daily basis, at the end of that lactation,
12% of the quarters were infected. The cows must be dry
treated every lactation, every quarter, to drop them back
down to 6%.

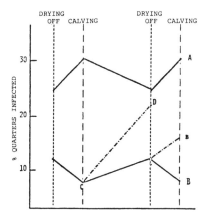

(A) HERDS NOT USING EITHER TEAT DIP OR DRYING-OFF THERAPY

(B) HERDS USING BOTH TECHNIQUES

EXPECTED TRENDS IN THE INFECTION LEVELS IN HERDS
THAT HAVE BEEN USING THE CONTROL SYSTEM BUT CEASE
TO TEAT DIP (CD) OR FAIL TO TREAT COWS AT
DRYING-OFF (AB).

Figure 11. Mastitis control

The researchers also kept track of all the people who
said that they were tired of paying $6.50 for teat dip
because they hadn't had a case of mastitis in years so had
quit teat dipping. They also kept track of the guys who
said they weren't going to pay a veterinarian $20 a cow or
$5 a cow, and they kept track of the guys who quit dry cow
treatment. In one year, one lactation, the percentage of
quarters infected in herds not teat-dipped went from their
low point they had worked towards for 3 years up to a higher

percentage. If the same dairyman said, "Well, I am not going to dry treat either," they got an increase in percentage of quarters infected. You add the percentages of the greater number of quarters infected, and the dairymen were back to where they were before the extension effort. The difference in the low percentage of quarters infected and the average percentage of quarters infected was 1,071 lb of milk!

41
THE EFFECT OF MACHINE MILKING ON MASTITIS IN DAIRY COWS

Jerry O'Shea

Research has not provided convincing evidence of a general relationship between machine milking factors and level of herd infection (Cousins, 1973), yet most researchers believe that machine milking methods are related to new infection rate; a similar opinion is held by dairy farmers (Cowhig, 1962).

NEW INFECTION AND INFECTION LEVEL

It is important to differentiate between new-infection rate and the level of infection in a herd. The level of infection in a herd at any time is determined by the dura- tion of infection and the rate of new infection. Since the duration of most infections is long, the rate of new infec- tion is a major determinant of infection levels.

PRE-1965 RESEARCH

The main machine milking factors investigated in rela- tion to mastitis were vacuum level, pulsation rate, pulsator ratio, liner bore, and overmilking. The results were conflicting and confusing and firm conclusions cannot be drawn from them. It is likely that the published experi- ments did not measure the important factors affecting new infection and that factors--other than those deliberately changed--also were changed. Even in present-day machines, changing one machine part or setting may affect several other settings. In the earlier experiments, it was often unclear if measurements were being made of new infection or of infection levels; therefore the definition of infection varied widely.

FLUCTUATING VACUUM

An Irish farm survey showed a significant negative regression of bulk milk cell count on the vacuum reserve of the milking plant (Nyhan and Cowhig, 1967). When vacuum reserve reached approximately 40 l/min/unit, the relationship between bulk milk cell counts and reserve disappeared (Nyhan and Cowhig, 1967; Le Du, 1980; Ollivier, 1974). Five experiments at Moorepark showed that unstable vacuum (a combination of large, irregular, and moderate cyclic fluctuations) caused higher new-infection rates than did stable vacuum (Nyhan, 1968; O'Shea and Walshe, 1970). Summary results are presented in table 1.

TABLE 1. SUMMARY RESULTS OF FLUCTUATING VACUUM EXPERIMENTS

Experiment Number	1	2	3	4	5
		No of new infections			
Stable vacuum	2	21	10	11	10
Unstable vacuum	13	41	21	24	23

Source: Nyhan (1968); O'Shea and Walshe (1970).

These differences in infection rates arose due to factors other than stimulation, role, i.e., completeness of milking, teat orifice eversion, teat canal hyperkeratization, and exposure time to the milking machine (O'Shea and O'Callaghan, 1980).

It may be concluded from these experiments that is is necessary to have a reasonable level of effective reserve, but it is not necessary to have effective reserves in excess of 40 l/min/milking unit. These reserve levels are largely in line with recently recommended vacuum pump sizes in the U.S. (Doane, 1980).

VACUUM STABILITY EXPERIMENTS

NIRD

A series of experiments of the NIRD, Reading, showed that large irregular fluctuations (F) or large cyclic fluctuations (C) per se did not increase new infection rates. However, a combination of F with substantial cyclic fluctuations caused a large increase in new infections. New infections were most likely to be initiated towards the end of milking and the "adverse" vacuum conditions only increased new infections when teats were connected to a common claw (as in most commercial machines). It may be concluded from these experiments that cyclic vacuum fluctuations should not be excessively large. Thus, it is important to have a claw air admission of 6 to 10 l/min; short milk tube bores should not be less than 8 mm, and the height of milk lift should not exceed 1.8 meters.

Moorepark

Five experiments were carried out (O'Shea and O'Callaghan, 1976 and 1980; O'Shea et al., 1976); the main treatment and results of 3 of these experiments are given in table 2. There were two main treatments in the experiments, i.e., FC and fc where F was irregular vacuum fluctuations, f was no irregular vacuum fluctuations, C was large cyclic fluctuations and c was small cyclic fluctuations.

TABLE 2. MAIN TREATMENTS AND RESULTS FOR THREE HALF-UDDER VACUUM STABILITY EXPERIMENTS

Experiment Number	1	2	3
Stable vacuum (fc)			
Liners supported	–	19	62
Liners not supported	9	–	–
Unstable vacuum (FC)			
Liners supported	–	–	48
Liners not supported	6	38	–

Source: O'Shea and O'Callaghan, 1976 and 1980; O'Shea et al., 1976.

In Experiment 1, the numbers of new infections were low with both FC and fc and did not differ significantly. Liner slip (air leakage between the teat and liner) was recorded in the fc half-clusters. In Experiment 2, the fc half-cluster was mechanically supported. There were more infections with FC than with fc. Experiment 3 was a repeat of Experiment 2, but the FC and fc half clusters were mechanically supported; there was not a significant treatment difference in new infections. Thus the difference in new infection rates appeared to arise from the presence or absence of liner slip, rather than from differences in vacuum stability per se.

LINER SLIP AND NEW INFECTIONS

Eight experiments at Moorepark to determine the importance of liner slip in new-infection rate showed that vacuum stability per se had little effect on new infection rate (O'Shea et al., 1976) and that the effect of generalized vacuum instability is mediated via liner slip. Liner slip is largely a function of liner design and also of vacuum stability. Results of five of these experiments are given in table 3.

TABLE 3.   TREATMENTS   AND   RESULTS   OF   FIVE   "LINER   SLIP"
           EXPERIMENTS

Experiment 8:   Half Udder Experiment With FC Conditions at
                All Quarters

Quarters	RF	LH	LF	RH
Supported liners	+	+	-	-
% quarters with new infections	60%	57%	7%	23%

Experiment Numbers	7	9	10	11
	Number (and %) quarters with new infections			
Quarters affected by slips*	46(77%)	17(43%)	23(84%)	13(41%)
Quarters not affected by slips	6(30%)	3(8%)	7(21%)	5(16%)

Source: O'Shea et al., 1976.
* At other quarters

    When slips occurred in one quarter, impacts of milk droplets were detected at other teat-ends connected via the claw to the teat in which the slip occurred and high new infection rates occurred in the "impacted" quarters. The results also showed that liner types differed in slip frequency. The results showed that stable liner types must be used and, even with high vacuum reserves, new infection rates are high with unstable liners. To minimize liner slip, it is necessary to have a reasonable vacuum reserve, adequate-sized pipelines, and reasonably low cyclic fluctuations to reduce the incidence of liner slip, even with liner design that generally slips infrequently.
    The results (table 3) suggested that liner slip was the main milking machine factor affecting new infection rates. The liner slip theory is compatible with many other results. Liner slips are most frequent at fore teats, and infections are most frequent in rear teats. Slip frequency is greatest at morning milkings and at high milk flowrates; this may partly explain the greater frequency of infection in cows with high milk flowrates. Slip frequency is increased with unstable vacuum, and especially 50:30 vacuum fluctuations as used by Cousins (1973). It may be postulated that the high infection rates measured by Cousins arose due to liner slip; this is particularly indicated by the low new infection rate associated with FC in quarter milking.

AIRBLAST EXPERIMENTS

Liner slip is a random event. To obtain further data on the effect of liner slip under different circumstances, slip was simulated in a series of experiments: airblasts were admitted every fifth pulsation cycle via a solenoid valve into the short milk tubes (O'Shea and O'Callaghan, 1980; O'Shea et al., 1981). The main treatments and results of 2 experiments are given in, table 4. In these experiments, airblasts occurred just after the liner was fully opened in the pulsation cycle.

TABLE 4. NEW UDDER INFECTIONS WITH STAPH. AUREUS AS A CONSEQUENCE OF AIRBLASTS DURING MILKING

Experiment Number	15				17			
Cow group	1		2		1		2	
Side of udder ·	RS	LS	RS	LS	RS	LS	RS	LS
Airblast (1/min)	150	30	150	0	150	150	150	150
Period of airblast[a]	0-100	0-100	0-100	-	0-100	0-20	0-100	80-100
New infections	9	6	11	0	9	6	8	6

Source: O'Shea and Callaghan (1980); O'Shea et al. (1981).
[a] 0 - 100% = Airblasts for total milking period.
0 - 20% = Airblasts for first 20% of milking.
80 - 100% = Airblasts for last 20% of milking.

In Experiment 15, there were almost an equal number of new infections with airflows of 30 1/min (almost inaudible) and 150 1/min (loud squeal). In Experiment 16, airblasts were at the start of milking, only at the end of milking, or during total milking period and caused almost equal numbers of new infections (table 4.) Thus inaudible liner slips and slips at any time during milking may cause new infections. Results of three further experiments are given in table 5. Challenge organisms, S. aureus (which colonize the teat canal), were used in two experiments and E. coli and Strep. dysgalactiae (which do not colonize the teat canal) were used in the third experiment.

In Experiment 18, five airblasts during overmilking greatly increased new-infection rates. Rough cluster removal also increased new-infection rate (Theil, 1974). Thus, it is important to remove clusters without airblasting the teat ends. With manual cluster removal, the vacuum to the cluster must first be turned off; then the vacuum is allowed to drop before removing the cluster. Automatic

cluster removers (ACRs) must be set to allow claw vacuum to
drop before the ram withdraws the cluster; ACRs, using an
airblast to close an automatic shut-off in the claw, should
not be used. Machine stripping should be avoided, this may
cause liner slips. In Experiment 19, airblasts into closed
liners did not cause new infections.

A simple system could be devised to close the liners at
the end of milkflow; the cluster pressure could increase to
atmospheric pressure before the cluster was removed.

In Experiment 20 (table 5), airblasts increased new
infections when teats were heavily contaminated with
environmental bacteria (E. coli and Strep. dysgalactiae)
although these organisms do not colonize the teat canal.
Liner slips increase new infection rates with any type of
mastitis pathogen contaminating the teat ends.

TABLE 5. EFFECT ON NEW-INFECTION RATES OF AIRBLASTS ON THE
TEAT-END DURING OVERMILKING (EXPERIMENT 18), WHEN
THE LINER IS COLLAPSED (EXPERIMENT 19) AND WITH A
CHALLENGE OF E. COLI AND STREP. DYSGALACTIAE
(EXPERIMENT 20)

Experiment Number	18		19		20	
Challenge organism:						
S. aureus	x		x		–	
E. coli and						
Strep. dysgalactiae	–		–		x	
Side of udder	RS	LS	RS	LS	RS	LS
Airblast location:						
Liner opened	x	–	–	–	x	–
Liner closed	–	–	x	–	–	–
Time of airblasts:						
Total milking period	–	–	x	–	x	–
Five airblasts						
during overmilking	x	–	–	–	–	–
Number of new infections	12	2	0	0	12	1

EFFECTIVE PULSATION

"Effective" pulsation reduces edema at the teat end
(Williams and Mein, 1980). The shear forces created by milk
flow through the teat canal can extrude the teat canal keratin; the vertical upward component of the force exerted by
the collapsing liner may push the keratin back into the teat
canal (Williams and Mein, 1980). "Effective" pulsation has
not been well-defined, but successful commercial liners

exert a compressive load ≥ 10 kPa on the teat when collapsed, and the duration of liner collapse also is important. Results of a trial with long and short liners (Mein
et al., 1982) are given in table 6. The lack of compressive
load on the teat-end (ineffective pulsation) increased
tissue damage and new-infection rates.

TABLE 6. EFFECT OF LINER LENGTH ON NEW-INFECTION RATES

| | | Percentage of Teats Preventing Full Liner Collapse | | | |
Effective Liner Length (mm)	Quarter	1 Minute From Start	3 Minutes From Start	End of Milking	New Infection
148	RF	0	10	14	1
148	LH	0	0	2	2
125	LF	100	100	100	10
110	RH	100	100	100	5

Source: Mein et al. (1982).

These and other results and observations show that the
d-value of the pulsation chamber waveform (British Standards
Institute) must be ≥ 15% of the pulsation cycle. Liners
(from the mouthpiece lip to the lowest collapsing part of
the barrel) should be long enough to allow liner collapse
beneath the teat end; the teat elongates by 40-50% under
vacuum (Mein, 1972). The actual length required depends on
teat lengths, and liner diameter determines the depth of
teat penetration. Pulsators with small extraction capacities may give unsatisfactory graphs with long liners; it
is always necessary to check the pulsation graphs.

OVERMILKING

Overmilking does not increase new infections (Natzke,
1978): front teats produce only 40% of the milk and are
consistently overmilked; yet there are more infections in
the hind teats (Rabold and Pichler, 1980). Moderate overmilking (40-60 sec) leads to lower herd infection levels as
indicated by cell count surveys (Brandsma and Maatje, 1980),
probably because the adverse effect of rough cluster removal
is less when the cluster has been drained of milk.

NONSTANDARD MILKING EQUIPMENT

- Deflector shields fitted in liners were designed to prevent impacts on the teat end; they give variable reductions in impacts (O'Callaghan and O'Shea, 1979; Mein, 1982). In a field trial in 21 herds with 2307 cows, shields reduced new infections by 10.5% (Griffin et al., 1980). Their general usefulness is questionable since a large proportion of the effect of shields was in herds with improperly installed milking machines.
- Valved claws are being investigated in the U.K. as a means of preventing new infections due to impacts (Griffin et al., 1980). Short-term bacterial challenge experiments at Moorepark suggest that the claw will cause a moderate reduction in new infections (O'Shea et al., 1982). More data are needed before definite recommendations can be made.
- The PPKME cluster (Hamman and Tolle, 1980) gives an unidirectional continuous milkflow. Short-term bacterial challenge experiments at Moorepark suggest that this cluster will cause similar new-infection rates as conventional claws (O'Shea et al., 1982). More data are needed before definite recommendations can be made.
- The bio-melke system appears to have no data yet on its effect on new-infection rate.
- Pressure equalization in the liner and pulsation chamber is a new system on the market in the past few years. There are no data available to show its effect on new-infection rate.
- The AHI claw has a vortex system designed to prevent impacts of milk between quarters and largely prevents contamination of the liners when the cluster is inverted (Phillips, 1982). While laboratory tests show theoretical advantages, there are no data available to show its effect on new-infection rate.
- The Duovac system has no data to show its effect on new-infection rate. Theoretically, it should not have an effect on new-infection rate. Trials over an 11-week period showed that there was no differences in cell count in two groups of cows--one milked with Duovac and the other milked with a standard Alfa Laval machine (O'Callaghan and O'Shea, 1982).

REFERENCES

Brandsma, S. and K. Maatje. 1980. Proc. Int. Wkp. Mach. Milk. and Mastitis. Moorepark. p 128.

Cousins, C. L. 1973. Ph.D. thesis, University of Reading, 157 pp.

Cowhig, M. J. 1962. Dairy husbandry survey (milking machines). An Foras Taluntais, Dublin. 25 pp.

Doane, M. K. 1980. M. S. thesis, Cornell University, Ithaca, N.Y.

Griffin, T. K., G. A. Mein, D. R. Westgarth, F. K. Neave, W. H. Thompson, and P. D. Maguire. 1980. J. Dairy Res. 47:1.

Hamman, J. and A. Tolle. 1980. Proc. Int. Wkp. Mach. Milk. and Mastitis. Moorepark. p 30.

ISO/DP 5707. Milking Machine Installations - Construction and Performance. In draft. British Standards Institute.

Le Du, J. 1980. Procs. Intl. Wkp. Mach. Milk. and Mastitis. Moorepark. p 82.

Mein, G. A. 1972. Mechanics of the teat and teatcup during milking. Ph.D. thesis, University of Reading, England. 107 pp.

Mein, G. A. 1982. Procs. Seminar - Milk production from pasture. Ruakura Research Station. p 79.

Mein, G. A., D. M. Williams, and M. R. Brown. 1982. Procs. Seminar - Milk production from pasture. Ruakura Research Station. p 83.

Natzke, R. P. 1978. Proc. Int. Symp. Mach. Milk. 17th Ann. Mtg. Nat. Mastitis Council, Inc. Louisville, KY. p 262.

Nyhan, J. F. 1968. Procs. Conf. Mach. Milk. Reading 1968. p 71.

Nyhan, J. F. and M. J. Cowhig. 1967. Vet. Rec., 81:122.

O'Callaghan, E. and J. O'Shea. 1979. Ir. J. Agric. Res. 18:237.

O'Callaghan, E. and J. O'Shea. 1982. Procs. Seminar - Milk production from pasture. Ruakura Research Station. p 116.

436

Ollivier, B.   1974.   Unpublished thesis.   ITEB, Paris.
    p 277-281.

O'Shea, J. and E. O'Callaghan.   1978.   Proc. Int. Symp. on
    Mach. Milk.   17th Ann. Mtg. Nat. Mastitis Council,
    Louisville, KY.   p 262.

O'Shea, J. and E. O'Callaghan.   1980.   Procs. Intl. Wkp.
    Mach. Milk. and Mastitis.   Moorepark.   p 6.

O'Shea, J., E. O'Callaghan, and W. J. Meaney.   1981.   Ir.
    J. Agric. Res.   20:163.

O'Shea, J., E. O'Callaghan, and W. J. Meaney.   1982.
    Unpublished data.

O'Shea, J., E. O'Callaghan, W. J. Meaney, and C. Crowley.
    1976.   Ir. J. Agric. Res.   15:377.

O'Shea, J. and M. J. Walshe.   1970.   Ir. J. Agric. Res.
    9:279.

Phillips, D. S. M.   1982.   Procs. Seminar - Milk production
    from pasture.   Ruakura Research Station.   p 81.

Rabold, K. and O. Pichler.   1980.   Proc. Int. Wkp. Mach.
    Milk. and Mastitis.   Moorepark.   p 121.

Thiel, C. C.   1974.   Bienn, Rep. Natn. Inst. Res. Dairying.
    p 35.

Williams, D. M. and G. A. Mein.   1980.   Proc. Intl. Wkp.
    Mach. Milk. and Mastitis.   Moorepark.   p 73.

42

MASTITIS CONTROL IN DAIRY COWS

Jerry O'Shea

Mastitis is inflammation of the udder due either to infection or physiological change; most mastitis is associated
with microbial infection. Normally the teat canal prevents
the entry of pathogens into the teat sinus, but occasionally
entry occurs and is followed by intramammary infection
(figure 1).

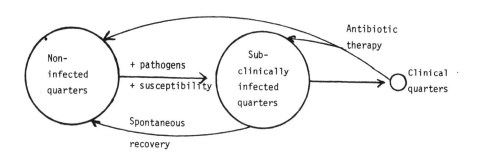

**Figure 1. The pattern of intramammary infection (Dodd et
al., 1964)**

The course of infection varies; often the early stage of infection is subclinical, i.e., the changes in milk composition and inflammation are so slight that they are not detected by the dairy farmer. In a large cow population,
about 1% of cows shows clinical symptoms of mastitis at any
one time, although about 50% will have one or more subclinical infected quarters.

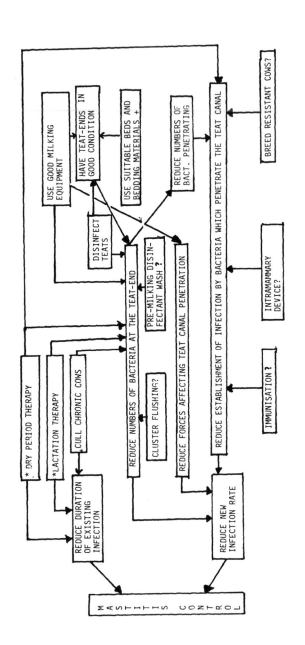

Figure 2. The main pathways of mastitis control

* ALWAYS RECOMMENDED   +   NOT RELEVANT IN SOME CLIMATES

LOSSES DUE TO MASTITIS

The losses in milk yield and lowered milk compositional quality of mastitic quarters has been extensively reviewed recently (Hoare, 1982). The average losses are as follows: quarters with subclinical mastitis produce 15% less milk than uninfected quarters and the milk yield of clinical quarters is depressed by up to 40%. Milk from mastitic quarters has a lower protein and fat percentage, a lowered nutritive value, and is not good for the manufacture of condensed products and cheese.

There are also further losses, i.e., quarters are lost; cows are culled for mastitis; antibiotics used for treatment of clinical mastitis are expensive as is discarding the milk from treated cows; labor costs for antibiotic treatment and discarding the milk from treated cows.

STRATEGY OF CONTROL

The average incidence of mastitis is a function of the percentage of cows infected and the duration of the infection as shown by the equation:

A	B	C
Average incidence	% cows infected at some period of time	Average duration cows are infected (%)

$$\text{Average incidence} = \text{\% cows infected at some period of time} \times \text{Average duration cows are infected (\%)}$$

To reduce the average incidence B and/or C must be reduced. Term B is reduced by reducing the number of new infections and by eliminating existing infection either by antibiotic treatment or culling. Term C is reduced by quickly eliminating existing infection. The strategy of control is shown diagrammatically in figure 2.

CONTROL

Any control system will be concerned with the rate of new infection and the elimination of existing infection. About 80% of infections are due to Staphylococcus aureus, Streptococcus dysgalactiae and Streptococcus uberis. While the primary reservoirs of staphylococcal and streptococcal pathogens are infected quarters, these bacteria can also be recovered from the teat skin, udders, coats, etc. (Neave et al., 1969). Staphylococcus agalactiae can easily be eradicated since it responds well to penicillin therapy and, in the absence of udder infection, it rarely persists in the environment for more than a few weeks.

There are five main factors involved in reducing the numbers of infections due to Staphylococcus aureus and Streptococcus dysgalactiae:
  - Correct milling machine function
  - Teat disinfection
  - Dry-cow therapy

440

- Attention to bedding of cows
- Lactation therapy

It should be noted (figure 2) that reducing the number of established infections also reduces the rate of new infection, presumably by reducing the bacterial challenge. Short term gains in mastitis control are largely achieved by reducing the level of existing infections by dry period therapy and to a lesser extent by lactation therapy and culling of chronically infected cows. In long-term control programs greater emphasis must be placed on reduction of the rate of new infection by using properly functioning milking equipment, postmilking teat disinfectant and, where applicable, using suitable cowbeds and materials. The reasons for this will become apparent in the paper.

DURATION OF INFECTION

Since farmers frequently take cognizance only of clinical infection, it is worth emphasizing that subclinical infections persist on average for about 70% of the lactation (Neave et al., 1969). Date in table 1 show that with all infection types, 55% of cows do not change state during a 12-month period. About 70% of cows infected with Staphylococcus aureus at the start of lactation are still infected at the end of lactation (Neave et al., 1969).

TABLE 1. CHANGES IN THE INFECTION STATUS OF 478 COWS IN 14 HERDS DURING THE 12 MONTHS

Infection status	No. of cows	% cows	Changes of state[a]
Infected for full year	185	39%	0
Not infected at any time	76	16%	0
Infected for part of the year	79	16%	
	68	14%	2
	70	15%	2

Source: (Neave et al., 1969)
[a]A single change from infected to noninfected or vice versa.

ROUTINES TO REDUCE DURATION OF INFECTION

The main factors to be considered under this heading are:
- Dry-period therapy.
- Lactation therapy.
- Culling of chronically infected cows.

Dry Period Therapy

Antibiotics. The dry period is the preferred time to treat subclinical infections. Such therapy is acompanied by a higher efficacy, and there is no risk of antibiotic contamination of salable milk. Dry-period therapy also permits regeneration of damaged tissue, prevents most dry-period new infections and significantly reduces the number of clinical cases of mastitis occurring at the beginning of the next lactation (Schalm and Ormsbee 1949; Neave et al., 1950; Neave et al., 1966; Smith et al., 1967; Meany, 1976).

With dry-period therapy, there is always the possibility that, because infections caused by certain strains of pathogens (e.g., some staphylococci) are difficult to eliminate, either because of antibiotic resistance or for other reasons not specific to the particular antibiotic used, the effectiveness of therapy will decline from the time a control program is started. Dry-period antibiotics give protection against new infections for a period of about 30 days. When dry periods are longer further infusions of antibiotics are theoretically desirable but in practice not worthwhile in most situations (Egan, 1982; Pankey, 1982).

TABLE 2. SUMMARY OF EXPERIMENTS WITH TEAT SEAL

Control — Osmonds D.C.
Treatment — Quick release antibiotic plus Osmonds D.C. plus teat seal
Cow no. — 513 divided between treatment and control

| | All cows | | New infections during dry period | |
	Quarters infected at drying off	No. of infections eliminated	All quarters in 347 cows that were infection-free at drying off	All quarters that were infection-free at drying off
Control	103 (10.3%)	48 (46.6%)	32 (4.6%)	52 (5.8%)
Treatment	113 (11.3%)	66 (58%)	17 (2.4%)	29 (3.8%)

Source: Meaney (1977)

There are large differences in the response (cure) for different dry-cow preparations and the response to all dry-cow preparations decreases as the period of sampling post-calving is extended (Meany, 1976; Meany and Nash, 1977; Pankey, 1982).

Teat seals. Teat seals were investigated as a means of preventing new infections during the dry period (Meany, 1977 and 1981). Summary results are given in table 2. The efficacy of the seal in preventing new infection was probably overestimated because of the low efficiency of the dry-cow therapy used. Good teat seals could be selectively used in quarters with a low CMT at drying off; quarters with high CMT values could be treated with dry-cow antibiotic preparations. The use of seals needs further investigation.

Lactation Therapy

Clinical mastitis must be identified and treated promptly. However the value of lactation therapy varies considerably with the product being used. The cure rates for different preparations varied from 18% to 53% (Meany, 1981). In practice lower cure rates may be expected since many of the infections may be of longer duration than those treated in the reported experiment and E. coli may be treated with inappropriate antibiotics.

Culling of Cows

Culling of cows that have more than 4 (sometimes 8) clinical cases per year is frequently recommended. However, there does not appear to be any experimental data showing the effect of such culling on infection levels or on new infection rates. It is possibly not worthwhle culling for mastitis alone, but many of these cows will be culled anyway for low yield. Culling has the greatest beneficial effect in herds with very low levels of mastitis.

REDUCTION OF NEW INFECTION RATES

In reducing infection rates, the main factors to be considered are:
- Properly functioning milking equipment
- Postmilking teat disinfection
- Bedding of cows
- Cluster flushing
- The IMD (intramammary device)
- Immunization of cows
- Breeding of cows resistant to infection.

Milking Equipment

The effect of machine milking on new infection rates has been dealt with in another paper (O'Shea, 1983) and will not be discussed here.

Postmilking Teat Disinfection

The practice of dipping teats in disinfectant after
milking to prevent intramammary infections has been investigated extensively and widely recommended (Kingwill et al.,
1970; Natzke and Bray, 1973; Shultze and Smith, 1972; Wesen
and Schultz, 1970; O'Shea et al., 1973). As a result, a
variety of postmilking teat dips have appeared on the
market. The lower new-infection levels associated with teat
dipping are achieved directly by reducing the overall
bacterial contamination at the tip of the teat (Kingwill et
al., 1970), and indirectly by improving the condition of the
teats.

Several products have been evaluated at least in
"simulated in-vivo" experiments (Philpot and Pankey, 1975).
However, 'simulated in-vivo' trials are not good indicators
of efficiency in new infection prevention (Bushnell et al.,
1978); products that perform badly in simulated in-vivo
trials are also poor in practice, but the converse is not
always true. Some inactivation of the bactericides may take
place due to milk contamination of the teat dips; in this
respect is is noteworthy that teat spraying is as effective
in preventing new infection as teat dipping (Meany, 1974).

Until recently Iodophor, PVP iodine, Iodine and Chlor-
hexidine (all at 0.5 to 1% concentration), and sodium hypo-
chlorite (with 4% available chlorine) have been the most
successful teat dips. Chlorine dioxide (0.04 to 0.4%) and
sodium dichloro-S-triazenetrione (0.3 to 1%) have performed
poorly. Water-based iodophor teat dips are generally four
to five times more effective than oil-based iodophors (Philpot and Pankey, 1975).

TABLE 3. COMPARISON OF 0.6% AVAILABLE IODINE IODOPHOR AND
0.3% GLUTARALDEHYDE

Treatment	No. of new infections		Causative organisms		
	Sub-clinical	Clinical	S. aureus	Streps	Other
Iodophor	19	12	26	2	3
Glutaraldehyde	11	8	12	5	2

Source: Meaney (1981).

Recent results have shown that activated gultaraldehyde
(0.3 w/v) is more effective than iodophor (0.6% iodine)
particularly for Staphylococcus aureus (table 3).

Bedding of Cows

The size of cow beds affects teat condition and presumably in this way affects infection (Schmidt-Madsen and Klastrup, 1980). Beds should be the correct length to prevent dunging on the beds and prevent teat treading. The type of bedding material used affects dry-period new infections and the incidence of new infection between calving (spring-calving cows) and going to grass (Meany, 1980). Summary results are given in table 4 and most new infections were due to E. coli and Streptococcus uberis. A thin layer of ground limestone on beds is recommended as a result of these experiments.

TABLE 4.   EFFECT OF TYPE OF BEDDING ON NEW INTRAMAMMARY IN-
FECTION IN THE DRY PERIOD AND BETWEEN CALVING AND
GOING TO GRASS

Bedding	No. of animals	% of quarters infected
Experiment 1		
Clay	156	13.0%
Concrete	170	24.0%
Experiment 2		
Concrete plus ground limestone	32	1.6%
Concrete plus mats + ground limestone	32	1.6%
Concrete	66	4.5%

Source: Meaney (1980).

Cluster Flushing

Cluster flushing between cows with 25 ppm iodine has reduced clinical infections considerably but its effect on the rate of new infection does not appear to have been quantified (Bushnel et al., 1978; Ruffo and Sangiorgi, 1980). Cluster pasteurization (which is more effective than cluster flushing) in addition to other hygienic measures did not significantly reduce new infection rates (Neave et al., 1969). The use of a disinfectant in the udder washing water did not reduce the rate of new infection due to common pathogens (Shedrake and Hoare, 1980) although it may play a role in controlling mycoplasma infection (Bushnell et al., 1978).

Immunization

It is unlikely that immunization will play a role in mastitis prevention in the immediate future (Anderson, 1978). Immunization generally involves raising the somatic cell count of the cow. Since a raised somatic cell count is defined as mastitis (whether or not pathogens are isolated), it is evident that cytological responeses cannot be used (whether or not enhanced by immunization) to prevent mastitis.

Intramammary Device (IMD)

The IMD is a small spiral of plastic inserted through the teat canal into the gland sinus to stimulate leucocyte production and/or increase the phagocytic activity of the polymorphs. The results obtained to date have not been consistent. The device is subject to the same cytological considerations as immunization.

Breeding Resistant Cows

Some lines of cows are more susceptible to mastitis than others, and thus it is theoretically possible to breed cows with a high resistance to mastitis. The probability of genetic progress is hampered by several factors. Resistance to mastitis is difficult to define, and a measurable parameter correlated with resistance has not been identified. The greatest genetic progress is made through sire selection, so the measurable parameter must be identifiable in bulls. Genetic progress in any trait is usually less when several traits are being selected for simultaneously. Parameters that increase mastitis resistance may be positively correlated with undesirable traits and so, even if there were an identifiable bull parameter, progress may still be slow.

REFERENCES

Anderson, J.C. 1978. Br. Vet. J. 134:412.

Bushnell, R.B., L. Brazil and D.E. Jasper. 1978. Proc. Int. Symp. Mach. Milk. 17th Ann. Mtg. National Mastitis Council Inc., Louisville, KY., p 400.

Dodd, F.H., F.K. Neave and R.G. Kingwill. 1964. J. Dairy Sci. 47:1109.

Egan, J. 1982. In press.

Griffin, T.K., A.J. Bramley and F.H. Dodd. 1980. Proc. Int. Wkp. Mach. Milk and Mastitis, Moorepark. p 19.

Hoare, R.T.J. 1982. Proc. Conf. on milk production from pasture, Ruakura, N.Z. p 85.

Kingwill, R.G., F.K. Neave, F.H. Dodd, T.K. Griffin and D.R. Westgarth. 1970. Vet. Rec. 87:94.

Meaney, W.J. 1974. Ir. J. Agric. Res. 13:157.

Meaney, W.J. 1976. Vet. Rec. 98:50.

Meaney, W.J. 1977. Ir. J. Agric. Res. 16:293.

Meaney, W.J. 1980. Procs. Moorepark Fmrs. Conf. p 42.

Meaney, W.J. 1981. An Foras Taluntais, Technical Bulletin. Agricultural Series Index 9, 1-81.

Meaney, W.J. and J.M. Nash. 1977. Vet. Rec. 100:492.

Natzke, R.P. and D.R. Bray. 1973. J. Dairy Sci. 56:148.

Neave, F.K., F.J. Dodd and E. Henriques. 1950. J. Dairy Res. 17:37.

Neave, F.K., F.H. Dodd and R.G. Kingwill. 1966. J. Dairy Res. 78:521.

Neave, F.K., F.H. Dodd, R.G. Kingwill and D.R. Westgarth. 1969. J. Dairy Sci. 521-696.

O'Shea J. 1983. Dairy Science Handbook 15:

O'Shea J., W.J. Meaney, O.H. Langley and J. Palmer. 1973. Ir. J. Agric. Res. 14:99.

Philpot, W.N. and J.W. Pankey. 1975. J. Dairy Sci. 58:205.

Philpot, W.N. and J.W. Pankey. 1975. J. Dairy Sci. 58:209.

Pankey, J.W. 1982. Personal communication.

Ruffo, G.F. and F. Sangiorgi. 1980. Proc. Int. Wkp. Mach. Milk. and Mastitis, Moorepark. p 103.

Schalm, O.W. and R.W. Ormsbee. 1949. J. Amer. Vet. Med. Ass. 115:464.

Schmidt-Madsen, P. and O. Klastrup. 1980. Procs. Int. Wkp. Mach. Milk. and Mastitis, Moorepark. p 137.

Schultze, W.D. and J.W. Smith. 1972. J. Dairy Sci. 55:426.

Sheldrake, R.F. and R.J.T. Hoare. 1980. J. Dairy Res. 47:253.

Smith, A., F.K. Neave and F.H. Dodd. 1967. J. Dairy Res. 34:47.

Wesen, D.P. and L.H. Schultz. 1970. J. Dairy Sci. 53:1391.

43

THE ESSENTIAL CRITERIA FOR THE COMPONENTS OF A MILKING MACHINE AND FOR COMPONENT TESTING

Jerry O'Shea

There are different standards for milking design and installations in various countries; in particular there are large differences in the recommended vacuum reserves and pipeline sizes. Additionally, "fashions" differ among countries; for example, individual pneumatic pulsators are used almost exclusively in some countries, while relay pulsators are used almost exclusively in other countries.

Researchers do not appear to have provided manufacturers with good guidelines for equipment design. In many cases designs have been advocated, although there were no experimental data to suggest that these designs were good. Some suggested designs were based on measurements of criteria that are not of practical importance to farmers; for example, maximum milk flowrate. In other cases, data obtained within a narrow range were extrapolated to multiples of the original range.

Dairy farmers need properly functioning, simple, milking equipment, and attempts are being made to achieve this objective in Ireland by means of a combination of four interdependent factors. Controlled testing of equipment under laboratory conditions provides guidelines in the selection of components that perform properly and are of good durability. Components are evaluated and are not recommended unless they have some measurable advantage. The standard of installations on farms is monitored. Milking machines on farms are tested for conformity with parts of the ISO Standard (British Standard Institute, 1980) and diagnosed faults are corrected. The frequency of faults on farms is highly correlated with the results of the laboratory performance tests. It is suggested that future standards should specify laboratory tests and acceptable performance criteria under these test conditions.

VACUUM PUMPS

Vacuum pumps are endurance tested over a period of at least 2,000 hours. In some earlier tests (Agricultural Institute, 1964 and 1969; Cowhig and O'Faherty, 1968; Walshe et al., 1970), the capacity of many vacuum pumps was found to be overstated by the manufacturers and some types were not durable. In more recent tests (Agricultural Institute, 1976; O'Callaghan and O'Shea, 1975; O'Shea and O'Callaghan, 1977), all vacuum pumps were found to have extraction capacities at least equal to the manufacturers rated capacities, and their durability was good.

Vacuum pumps should be capable of extracting their rated capacity for a test period of at least 6,000 hours. Recommended vacuum pump capacities provide for a plant effective reserve equal to ISO Standards; farm experience has indicated that these effective reserve standards are satisfactory when liners with low slip levels are used. Liner slips can admit up to 170 l/min without the cluster falling off the cow. With high liner slip levels, very high effective reserves are necessary. The approximate vacuum pump capacities used in Ireland and the minimum effective reserves needed are shown in table 1.

TABLE 1. APPROXIMATE VACUUM PUMP CAPACITIES (l/MIN) AND MINIMUM EFFECTIVE RESERVES (l/MIN) FOR DIFFERENT NUMBERS OF UNITS IN IRELAND

Number of units	2	4	6	8	10
Vacuum pump capacity	330	390	510	630	750
Effective reserve	150	200	250	300	350
Number of units	12	14	16	18	20
Vacuum pump capacity	840	930	1020	1110	1200
Effective reserve	370	390	410	430	450

VACUUM REGULATORS

Vacuum regulators are tested for broad conformity to an Irish Standard (O'Shea, 1973). Regulators are expected to meet four main criteria when tested new and after 2.5×10^5 cycles of operation. Regulators should have stated rated capacity and rated vacuum. The rated capacity must be attained between rated vacuum and rated vacuum plus 1.7 kPa; the operating vacuum, i.e., the vacuum at half the rated capacity must not differ by more than 1.7 kPa from the rated vacuum; and the leakage airflow at rated vacuum minus 3.4 kPa should not exceed 28 l/min or 5% of rated capacity whichever is lesser. These standards were applied in earlier tests (O'Shea, 1973; O'Shea and O'Callaghan, 1980); in more recent tests (O'Shea and O'Callaghan, 1980), the 1.7 kPa and 3.4 kPa have been changed to 2 kPa to parallel the ISO Standards (British Standard Institute, 1980). The test rig used is diagrammatically represented in figure 1.

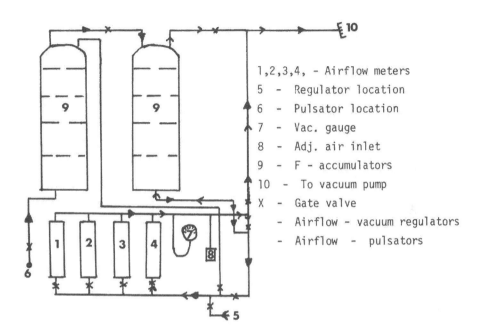

1,2,3,4, - Airflow meters

5 - Regulator location

6 - Pulsator location

7 - Vac. gauge

8 - Adj. air inlet

9 - F - accumulators

10 - To vacuum pump

X - Gate valve

— - Airflow - vacuum regulators

— - Airflow - pulsators

Figure 1. Diagrammatic representation of test rig for vacuum regulators

Regulators then are mounted on or adjacent to a vacuum pipeline in accordance with manufacturers' recommendations, and simulated wear is begun. Regulators are opened and closed 20 times per minute; the vacuum is varied from 56 to 40 kPa by admitting air through solenoid valves. Regulator inlets are reduced to 10 mm bore to reduce the vacuum capacity needed to attain a vacuum level of 56 kPa. The measurements made on the new regulators are repeated after the regulators have completed 2.5×10^5 cycles of operation.

This test method gives a good indication of the performance of regulators on farms and could be incorporated in a future ISO Standard. Results of trials to date have shown that about 30% of regulators conform to the standard, another 30% are near conformity, and the performance of the remaining 40% is well below the standard. Although the permissible level of vacuum fluctuations has not been quantified, it is not logical to introduce vacuum fluctuations by using sticking vacuum regulators or to waste vacuum reserve by using regulators with large leakage levels. Furthermore, the cost difference is negligible between the best and worst vacuum regulators on the Irish market.

PULSATORS AND PULSATION

Pulsators have been performance tested in Moorepark on several occasions (Agricultural Institute, 1968, 1969, and 1976; O'Shea and O'Callaghan, 1977 and 1980). Four to six samples of each pulsator type have usually been tested with some earlier tests over a 2,000 hour period, but with most recent tests for 5,500 hours. Table 2 shows the main measurements made on pulsators. Measurements and endurance testing are done at a vacuum of 50.8 ± 0.2 kPa. Standard simulated clusters of 0.8l (two 0.4l volumes for alternate pulsators) are used at the end of 2 m pulse tubes for the measurement of air consumption, pulsation rate, and pulsation chamber waveform. Pulsation chamber waveforms are recorded using a pressure transducer and a recording oscillograph (O'Callaghan, 1974); the paper speed is 120 mm/sec and trace deflection is 2 mm/kPa.

TABLE 2. MEASUREMENTS MADE ON PULSATORS DURING THE TEST PERIOD ARE MARKED "x"

	Period During Test		
	Start	Every 500 hours	End
Pulsation rate	x	x	x
Pulsation chamber waveform	x	x	x
Pulsator air consumption	x		x
Reference graphs for pulsator controllers	x	x	x

Ideal pulsation characteristics have not been determined, but pulsators should fulfill the following requirements over the test period: (1) they should not break down; (2) pulsation rate should have maximum range of 6 c/min for all samples of each make; (3) pulsator ratio should have a maximum range of 8% and should not be less than 50%, and the "b" and "d" values of the pulsation chamber waveform should not be less than 20% and 15%, respectively; (4) limping in alternate pulsators should not exceed 6%; and (5) limiting pulse volume should be at least 40% greater than cluster-pulsed volume.

Pneumatic relays with a mechanical master have met these requirements over a 5,500 hour test period (O'Shea, 1983). Mechanical slide relays and the better electronic relays have met these requirements in 2,000 hour test periods (Agricultural Institute, 1976; O'Callaghan and O'Shea, 1975, O'Shea and O'Callaghan, 1977) and possibly could also meet the requirements up to 5,500 hours. Individual pneumatic pulsators generally have not been as satisfactory. About 25% of all makes of individual pulsators were able to meet these requirements at 2,000 hours (O'Shea and O'Callaghan, 1977 and 1980; Agricultural Institute, 1976), and in a recent test only two out of ten makes tested could meet such requirements at 3,000 hours, and none met the requirements at 5,000 hours. Thus, relays generally are preferred to individual pneumatic pulsators. Relay pulsators are also much cheaper than individual pneumatics (table 3).

TABLE 3. COMPARATIVE FEATURES OF RELAYS AND INDIVIDUAL PULSATORS

RELAYS
Generally cheaper
Less maintenance
Cheaper maintenance
Maintenance can be done by farmer
No pulsation rate variation between pulsators
Not as greatly affected by dust

INDIVIDUAL PNEUMATICS
Breakdown of one does not affect others
Individuals (or small electronic relays) most suitable for
    bucket and RTS plants
Not as subject to incorrect installation

The standards applied on farms are those for laboratory tests, and in addition, pulsation rate should be within the range 53 to 63 cc/min. Slower rates increase milking times, while faster pulsation rates do not decrease milking times (O'Shea, 1973) but do increase pulsator air consumption. Increasing the pulsator ratio decreases milking time (O'Shea, 1983), but the requirement of a d-value of at least 15% puts an upper limit of about 66% on pulsator ratio. There are negligible differences in milking characteristics between 2 x 2 and 4 x 0 pulsation; 4 x 0 pulsation is usually used with relays, since plumbing arrangements are simpler.

Little emphasis is placed on the a- and c-values of the pulsation chamber waveform, as these have little effect on milking characteristics (O'Shea, 1983). However, some alterations of a- and c-values may be necessary to get the desired pulsator ratio and d-value (O'Shea, 1983).

Satisfactory pulsation requires air pipelines of adequate size. The size of air pipelines being recommended in Ireland (for herringbones) are those recommended by ISO (British Standard Institute, 1980), as shown in table 5; these sizes are somewhat larger than those necessary for satisfactory pulsation graphs (O'Shea and O'Callaghan, 1977 and 1978).

454

TABLE 4. SUMMARY RECOMMENDATIONS OF PULSATION SETTINGS ON
FARMS

Pulsation rate should be 53 to 63 c/min with a maximum range
of 6 c/min in any plant.

Pulsation ratio should be 55 to 70% with a maximum range of
8 percentage points in any plant.

The d-value should be at least 15%.

The milking performance difference between 4 x 0 and 2 x 2
pulsation is negligible.

TABLE 5. SIZES OF AIR PIPELINES RECOMMENDED IN IRELAND

	Number	of	Units
	<4	5 to 14	15 to 20
Diameter of main air pipeline to branch of pulsation line (mm)	31	31	50
Air pipeline (pulsation) diameter (mm)	25	31	38

TYPE OF PLANT

    Milking pipeline machines are recommended rather than
recorder plants (O'Callaghan and O'Shea, 1978) (table 6).
Pipeline plants are simpler and cheaper, and the chore of
jar emptying is avoided.
    It has not been established that recorder jars have a
beneficial effect in damping vacuum fluctuations. There is
minimal interunit crosstransfer of milk when entries into
the milking pipelines are vertical and downward (O'Callaghan
and O'Shea, 1978). Milking pipeline sizes recommended by
ISO are somewhat larger than necessary (O'Callaghan and
O'Shea, 1978), but ISO sizes are accepted in Ireland to conform with the Standard. The recommended bores of milking
pipelines and washlines are given in table 7.
    ISO reservation about highline milking plants are
ignored. Mid-level milking pipelines, about 1.4 m above cow
standings, are used; these reduce the effect of milk lift on
vacuum fluctuations, have minimal effects on FFA levels
(Doody et al., 1975; Judge et al., 1977), and allow the use
of one unit per two stalls (table 8).
    The number of cows milked per cluster per hour in one
stall per unit herringbone parlor is only 60% of the number

in two-stall-per-unit parlors (O'Callaghan et al., 1982) (table 9).

TABLE 6. ADVANTAGES OF PIPELINE PLANTS OVER RECORDER PLANTS

Pipeline Plants

Are cheaper to install
Are cheaper to maintain (less rubberware)
Are cheaper to clean (20% less solution needed)
Do not clutter the milking pit
Lead to more efficient cooling with plate coolers
Make it possible to use side-by-side parlors

TABLE 7. DIAMETER (BORE) OF MILKING PIPELINES AND WASHLINES FOR DIFFERENT NUMBERS OF UNITS IN HERRINGBONE MILKING SHEDS IN IRELAND

| | Number | of | Units |
	4 to 8	9 to 16	17 to 18	19 to 20
Milking pipeline bore (mm)	31	2 x 38 or 50	50	50
Washline bore (mm)	31	2 x 31 or 38	38	50

TABLE 8. RELATIVE ADVANTAGES OF HIGH-LEVEL AND LOW-LEVEL PIPELINE PLANTS

Highline Plants

Allow the use of 2 stalls per unit
Are easily cleaned externally
Are easily installed and dismantled
Have more readily accessible joints
Simplify milk recording
Are not likely to result in milk contamination due to holes in rubberware

Lowline Plants

2% to 8% faster milking
Lower cyclic fluctuations (but not lower mastitis)

TABLE 9. MAIN RECOMMENDATIONS FOR PLANT DESIGN

Use milking pipeline rather than recorder plants
In herringbone parlor, use 2 stalls per unit, e.g., 10
    stalls per 5 units
Use mid-level pipelines, i.e., about 1.4 m above the cow
    standing

LINERS AND CLUSTERS

Over the past five years, our main emphasis in milking
research has been on liner performance; this work has been
reported in another paper (O'Shea, 1983). Special emphasis
is placed on the selection of the better liner types for use
on farms, although practical tests for liner function on
farms are not available.

Specific standards cannot be set for liner performance. Liners are considered satisfactory when statistical
analyses show that their performance for five criteria is
not significantly less satisfactory than that of the most
satisfactory liners. The five criteria are: time to stripping, machine and strip yield, slips 12 l/min, and fall-
offs. Tests on liners should be done on a regional basis
because of differences between breeds of cows and teat
sizes. The above test method has been useful for the initial screening of liner performance. As the differences
between liners on the market have decreased, however, the
method has shown disadvantages, particularly for measuring
slip levels.

A new method is being developed to measure slips and
milking times in an 11-unit herringbone parlor in which
approximately 130 cows and heifers are being milked. This
method will be used for further segregation of liners shown
to be satisfactory with the original test method. It is
also hoped that the method will be fairly easily used on
problem farms. The method involves the use of a milk-air
separator at each milking unit and measuring the pressure
differential across an orifice plate to estimate airflow
(slips); milking times will be recorded using a magnetic
float, as in the previous method. All data will be processed by microcomputer and then teleprinted.

MILK RECORDING

Milking meters are used for milk recording on farms and
have been tested for accuracy of recording and sampling
(O'Shea, 1983; Connolly and O'Shea, 1973; O'Shea and
O'Callaghan, 1980; O'Shea et al., 1973).

Summary results are given in table 10 for the latest
models tested. For culling and breeding purposes, relative

--rather than absolute--yields are adequate. Thus, all milk meter types, except Bodmin should be satisfactory for farm use. However, it could be argued that the Milkoscope is best, because it is the most accurate. The cost and robustness of the meters also should be considered.

TABLE 10. SUMMARY RESULTS OF TESTS ON MILK METERS

Milk Meter Yields Expressed as a Percentage of True Yields

Milk	Fat Yield		Milk Yield	
	am	pm	am	pm
Bodmin	105.8	103.0	99.5	104.0
Milkoscope MK II	99.8	99.9	100.9	100.8
Tru-Test	99.2	102.9	98.6	96.0
Waikato	98.9	99.9	102.3	100.0

Percentages of observations which differed by more than 5% from the true values

Meter	Fat Yield			Milk Yield		
	am	pm	am/pm	am	pm	am/pm
Bodmin	3.21	22.6	27.2	50.6	65.0	57.9
Milkoscope	9.80	10.0	9.9	3.8	2.4	3.1
Tru-Test	49.3	24.2	35.7	15.6	33.8	24.7
Waikato	18.8	3.9	11.4	29.9	35.3	32.5

NONESSENTIAL AND NONSTANDARD EQUIPMENT

Automatic cluster removers (ACRs) have been tested and most types have operated mechanically satisfactorily (O'Shea and O'Callaghan, 1980a). However, ACRs are not recommended in two-stall-per-unit herringbone parlors. Limited overmilking does not increase new-infection rate. ACRs do not reduce the work routine time for operators, unless they have been wasting time moving up and down the pit to remove clusters as soon as cows have stopped milking. ACRs improve throughput in rotary parlors, provided teat disinfection is automated; in this case, one line operator can be dispensed with.
In any case, cluster remover types should not be used that admit airblasts into the claws (to activate the automatic shut-off) at cluster removal. With nonairblast ACRs, the ram for removal should not begin cluster removal until vacuum in the cluster has dropped.
Equipment must have some measurable advantage if it is to be recommended. Nonstandard equipment must be at least

as good as standard equipment. Tests on Alfa Laval Duovac and Tru-Test LDS Units showed that these units have no advantages over standard vacuum-pulsation systems (O'Shea, 1983) as shown in tables 11 and 12. Tests on the old type Nu-Pulse claw-cum-pulsator have shown that it has some disadvantages compared with standard milking units (O'Shea and O'Callaghan, 1980b), as shown in table 13.

TABLE 11. SUMMARY RESULTS OF TEST ON DUOVAC SYSTEM

	Standard	Duovac	LDS	Sig.
Milk yield (1/cow /milking)	4.610	4.470	0.416	NS
Fat yield (kg/cow /milking)	0.187	0.181	0.021	NS
Protein yield (kg /cow/milking)	0.168	0.167	0.026	NS
Time to stripping (min/cow/milking)	4.860	4.960	0.450	NS
Log_{10} cell count	2.440	2.480	0.147	NS

TABLE 12. SUMMARY RESULTS OF TEATS ON TRU-TEST LDS UNIT

	Standard Pulsation	LDS Unit	LDS	Sig.
Milk yield (1/cow /milking)	6.250	6.630	0.594	NS
Fat yield (kg/cow /milking)	0.186	0.209	0.027	NS
Protein yield (kg /cow/milking	0.205	0.222	0.019	NS

TABLE 13. MAIN RESULTS OF TESTS ON NU-PULSE SYSTEM

	Standard	Nu-Pulse	LDS	Sig.
Milk yield (1/cow /milking)	6.430	5.140	0.513	NS
Fat yield (kg/cow /milking)	0.225	0.222	0.023	NS
Protein yield (kg /cow/milking	0.213	0.213	0.023	NS
Time to stripping (kg/cow/milking)	5.890	6.670	0.619	*
Log_{10} cell count	2.422	2.643	0.196	*
FFA (meq/1)				

MANUFACTURERS COOPERATION

Based on test data, manufacturers quickly improve components that have not performed satisfactorily. In many cases, particularly among smaller manufacturers who do little research, the manufacturers have not been aware of the malfunctions of their equipment. Thus, these tests provide guidelines to farmers in selecting suitable, durable equipment to manufacturers in improving their equipment, and to those setting standards in preparing performance standards.

MILKING RESEARCH IN THE U.S.

Although there is no testing program for milking-machine components in the U.S. equivalent to the Irish program, the findings in Ireland on endurance and performance tests on pulsators, vacuum regulators, milk meters, and automatic cluster removers should be applicable in the U.S. However, many of the components being sold in the U.S. are not sold in Ireland and have not been performance tested there. Thus, at least some testing is needed in the U.S. Liner performance data from Ireland may not be applicable in the U.S. because cow teat sizes may vary between the two countries. Local data are needed on liners in the U.S.

Recent studies have determined vacuum pump and pipeline sizes needed in U.S. conditions; further work is needed in this area, since it appears that these components in the U.S. are grossly oversized. Further work is also needed on milking shed design.

The end result of these studies should be the production of a simple booklet that would be frequently updated. This booklet should contain standards, performance data for components, milking shed design data and drawings, and also general information on milking routines, machine milking and mastitis, and maintenance and testing of milking equipment. Such a booklet is available in Ireland (O'Callaghan et al., 1982).

460

REFERENCES

Anon. 1968. Test Report 38. Milking Machines. An Foras
Taluntais, Dublin, Ireland.

Anon. 1969. Test Report 52. Milking Machines. An Foras
Taluntais, Dublin, Ireland.

Anon. 1976. Test Report 59. Milking Machine Components.
An Foras Taluntais, Moorepark, Fermoy, Ireland.

Anon. 1976. Test Report 60. An Foras Taluntais, Moorepark, Fermoy, Ireland.

Anon. 1980. ISO/DP 5707. Milking Machine Installations-
Construction and Performance (In draft). British
Standards Institute, London.

Connolly, J. and J. O'Shea. 1973. Ir. J. Agric. Res.
12:83.

Cowhig, M. J. and T. O'Flaherty. 1968. Ir. J. Agric. Res.
7:161.

Doody, K., J. O'Shea, and T. F. Raftery. 1975. IDF Document 86, FIL-IDF 41 Square Vergote, Brussels, Belgium,
p. 146.

Judge, F. J., M. G. Fleming, J. O'Shea, and T. F. Raftery.
1977. Ir. J. Agric. Res. 16:115.

O'Callaghan, E. 1974. Ir. J. Agric. Res. 13:323.

O'Callaghan, E. and J. O'Shea. 1975. Ir. J. Agric. Res.
14:107.

O'Callaghan, E. and J. O'Shea. 1978. Ir. J. Agric. Res.
17:79.

O'Callaghan, E., J. O'Shea, A. J. Kavanagh, and H. J.
Doyle. 1982. Machine Milking and Milking Facilities.
ISBN O - 905442-59-8. An Foras Taluntais, Dublin,
Ireland.

O'Shea, J. 1973. Ir. J. Agric. Res. 12:111.

O'Shea, J. 1983. Dairy Science Handbook 15.

O'Shea, J. and E. O'Callaghan. 1980. Experiments on
milking machine components at Moorepark. 1976-1979.
An Foras Taluntais, Moorepark, Fermoy. Ireland.

O'Shea, J. and E. O'Callaghan. 1980. Experiments on milking machine components at Moorepark. 1976-1979, p. 1. An Foras Taluntais, Moorepark, Fermoy, Ireland.

O'Shea, J. and E. O'Callaghan. 1980. Experiments on milking machine components at Moorepark. 1976-1979, p. 11. An Foras Taluntais, Moorepark Fermoy, Ireland.

O'Shea, J. and E. O'Callaghan. 1980. Experiments on milking mahine components at Moorepark. 1976-1979, p. 27. An Foras Taluntais, Moorepark, Fermoy, Ireland.

O'Shea, J. and E. O'Callaghan. 1980. Experiments on milking machine components at Moorepark. 1976-1979, p. 34. An Foras Taluntais, Moorepark Fermoy, Ireland.

O'Shea, J. and E. O'Callaghan. 1977. Ir. J. Agric. Res. 16:105.

O'Shea, J. and E. O'Callaghan. 1978. Ir. J. Agric. Res. 17:97.

O'Shea, J., J. Palmer, and J. Connolly. 1972. Ir. J. Agric. Res. 11:159.

Walsh, J. P., P. V. Kinsella, J. M. A. Palmer, and J. Connolly. 1970. Ir. J. Agric. Res. 9:127.

44
THE RANGE AND PERFORMANCE OF CONVENTIONAL MILKING-MACHINE CLUSTERS

Jerry O'Shea

The most important milking characteristics are machine yield, strip yield, time to stripping, liner slips, and cluster fall-off. The relative importance of these characteristics varies with herd size and labor available.

Machine stripping increases milking work-routine time. Machine milking time is usually of less importance since its effect on herd milking rate is largely obviated by varying the number of clusters used. Cluster slipping and cluster fall-off reduce herd milking rates, and liner slips have been shown to increase the rate of new infection (O'Shea et al., 1976, 1976[a], 1979, 1981; O'Shea and O'Callaghan, 1978, 1980).

There have been several studies (Cowhig, 1968, 1968a) on the effect of milking performance of pulsator ratio, pulsation rate, cluster weight, and vacuum level. There have been few published reports on the effect of liner design on milking performance (McGrath and O'Shea, 1972; McD.Gibb and Mein, 1976).

The present paper describes the range of milking performance of liners of different designs. These are compared with the range arising from altering of other milking machine factors.

EXPERIMENTAL METHODS

The data presented are selected from several experiments (O'Shea and O'Callaghan, 1980; Kelly et al., 1982; O'Shea et al., 1982, 1982a). Although the results are not presented for all treatments in some experiments, the Least Significant Difference (LSD) values quoted are those for all treatments. The data are selected to demonstrate the range of differences measured.

Most of the experiments were Latin Square design experiments that compared 12 clusters over a 36-day period; there were 12 groups of 3 cows each and there was a changeover of treatments after 3 days. A few experiments were 8 x 8 Latin Square or 10 x 10 Latin Squares. There were also incomplete block changeover experiments.

The experimental methods have been described in detail
(O'Shea and O'Callaghan, 1980) and only the main details are
repeated here for clarity. The main characteristics measured were (a) machine yield (yield before machine stripping), (b) strip yield, (c) time to stripping, (d) liner
slips, and (e) cluster fall-offs. Slips were defined as
airflows ≥ 12 l/min between the teats and the liners. Slips
were measured using hot wire anemometers. A UV-recorder was
used to measure milking times and slips. Milk yields were
recorded to ± 8 ml using calibrated tall cylinders. Liners
were pulsed in their shells for 18 hours prior to being used
on experiment; this allowed liners to reach equilibrium tension.

The test method used had some limitations; these limitations have been discussed (O'Shea and O'Callaghan, 1980).
Generally the differences between treatments (clusters) were
minimized, since interactions between adjacent units were
eliminated to use the hot wire anemometers. The measurement
equipment is very expensive, thus only four milking units
were operated. The plant used was effectively four one-unit
machines. The four units limited cow numbers to about 40
cows per experiment. Heifers were not used in the experiments due to the danger of breaking some of the equipment;
in practice, slip problems are likely to be greatest with
heifers. The fact that a cluster did not interact with adjacent clusters was not typical of practical milking. Airflows up to 170 l/min have been recorded during liner slip;
vacuum stability would be affected by several simultaneous
slips in a commercial plant installed to ISO Standards and
slip. Cows used were Friesian type, spring calving, and
yielded 3,500 to 5,000 l per annum. Milking was at 17/7 hr
(night/day) intervals. Mastitis control was by sampling at
14-day intervals; clinical and subclinical infections were
treated with appropriate antibiotics. A minimal wash with
cold running water followed by immediate cluster application
was used in all experiments.

MILK YIELD

There are interactions between stage of lactation and
most milking characteristics (Smith et al., 1974). The
stage of lactation effect is largely due to differences in
milk yield. In the present experiments, the effect of milk
yield was apparent from comparisons of a.m. and p.m. data
(table 1). Attention is drawn to the misleading use of
strip yield as a percentage of total yield in the measurement of liner performance. As shown in table 1, strip yield
as a percentage for total yield varies with milk yield.

TABLE 1. COMPARISONS OF A.M. AND P.M. DATA WITH TWO REPRE-
SENTATIVE CLUSTERS

Cluster	A	B	LSD	A	B	LSD
Milking	am	pm		am	pm	
Machine yield (1)	10.28	10.05	0.231	4.77	4.33	0.146
Strip yield (1)	0.13	0.43	0.231	0.17	0.44	0.036
Slips	40.0	200.0	67.2	11.2	73.7	45.2
Fall-offs	3.8	10.1	8.8	0.0	5.0	3.2

RANGE OF PERFORMANCE WITH CLUSTERS

Data are presented in table 2 to demonstrate the range
of performance (sums of a.m. and p.m. data) measured with
different commercial clusters using standard vacuum (51 kPa)
and pulsation characteristics. The data show that clusters,
per se, cause a wide variation in milking characteristics.
The data are presented as percentages of the best value.

TABLE 2. RANGE OF MAIN PERFORMANCE VALUES FOR ALL TRIAL-
VALUES EXPRESSED AS A PERCENTAGE OF THE BEST VALUE

	Best value	Worst value
Machine yield	100	92
Strip yield	100	600
Time to stripping	100	133
Slips $\geqslant$ 12 1/min	100	800
Fall-offs	100	800

CLUSTER WEIGHT

In general, milking characteristics are more critically
related to liner design than cluster weight (table 3).
Strip yield decreased and slips and fall-offs increased with
increasing cluster weight with a given type of liner. The
differences between two clusters of the same weight with two
liner types was much greater than the differences between
two weights of clusters with the same liner type.

TABLE 3.   EFFECT OF CLUSTER WEIGHT AND LINER TYPE ON MILKING
           CHARACTERISTICS.   YIELDS AND TIME ARE PER COW PER
           MILKING; SLIPS AND FALL-OFFS ARE PER 100 COW MILK-
           INGS

Liner type	A	A	B	C	LSD
Cluster weight (kg)	2.88	2.38	2.35	2.98	
Machine yield (1)	7.44	7.51	7.19	7.43	0.144
Strip yield (1)	0.26	0.29	0.44	0.15	0.029
Time to stripping (min)	6.90	6.68	7.05	7.03	0.464
Slips ⩾ 12 1/min	140.0	106.9	136.9	5.0	45.4
Fall-offs	5.7	5.6	6.5	1.9	4.8

LINER MATERIAL

     Comparisons of several liner designs (made in the same
mold) showed that the materials from which liners were made
had similar effects on milking characteristics (table 4).

TABLE 4.   EFFECT OF LINER MATERIAL ON MILKING CHARACTERISTICS; YIELD AND TIME
           DATA ARE PER COW PER MILKING AND SLIPS AND FALL-OFFS ARE PER 100 COW
           MILKINGS

Liner type (and material)	A (1)	A (2)	LSD	B (1)	B (2)	LSD
Cluster weight (kg)	2.92	2.92		3.05	3.05	
Machine yield (1)	6.03	5.96	0.209	6.29	6.19	0.289
Strip yield (1)	0.20	0.25	0.049	0.20	0.19	0.025
Time to stripping (min)	4.43	4.57	0.437	5.42	5.48	0.399
Slips ⩾ 12 1/min	50.7	67.4	31.2	16.7	31.2	15.7
Cluster fall-offs	1.4	3.5	4.8	6.3	10.4	9.3

PULSATION RATE AND PULSATOR RATIO

     A series of experiments were carried out in which dif-
ferent liner types were used with different pulsation settings. These experiments determined the effect of pulsation
settings on milking characteristics, especially liner slip,
and also established that there was not an interaction between liner type and pulsation settings (table 5). This
means that liners that perform well at one pulsation setting
also perform well at other pulsation settings and vice
versa. In table 5, LSD I refers to the same pulsation settings and LSD 2 refers to differences between pulsation settings. Pulsation rate (c/min) in the range 40 to 170 c/min
had little effect on milking characteristics.   Time to

stripping was decreasd significantly by increasing pulsation rate from 50 to 65 c/min; further increases in pulsation rate had virtually no effect on milking characteristics (table 6).

Summarized results presented in table 7 are from an experiment in which two liner types were each used with four different pulsation settings. Liner design affected most milking characteristics significantly, but pulsation rate and pulsation phase did not; pulsator ratio significantly affected milking times only. There were no significant interactions between liner type and pulsation settings for any of the milking characteristics measured.

TABLE 5. SAMPLE RESULTS SHOWING THAT THERE WAS NOT AN INTERACTION BETWEEN LINER DESIGN AND PULSATION SETTINGS. YIELD AND TIME DATA ARE PER COW PER MILKING AND SLIPS AND FALL-OFFS ARE PER 100 COW MILKINGS

Liner type	A	A	B	B	LSD 1	LSD 2
Pulsation rate (c/min)	50	65	50	65		
Pulsator ratio	53	73	53	73		
Machine yield	6.34	6.36	6.47	6.41	0.156	0.126
Strip yield	0.16	0.13	0.20	0.19	0.108	0.054
Time to stripping	4.96	4.86	5.30	5.19	0.344	0.170
Slips $\geqslant$ 12 l/min	0.0	3.9	4.4	5.5	12.0	6.0
Fall-offs	0.0	0.0	2.2	3.2	8.2	4.0

TABLE 6. SAMPLE RESULTS SHOWING THE EFFECT OF PULSATION RATE ON THE MAIN MILKING CHARACTERISTICS. VALUES ARE EXPRESSED AS PERCENTAGES OF THE CONTROL VALUES

	Pulsation rate (c/min)					
	40	50	65	100	135	170
Time to stripping	116.1	115.9	100	93.0	94.0	92.0
Strip yield	78.0	70.0	100	108.3	116.7	133.3
Slips $\geqslant$ 12 l/min	60.2	170.0	100	29.4	50.0	89.7

INCREASING AND DECREASING VACUUM OF PULSATION CYCLE

The a-value (increasing vacuum) and c-value (decreasing vacuum) of the pulsation chamber waveform had little effect on milking characteristics. There was an increase in time

TABLE 7. PULSATION SETTINGS USED WITH TWO LINER TYPES AND
THEIR EFFECTS ON MILKING CHARACTERISTICS. THE
EFFECTS OF PULSATION RATE AND PHASE WERE NOT SIGNIFI-
CANT AND DATA ARE OMITTED

Pulsation treatment codes	PI	P2	P3	P4
Pulsation rate (c/min)	50	65	65	65
Pulsation ratio (%)	53	74	74	53
Pulsation phase	4 x 0	2 x 2	4 x 0	4 x 0

	Effect of liner types			
	A	B	LSD	Sig.
Time to stripping (min/cow/milking)	4.47	5.08	0.152	***
Machine yield (l/cow/milking)	6.22	6.28	0.140	NS
Strip yield (l/cow/milking)	0.22	0.15	0.032	***
Slips ⟩ 12 l/min/100 milkings	31.4	2.1	6.0	***

	Effect of pulsator ratio			
	74%	53%	LSD	Sig.
Time to stripping (min/cow/milking)	4.55	5.01	0.214	***
Machine yield (l/cow/milking)	6.28	6.30	0.200	NS
Strip yield (l/cow/milking)	0.19	0.21	0.046	NS
Slips ⟩ 12 l/min/100 milkings	13.3	17.6	8.6	NS

***Significant p<0.001 NS = not significant

TABLE 8. SAMPLE RESULTS SHOWING THE EFFECT ON THE MAIN
MILKING CHARACTERISTICS OF THE A- AND C-VALUES OF
THE PULSATION CHAMBER WAVEFORM. VALUES ARE EX-
PRESSED AS PERCENTAGES OF THE CONTROL VALUES

	a-values (%)			
	10	25	40	LSD
Time to stripping	100	100.4	105.2	3.06
Strip yield	100	100.0	111.0	31.8
Slips ⟩ 12 l/min	100	110.6	135.9	18.1
	c-values (%)			
	10	25	40	LSD
Time to stripping	100	96.8	96.5	2.98
Strip yield	100	80.0	80.0	27.0
Slips ⟩ 12 l/min	100	107.8	95.5	15.8

to stripping, strip yield, and slips with increasing a-
values; thus it is desirable to have a-values fairly close
to 10%. As the c-value was increased, time to stripping was
reduced but other milking characteristics were not altered
(table 8). It is important to remember that if the shape of
pulsation graphs is changed, the d-value (atmospheric phase)
must be at least 15%.

469

SUMMARY: LINERS AND PULSATION

The changes in milking characteristics measured with different liner types and different pulsation settings are summarized in table 9. Liners caused changes of much greater magnitude than the changes arising from pulsation.

TABLE 9. CHANGES IN MILKING CHARACTERISTICS BY USING DIFFERENT LINER TYPES AND DIFFERENT PULSATION SETTINGS

	Machine yield	Strip yield	Milking time	Liner slips & fall-offs
Bad liner (vs. good liner) causes:	7% loss	600% inc.	33% inc.	800% inc.
2 x 2 (vs. 4 x 0) causes:	No change	5% inc.	5% dec.	No change
50 c/min (vs. 65 c/min) causes:	No change	30% dec.	No change	No change
40 c/min (vs. 170 c/min) causes:	No change	70% dec.	24% inc.	No change
53% ratio (vs. 74% ratio) causes:	No change	No change	11% inc.	No change
a-value 10% (vs a = 40%) causes:	No change	No change	5% dec.	40% dec.
c-value 10% (vs c = 40%) causes:	No change	No change	5% inc.	No change

LINER DESIGN FACTORS AFFECTING MILKING CHARACTERISTICS

The results of several milking experiments with standard pulsation were statistically combined. Liner types used in these experiments were x-rayed while hanging free and also in a transparent shell with a plastic teat, loaded with an 0.5 kg weight, inserted in the liner. Twenty-five measurements were made on each liner type from these x-rays and 10 further variables were derived from these measurements. Milking characteristics data were correlated with and regressed on the liner measurements for individual experiments, for groups of similar liner types, and for data from combined experiments. The main factors that appeared to affect milking characteristics are shown in table 10 and figure 1. These data should be treated only as guidelines since the analyses may not have indicated the causative factors and may only have highlighted factors that were positively correlated with the causative factors.

TABLE 10. FEATURES OF LINER DESIGN SHOWN BY CORRELATION AND REGRESSION ANALYSES TO AFFECT MILKING CHARACTERISTICS – DATA ARE GUIDELINES ONLY

Liner features	Effect of liner feature on milking characteristics			Apparent optimum for liner features
	Strip yield	Slips 12 l/min	Time to stripping	
Increasing bore at lower barrel	No effect	Decrease	No effect	25 mm
Increasing mouthpiece bore	Decrease	Increase	No effect	22 mm
Increasing cavity size	Decrease*	No effect	No effect	300 mm^2
Increasing cavity height	Decrease**	Decrease*	No effect	30 mm
Decreasing mouthpiece deflection at 0.5 kg	No effect	Decrease	No effect	1 to 3 mm
Increasingly gentle slope at shoulder	Decrease	Decrease**	No effect	Very gentle
Increasing liner length	No effect	Decrease	Decrease	150–160 mm

**Largest effect *Fairly large effect

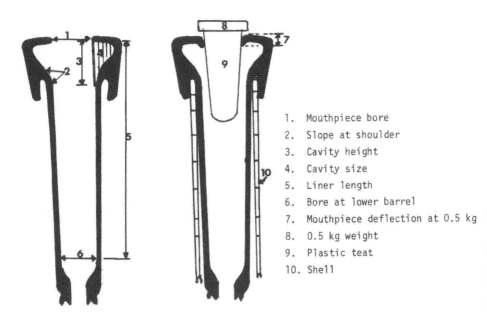

1. Mouthpiece bore
2. Slope at shoulder
3. Cavity height
4. Cavity size
5. Liner length
6. Bore at lower barrel
7. Mouthpiece deflection at 0.5 kg
8. 0.5 kg weight
9. Plastic teat
10. Shell

Figure 1. Liner Features Affecting Milking Characteristics

REFERENCES

Cowhig, M.J. 1968. A study of the performance of eight milking machines. M.Sc. Thesis, N.U.I., Dublin, 85 pp.

Cowhig, M.J. 1968. Procs. Symp. Mach. Milk., Reading, 1968 p 15.

Kelly, T.G., J. O'Shea, E. O'Callaghan, and B. McKenna. 1982. Unpublished data.

McD. Gibb, I. and G.A. Mein. 1976. Aust. J. Dairy Techol. 31:148.

McGrath, D., and J. O'Shea. 1972. Ir. J. Agric. Res. 11:339.

O'Callaghan, E., J. O'Shea, W.J. Meaney, and C. Crowley. 1976. Ir. J. Agric. Res. 15:401.

O'Shea, J. and E. O'Callaghan. 1978. Proc. Int. Symp. Mach. Milk. 17th Ann. Mtg. Nat. Mastitis Council, Inc., Louisville, KY, p 262.

O'Shea, J., E. O'Callaghan, and W.J. Meaney. Ir. J. Agric. Res. 18:225.

O'Shea, J., and E. O'Callaghan. 1980. Proc. Int. Wkp. Mach. Milk. and Mastitis, Moorepark, p 6.

O'Shea, J., and E. O'Callaghan. 1980. Experiments 6 to 17 "Experiments on Milking Machine Components at Moorepark, 1976-1979" pp 40-191.

O'Shea, J., E. O'Callaghan, and W.J. Meaney. 1981. Ir.J. Agric. Res. 20:163.

O'Shea, J., E. O'Callaghan, and W.J. Meaney, and C. Crowley. 1976. Ir. J. Agric. Res. 15:377.

O'Shea P., E. O'Callaghan, and J. O'Shea. 1982. Unpublished data.

O'Shea, P., J. O'Shea, E. O'Callaghan, and A. Gonzalez. 1982. Unpublished data.

Smith, J.W., R.H. Miller, N.W.Jr. Hoover, and E.D. Moore. 1974. J. Dairy Sci. 57:1364.

Part 9

COMPUTER TECHNOLOGY

45
AGNET: A NATIONAL COMPUTER SYSTEM FOR CATTLEMEN

Harlan G. Hughes

In the early 1970s, two professors at the University of Nebraska conceived the idea of an agricultural computer system designed specifically for farmers and ranchers. They developed the computer system now known across the country as AGNET--The Agricultural Computer Network. In 1977, the governors of five states (Nebraska, South Dakota, North Dakota, Montana and Wyoming) jointly funded a pilot project to test if farmers and ranchers in their respective states would use a computer system to make better management decisions. AGNET has now developed so that over 400,000 calls a year are being made to the AGNET computer. AGNET is, indeed, a management tool for agriculture.

Wyoming now has computer terminals in all 23 county extension offices, and county extension agents are now receiving training on how to use these terminals with their farmer and rancher clientele.

AGNET is one of three computers in the Wyoming computer center. This operator controls the AGNET computer from the central station. If we have done the job right, the operator should not have to do much. Due to the speed of the computer, we prefer that the machine do as much of its own operation as possible. This operator, however, can and does take over control of the machine whenever necessary.

AGNET is a mass storage system. Behind the dark windows in AGNET storage units are stacks of phonograph-like records used for storage of data and programs. All of AGNET's programs are stored on disks so that when you type in the name of a program, the computer can immediately go to the appropriate disk and find the requested program. We do not have to wait for an operator to mount a tape or to do any manual intervention. AGNET is one of the largest mass storage systems in the world.

Farmer advisors on the AGNET payroll are very special persons to AGNET. George, one of the advisors, is a real character whom I wish everyone could met. George's role is to help make sure that what we have on AGNET will work for farmers and ranchers. I have heard George say, "Harlan, that is the dumbest #%&"* thing I have ever heard!" Or I have heard George say, "That may be well and fine in your ivory tower, but out on the farm we do not have that kind of data." We have two half-time farmers on the AGNET payroll and they play a very important and unique role in the design and operation of the total AGNET system.

AGNET is equipped so that we can have over 200 tele-phone calls coming in to the computer at one time. We are now averaging a phone call into AGNET every four minutes, seven days a week, 24 hours a day. That is over 400,000 phone calls a year.

The AGNET computer in Nebraska is located in the base-ment of the State Capital. By design AGNET is not on a uni-versity campus computer (and probably never will be on a campus computer) because of our computer needs and demands. A university computer is set up for research and administra-tive data processing. We need a service-oriented computer center that can consult us before the system is changed or shut down. Our users are not computer science PhD's and be-come frustrated with computer down time or off time. AGNET is often our user's first contact with a computer and since they are paying for the computer time, we place some strin-gent demands on the computer center.

THE FULL PARTNER STATES AND THE STAFF

In 1977, five states previously mentioned became full partners in the AGNET system. The best way to describe a full partner state is to say that each state has a member on the AGNET Board of Directors.

In July of 1980, the state of Washington joined as a full partner state and in October 1980 the state of Wiscon-sin joined AGNET. In July of 1981, Wisconsin withdrew, which leaves six full-partner states in the AGNET system, but other states currently are considering partner status.

The concept behind AGNET is to share the development and operating costs among the full-partner states. There are approximately 17 people on the AGNET payroll. Of the 17 people, Wyoming is paying for two. Each state pools its re-sources with the other states so that each state can take advantage of the total efforts of the total 17 people.

I have a goal in life and it is to dissolve state boundaries when it comes to information dissemination and use. We are proving that states have information and com-puter programs to share across state lines, and as we share our extension resources, the winners are our clientele.

AGNET PROGRAM LIBRARY

The six partner states in the AGNET system have de-veloped the world's largest agricultural and home economics computer-program library in the world. Today there are over 200 programs available to AGNET users. With a library of this size, no one is expected to use or even know how to use all the programs.

Our goal is not to have users able to use all the programs in the library, but rather to have a large enough
library so that every user can find at least one program of
interest. This large smorgasbord of programs means that
users should be able to find several programs of special interest. Appendix A provides a partial list of the programs
available on AGNET. I have grouped the programs by subject
matter to facilitate user interests.

The AGNET library has been put together with approximately 35 man-years of programming effort. In addition,
each program development is supervised by a subject matter
extension specialist who is responsible for the content of
the program. Each subject-matter program is owned by the
subject matter specialist and not by AGNET.

AGNET is exceptionally well equipped for the livestock
producer. There are livestock ration-formulation programs
available for range cattle, feedlot cattle, hogs, sheep, and
poultry. There are programs available that will let you
simulate on paper what your cattle will do in the feedlot
given a description of your cattle and the ration that you
are going to feed them. There are livestock budgets and
planning prices stored in selected programs.

AGNET also has programs for the crop farmer. Machin-
ery-cost calculators and crop budgets are available. In
addition, there are whole farm or ranch budgeting programs
designed to help you make long-run business investment decisions. There are many, many more programs designed to
help you make better management decisions.

HARDWARE USED TO ACCESS THE AGNET LIBRARY

Touch-tone telephone. The first computer terminal that
I installed in a county extension office in 1972 was a
touch-tone telephone. It cost us $14 per month. We used
the number pad to send information to the computer and the
computer sent back the information over the special loud
speaker attached to the phone. We would send in the input
numbers by typing them into the telephone. The computer
would talk back and say, "Answer number 1 is 420." We
printed the answer onto a preprinted form that explained the
interpretation of the number. This touch-tone terminal
served us very well as a low-cost computer terminal. Industry still uses this type of small, low-cost terminal.

Execuport terminal. It soon became evident that we
would like to have terminals in our county extension offices
that would print out the computer information. We now have
five of the Execuports in the Wyoming AGNET inventory.
These cost $1,400 for reconditioned terminals.

Texas Instrument 745 Terminal. We have installed small
portable TI-745s in most of our extension offices. The TI-
745 weighs 13 pounds, has a clamp-on lid and a handle. It

is the size of a small briefcase and weighs about half as much. Wyoming agents transport their terminals all over their counties. The TI-745 costs approximately $1,400 new.

North Dakota's CRT Terminal. Terminals come in all sizes and shapes. The Animal Science Department at North Dakota has a CRT terminal with a TV screen where one can read the data. It also has a printer that can be used to generate a printed copy of the output when desired. These dual-purpose units cost more money, but the flexibility is convenient and does reduce paper costs.

A Decwriter terminal. A Decwriter terminal is used by the Department of Agriculture in Alberta, Canada. I used their terminal to check my electronic mail. Alberta Agriculture subscribes to the AGNET system. This terminal costs around $2,000.

Teletype 43 terminal. My secretary and I use Teletype 43 terminals. Obviously this is the terminal that I like best. The TT-43 gets used more hours than any of our terminals and is virtually a maintenance-free terminal. The only problem is that it is not portable. It weighs 45 pounds and has the terminal plus the telephone coupler and the paper to move. This terminal also costs approximately $1,400.

Terminal with TV screens. We have one special terminal that drives 23-inch TV screens for demonstration and teaching purposes. These are the same TV screens that you see in airports with flight schedules. We use these screens so that clientele and students can see exactly what we type on the terminal and exactly what the computer sends back to the screens. These screens have helped to promote AGNET in Wyoming. The screens work so well that I will not give a demonstration without these screens. The special terminal and the two screens cost approximately $4,000; therefore, we have only one in Wyoming.

MICROCOMPUTERS FOR FARM AND RANCH

Let's now boil this all down. What does it mean for you on the farm or ranch?

Agriculture is going to have some serious challenges in the 80s. During the 60s and 70s your challenge was production, but the challenge in the 80s is going to be financial management.

Yes, the computer has the potential to improve your financial management. Let me make a prediction. Those of you that will be farming in 1990 will be using computers. Those of you that do not want to use a computer will not be farming in 1990. I often hear, "No damn computer is going to tell me how to run my ranch!" I predict that that person

won't be farming in 1990. Many will have retired and others will have gone out of business. Computers are going to become commonplace on U.S. farms and ranches during the 1980s. Producer owned microcomputers can be useful in relation to AGNET. As I travel around the country talking to farmers and ranchers, I hear them expressing interest in three applications of microcomputers. The three applications are:
- Business accounting.
- Herd performance reporting.
- Financial management.

In an accompanying paper, I have discussed microcomputers, their use, and purchase.

Information Networking on AGNET

If you have a telephone coupler for your micro, you can access the following from AGNET:
- Current commodity market prices.
- Current USDA, Foreign Ag and Wyoming news releases.
- Agricultural outlook and situation reports.
- Western Livestock Market Information Project livestock analyses.
- Hay for sale.
- Sheep for sale.
- Certified pesticide applicators in Wyoming.
- People interested in judging county and state fairs.
- Horticultural tips during the summer.
- Home-canning tips during the canning season.
- Emergency information such as drought tips, Mount St. Helen's emergencies, etc.

You can even use your micro to access the UPI and AP news services for news stories dealing with, for example, the Farm Bill or "beef." The AP and UPI news services are available from two commercial time-sharing companies. You can do all this today with your micro if it has a telephone coupler on it.

Marketing Information On AGNET

We are putting about 17 different daily market-price files on AGNET, including the futures opening and closing prices. We have Chicago, Kansas City, and Minneapolis futures going onto AGNET. In addition, we have both national and selected local cash markets. We are reporting local feeder-cattle sales in Wyoming, Northeastern Colorado, and Western Nebraska. Local grain and cattle markets are being put on weekly for Nebraska. Feedlot reports for the major cattle feeding states are going periodically. Export data is also going on weekly. AGNET is becoming a major source of market information for agricultural producers.

This appears to be the major reason for most of our producer subscriptions to AGNET. They want current market information.

SUMMARY

In Wyoming we are using the AGNET system to provide Wyoming farmers and ranchers with their first contact with computers for:
- Record keeping such as beef herd performance.
- Problem solving for computer-aided decision making.
- Information networking such as daily market information.
- Electronic mail to speed up the delivery of research and extension information to clientele.

Computerized Management Aids (CMAs) are not new to agriculture. They are just new to the west. Leading midwest farmers have been using CMAs for over 10 years.

481

APPENDIX A

PARTIAL LIST OF AGNET PROGRAMS AVAILABLE

Livestock Production Models on AGNET:

BEEF   Simulation and economic analysis of feeder's performance.
BHAP/BHPP   Beef herd performance program and beef herd analysis program.
COWCULL   Package to help determine which dairy cow to cull and when.
COWGAME   Beef genetic selection simulation game.
CROSSBREED
            Evaluates beef crossbreeding systems & breed combinations.
FEEDMIX   Least cost feed rations for beef, dairy, sheep, swine, & poultry.
FEEDSHEETS
            Prints batch weights of rations including scale readings.
RANGERATION
            Ration balancer for beef cows, wintering calves, horses & sheep.
SWINE   Simulation and economic analysis of feeder's performance.
TURKEY   Simulation and economic analysis of turkey's performance.
VITAMINCHECK
            Checks the level of vitamins & trace minerals in swine diet.
WEAN   Performance testing of weaning-weight calves.
YEARLING   Performance testing of yearling-weight calves.

AG Engineering Models on AGNET:

BINDRY   Predicts results of natural air & low temp. corn drying.
CONFINEMENT
            Ventilation requirements & heater size for swine confinement.
DRY   Simulation of grain drying systems.
DUCTLOCATION
            Determines ducts to aerate grain in flat storage bldg.
FAN   Determination of fan size and power needed for grain drying.
FUELALCOHOL
            Estimates production costs of ethanol in small-scale plants.
GRAINDRILL
            Calculates the lowest cost width for a grain drill for your farm.

PIPESIZE   Computes most cost-effective size irrigation pipe to install.
PUMP   Determination of irrigation costs.
SPRINKLER   Examines feasibility of installing sprinkler irrigation.
STOREGRAIN
            Cost analysis of on-farm and commercial grain storage.
TRACTORSELECT
            Assists in determining suitability of tractors to enterprise.

Crop Production Models on AGNET:

BASIS   Develops "historical basis" patterns for certain crops.
BESTCROP   Provides equal return yield & price analysis between crops.
CROPINSURNACE
            Analyzes whether to participate in crop insurance program.
FLEXCROP   Forecasts yields based on amount of water available for crop growth.
IRRIGATE   Irrigation scheduling.
RANGECOND   Calculates the range condition and carrying capacity.
SEEDLIST   Lists seed stocks for sale.
SOIL LOSS   Estimates the computed soil-loss (tons/acre/year).
SOILSALT   Diagnoses salinity & sodicity hazard for crop production.
SOYBEANPROD
            Demonstration soybean production management model.

Home Economic Models on AGNET:

BEEFBUY   Comparison of alternative methods of purchasing beef.
BUSPAK   Package of financial analysis programs.
CARCOST   Calculates costs of owning & operating a car or light truck.
DIETCHECK   Food intake analysis.
DIETSUMMARY
            Summary of analysis saved from DIETCHECK.
FIREWOOD   Economic analysis of alternatives available with wood heat.
FOODPRESERVE
            Calculates costs of preserving foods at home.
MONEYCHECK
            Financial budgeting comparison for families.
PATTERN   Helps select a commercial pattern size & type for figure.
STAINS   Tells how to remove certain stains from fabrics.

4-H and Youth Models on AGNET:

CARCASS   Scoring & tabulation of beef or lamb carcass
          judging contest.
FAIR   Scoring and tabulation of judging contests.
JUDGELIST List of judges available for fairs and con-
          tests.
PREMIUM   Compiles and summarizes fair premiums.

Farm and Ranch Planning Models on AGNET:

BUSPAK   Package of financial analysis programs.
CALFWINTER
          Analyzes costs and returns associated with
          wintering calves.
COWCOST   Examines the costs and returns for beef cow-
          calf enterprise.
CROPBUD   Prints out select Wyoming crop budgets.
CROPBUDGET
          Analyzes the costs of producing a crop.
DAIRYCOST Analyzes the monthly costs and returns with
          milk production.
EWECOST   Analyzes the costs & returns of sheep produc-
          tion enterprise.
FARMPROGRAM
          Analyzes USDA Acreage Reduction Program.
GRASSFAT Analyze costs and returns associated with
          pasturing calves.
LANDPAK   Package of programs to assist in land manage-
          ment decisions.
LSBUDGETS Designed to print out stored livestock bud-
          gets.
MACHINEPAK
          Machinery analysis package.
PLANPAK   Package of programs designed to help analyze
          and plan aspects of the business.
PLANTAX   Income tax planning/management program.

Information Networking on AGNET:

CONFERENCE
          A continuing dialogue among users on a speci-
          fic topic.
EWESALE   Lists sheep for sale.
FAS   Prints trade leads & commodity reports pro-
          vided by USDA-FAS.
GUIDES   Prints available reports of reference materi-
          al information.
HAYLIST   Lists hay for sale.
MAILBOX   Used to send and receive mail.
NEWS   Latest notifications about programs and user-
          related information.
NEWSRELEASE
          A program for rapid dissemination of news
          stories.

WHO IS   Retrieves name and company affiliation of individual users.

WYOPROGS List of specialized Wyoming programs available only to Wyoming users.

Market Price Retrievals, Plotting, and Forecasting Models on AGNET:

CASHPLOT Prints a plot of selected cash prices.

CORNPROJECT
     Projects avg U.S. corn price for various marketing years.

MARKETCHART
     Prints various charts on selected future and cash prices.

MARKETS   Various market reports and specialists' comments.

PRICEDATA Prints selected historic cash and/or futures prices.

PRICEPLOT Designed to plot market prices in graphic form.

Miscellaneous Programs on AGNET:

EDPAK   Demo programs illustrating computer-assisted instruction.

FILLIN   A "fill in the blank" quiz routine.

GAMES   Package of game programs.

INPUTFORMS
     Prints available input forms.

JOBSEARCH Matches abilities and interests to occupations.

MC   A multiple choice quiz routine.

MICROPROGRAM
     Lists programs for microcomputers.

TESTPLOT Standard analysis of variance.

TREE   Summarization of community forestry inventory.

46

RANCHER-OWNED MICROCOMPUTER SYSTEMS:
WHAT'S AVAILABLE

Harlan G. Hughes

In the fall of 1977, Radio Shack started advertising the TSR-80 microcomputer for Christmas. This was the beginning of general-public awareness of the personal microcomputer. Another highly advertised microcomputer is the Atari which can be hooked up to a regular TV set, but the Atari is a game computer and, to my knowledge, has no agricultural programs available yet.

CURRENT MICROCOMPUTERS FOR RANCH AND FARM USE

There are two levels of microcomputers being considered by farmers. For the lack of any other terminology, I will use Level I and Level II as the classifications. Level I micros are the lowest cost and most popular systems. The three most common Level I micros in agriculture are the Radio Shack, Apple, and Pet Commodore.

Level I Hardware

Radio Shack Models I, II, & III. The Animal Science Division at the University of Wyoming has a Model I Radio Shack microcomputer. As is typical of most microcomputers, it has a keyboard, a TV screen, a disk unit, and a printer. Dr. Schoonover from Wyoming has developed a herd performance program for the Radio Shack Model I and III microcomputers. This program keeps track of the cow/calf information that ranches have been keeping on 3- x 5-inch cards. Once the data is inside the computer, management reports can be quickly printed out to help the rancher determine the cows to keep and the cows to cull. The same herd performance program that Dr. Schoonover has on the Radio Shack microcomputer is also on the AGNET system. We have several Wyoming ranchers currently using these herd performance programs.

The Radio Shack Model II has the disk drive built into the unit. Radio Shack refers to this as their small business machine.

The Radio Shack Model III has two disk drives built into the unit and presents pictures and graphs of your data.

The Model III can present a bar graph to show how ranch profits have changed the last 5 years. It has been suggested by some ranchers that a graph is purely academic if it represents ranch profits since profits have disappeared rather than changed.

Apple Computers. AGNET has an Apple computer with which we have one of our Teletype 43 AGNET terminals as a slave terminal. With proper connections, you can use your existing terminal as a slave printer on your microcomputer. Also, if you have a black and white TV you can back it up as the CRT on the Apple (and other brands as well). The resolution is not quite as clear as a regular monitor, but it is a cheaper way to get set up with a microcomputer.

Dr. Menkhaus, at the University of Wyoming, uses his Apple microcomputer in his Price Analysis class to teach undergraduate students how to use microcomputers.

The newest Apple is the Apple III. It has been out for about a year, but has had some technical troubles that has set its acceptance back. The Animal Science Division at Wyoming cancelled its order for the Apple III and ordered the Apple II Plus. This fast-growing company moved into a new product and forgot something called "quality control."

Pet Commodore Microcomputer. The Pet Commodore is being used by Alberta Agriculture in Canada and the Ag Economics Department at Wyoming. The Canadians have written a fair amount of agricultural software for the Pet and have been willing to share it with Wyoming so that we do have several decision aids for our Pet Commodore.

Word Processor on Screen. Micros also can be used for word processing. You can buy word processing programs for almost all micros that will let you use your micro to generate printed materials like letters and reports.

Word processing allows you to electronically add words, delete words, add paragraphs, move paragraphs, etc. When you have your paper like you want it, you can print out the letter or paper on the computer's printer. I now write all my papers on the word processor.

While word processing will not be a big thing for many farmers or ranchers, it might be of value to those of you that are officers of farm organizations. Dave Flintner, President of Wyoming Farm Bureau, could surely use word processing in his Farm Bureau business.

Level II Computers

The more common level II microcomputers that farmers are considering are: Northstar, Vector Graphics, Superbrain, Hewlett Packard, and Cromenco. There are also other brands but they tend to be less popular.

Northstar Microcomputer. One Level II microcomputer that is fairly popular is the Northstar. Country Side Data out of Utah is selling agricultural software for the Northstar computer.

Vector Graphics Microcomputer. Another Level II microcomputer is called the Vector Graphics. Homestead Computers out of Canada has several software packages for the Vector. In addition, Loren Bennett in California has a dairy-ration package for the Vector.

This microcomputer and others can be equipped with a "professional" printer that is used for word processing. If we had a letter typed with this type of printer, I could convince you that the letter was typed by my secretary on her IBM electric typewriter. Professional printers sell for around $3,000; however, if you are going to do word processing, a professional printer is preferred.

One purebred cattleman has a professional printer on his micro. He uses the word processor to write individual letters to his purebred cattle customers. He keeps a list of potential customers inside his computer. When he has a bull for sale, he then uses the word processor to generate and address personal letters to each customer. Each customer thinks the cattleman personally typed the letter to them. In reality, his microcomputer merged the names into the standard letter stored in the micro. This cattleman argues that this is a very cost-effective way to advertise his purebred cattle. The key is the professional printer and the word processing software.

Superbrain microcomputer. A Superbrain is used by South Dakota AGNET with disk drives that are built into the cabinet. This is extremely nice when you move the microcomputer around.

Hewlett Packard. Hewlett Packard recently announced the HP-125 as their small business machine. HP long has a reputation of producing high quality products, and we believe this is also true for their microcomputers. To date, I am not aware of any agricultural software available for the HP machines.

Cromenco Computer. Cromenco microcomputer is configured to be a fairly powerful microcomputer, yet there are several empty slots for future additions to meet your expanding needs. The Level II machines are considerably more flexible than the level I machines.

Comparing Level I and II Microcomputers

There are several differences in the Level II micros as compared to the Level I micros. The key differences are: 1) basic language compilers that are faster than Level I interpretors, 2) 80 character screens that make VISICALC

and communications easier to use, 3) more standard operating
systems such as CP/M (this means it is easier to exchange
programs from one machine to another), 4) more error diag-
nostics for software and hardware, 5) and the S-100 buss
(for more hardware exchangeability).

Hardware Accessories

Data cassette. In the past, we used cassettes for data
and program storage. In fact, you can use your kids' cassettes and their tape recorder on your micro to record data
and programs. While this is a very cheap storage device, by
today's standards it is too slow and inflexible.

Floppy disk. The technology that has made microcomputers of value to agriculture is the floppy disk--a phono-
graph record with a paper covering around it. Instead of
recording music on the disk, the micro records data and computer instructions on the disk. The floppy disk now provides the microcomputer with mass storage capability. Dr.
Schoonover can store data for 500 cows in the beef program
on one of these disks. If you have 1,000 cows, you simply
use two disks. In fact, you can have as many of these disks
as you want on the shelf. You just pull off the shelf the
disk that you want and put it into your microcomputer.

Hard disk. The newest storage technology is the hard
disk. Inside this little box is the ability to store 5
million characters of data. You could store all the management information that you would ever need or generate on
your farm or ranch on one hard disk. Most farmers or ranchers do not have this kind of data storage need. The purebred cattleman I know with a Vector Graphics machine keeps
all his pedigree information for his cow herd on the hard
disk. He can go back to 1932 with his pedigree searches.
He feels that the microcomputer has helped his purebred
business out considerably.

Instructional Aids

A Radio Shack Teaching Center on our campus has 15
microcomputers hooked up to a sixteenth computer. The
sixteenth computer can monitor the other 15 computers.
Wyoming's Agricultural Extension Service needs one of these
to bring 15 ranchers or farmers in for computer training.
You learn more about microcomputers by hands-on experience
than from lecture or books. Many high schools and vocation-
al technical schools have such instructional centers but the
university extension services are behind.

MICROCOMPUTERS FOR FARM AND RANCH

Let's now boil all this down--what do microcomputers mean for you on the farm or ranch? Agriculture is going to have some serious challenges in the 80s. During the 60s and 70s your challenge was production, but the challenge in the 80s is going to be financial management. And the computer has the potential to improve your financial management. Let me make a prediction. Those of you that will be farming in 1990 will be using computers, and those of you that do not want to use a computer will not be farming in 1990. I often hear, "No damn computer is going to tell me how to run my farm!" I predict that that person won't be farming in 1990. Many will have retired and others will have gone out of business. Computers are going to become commonplace on U.S. farms and ranches during the 80s.

As I travel around the country talking to farmers and ranchers, I hear them expressing interest in three applications of the microcomputer. The three applications are:
- Business accounting.
- Herd performance reporting.
- Financial management.
Top producers are recognizing that they need to keep better books. They are looking to the microcomputer as a means to make bookkeeping easier and more flexible. They want current cashflow situations several times during the year. Today's profit margins do not allow the management errors that you could get by with in the 70s.

Top ranchers know the benefits of good cow-calf records and they have been keeping them on the 3- x 5-inch cards; however, it takes a lot of time to sort them into useful management reports. A herd performance system fits well onto a microcomputer and once the data is in the computer, management reports can easily be printed out. We even know of one rancher that takes his micro right out to the scales and enters the calf weights as they are weighed. When the last calf is weighed, he pushes the button and identifies the cows to be immediately culled. By not having to wait for culling data, this rancher argues that the dollar amount saved from not rounding up cattle the second time will pay for his microcomputer.

Bankers are requiring more and more financial information before they will make loans to producers. Top producers are starting to see the potential of being able to use the microcomputer to help generate these needed reports: financial statements, profit and loss statements, cash flow projections, five-year plans, etc.

VISICALC - a financial management tool. One of the most powerful financial management tools available is VISICALC. It is designed so that you don't have to be a programmer to program your own financial management programs. There is nothing equivalent on AGNET! Since I don't know how to describe in words what VISICALC can do, I suggest that you stop into a computer store and ask for a VISICALC demonstration.

Disk oriented system. In order to have sufficient capacity to handle your agricultural applications, producers should buy a disk-oriented system. It should contain:
- Dual-disk drives.
- A good 80-column printer.
- 32K to 48K memory (the horsepower of the computer).
- 80-column screen (preferred over a 40-column screen).
- Telephone coupler.

The system will cost between $4,000 to $5,000 for the hardware and about $2,000 to $3,000 for programs (software) for your farm or ranch.

Telephone coupler. One of the extremely useful attachments that you can purchase for your microcomputer is a telephone coupler. This will allow you to use your micro as a terminal to large mainframe computers such as AGNET, TELEPLAN, and CMN. You can call the mainframe on the telephone and type in your information on your micro's keyboard and have the output printed out on your micro's printer. The cost of a phone coupler is around $300 and you can access:
- Current commmodity market prices.
- Current USDA, Foreign Ag, and Wyoming news releases.
- Agricultural outlook and situation reports.
- Western Livestock Market Information Project livestock analyses.
- Hay for sale.
- Sheep for sale.
- Certified pesticide applicators in Wyoming.
- People interested in judging county and state fairs.
- Horticultural tips during the summer.
- Home-canning tips during the canning season.
- Emergency information such as drought tips, Mount St. Helen's emergencies, etc.

You can even use your micro to access the UPI and AP news services such as news dealing with the Farm Bill and "beef." The AP and UPI news services are available from two commercial time-share companies. You can do all this today with your micro if it has a telephone coupler on it.

HOW TO BUY A SMALL COMPUTER

What should a farmer and rancher do if he is thinking about buying a small computer?
There are two newsletters that I recommend that you subscribe to on computers in agriculture. Successful Farm-

ing publishes one newsletter for $40.00 per year. They make useful evaluations of hardware and agricultural software. The second newsletter is published by Doane-Western Agricultural Service out of St. Louis, Missouri. Their subscription rate is $48.00 per year. If you are seriously considering a microcomputer, I strongly recommend that you subscribe to one or both of these newsletters.

The second thing I recommend that you do if you are considering purchasing a computer is attend one of the computer seminars that are being held around the country. Almost every state extension service is holding these seminars specifically for farmers and ranchers interested in learning more about microcomputers and the potential agricultural applications. Contact your local county extension agent or extension advisor for information on these seminars.

Books and magazines on microcomputers and how to use and program them are also helpful when selecting a microcomputer. I strongly encourage farmers and ranchers who are thinking seriously about purchasing a computer to get one or two magazines or books on microcomputers. Farmers and ranchers read several agricultural-related magazines, so why not read at least one computer-related magazine.

I personally subscribe to BYTE. It is a good magazine to read to find out what kind of hardware is available and to learn the jargon of computers.

I also subscribe to the Personal Computing magazine. It has stories written by people who are familiar with microcomputers for people like you and me who are not familiar with microcomputers.

SUMMARY

Microcomputers are the new farm- and ranch-management tools and innovative producers are buying them. More and more farmers and ranchers are going to own one or more microcomputers.

If you buy a microcomputer, be sure and buy the telephone coupler so that you can access the agricultural information networks being set up across the country. You will need to spend around $4,000 to $5,000 for a microcomputer with enough horsepower and flexibility to do your farm or ranch applications. I assure you that we are going to see considerably more farm and ranch purchases in the next five years.

47
SIX STEPS FOR A CATTLEMAN TO TAKE IN BUYING A COMPUTER

Harlan G. Hughes

INTRODUCTION

Today's low profit margins and high interest rates place a premium on a cattleman's management-information system. Automation of that system lends itself to the microcomputer. Microcomputers represent a relatively new farm and ranch-management tool that farmers and ranchers are investigating. Purchasing one may prove to be one of the few profitable equipment purchases of the 1980s. One study indicates that as high as 64% of the producers interviewed were planning to buy a microcomputer as a management tool in the next five years. Twenty-seven percent indicated they would purchase a microcomputer in one to two years. These producers ranked business record keeping as the number one management function they wanted to perform on the microcomputer. The preparation of financial balance sheets and income and cash-flow statements ranked second. Breakeven analysis of individual enterprises and crop-production records ranked as the third and fourth management functions, respectively.

An Alberta, Canada, study of producers owning microcomputers indicated they were using the microcomputers for 1) farm planning, 2) financial record keeping, 3) physical record keeping, and 4) analysis of records (cash flow, breakeven analysis, and costs of production).

What kind of microcomputers do producers own? Sixty percent of the Canadian producers owned Radio Shack and the rest owned Apple, Pet Commodore, Vector Graphics, and others.

A recent Successful Farming magazine survey indicated that 46% of the respondents owned Apples, 34% owned Radio Shacks, 4% owned IBMs, 4% owned Commodore or Pet and the remaining percentage covered all other brands.

SIX STEPS FOR A COST-EFFECTIVE INVESTMENT

A producer-owned microcomputer should pass the same cost/benefit analysis as any other machinery investment.

Costs can be easily identified and documented; however, the benefit of improved management is considerably more difficult to document. What is clear, however, is that benefits received depend heavily on the preparation that the cattleman makes before purchasing the microcomputer.

Step 1

Before purchasing a microcomputer, study your management-information needs. Collection and analysis of management information requires time and money. You cannot afford to collect management information that you do not use or need. Some questions that you should ask are: What are the most important and significant decisions that I need to make? What information is needed to make these decisions? Can the generation of the needed information be scheduled? Can a microcomputer make this information collection easier? Studying your information requires some time and effort. It may well be worth your time to hire a consultant or visit with your university extension service and get a second opinion.

Step 2

Identify computer programs (software) that are available that might meet your management-information needs. As a cattleman you have four potential ways that you can obtain needed software. You can (1) buy it from a commercial vendor, (2) obtain it from the extension service, (3) hire it custom programmed, or (4) program it yourself.

If the software needed is available from a commercial vendor, this may well prove to be the most satisfactory method of acquiring software. Sometimes, however, you'll need software that is not available from a commercial vendor. The local extension service may have what is needed. Occasionally the only viable alternative is to hire a program custom-programmed or to program it yourself. Unless you have special training or a lot of spare time, I cannot recommend that you program the software on your own. Obtaining software tailored to your specific needs will be the most difficult and time-consuming task.

Step 3

Determine the hardware specifications required to execute the needed software. The size of the business affects the volume of management information needed and this, in turn, determines the size of hardware needed. Microcomputers come in different sizes (memory units), have different storage capabilities on the diskette (floppy disk), and have different add-on capabilities (80 column screens, upper and lower case characters, computer languages, CP/M operating systems, telephone modems, word processing software, etc.) Again, it is recommended that you contact a consultant or the extension service. Computer dealers are not necessarily the best information sources for determining specific hardware needs. Generally, they promote what they have to sell.

Step 4

Contact local hardware dealers and determine the viable hardware alternatives. Cattlemen should use the same criteria that they would use for any other equipment purchase: dealer knowledge of his own hardware, quality of the service department, apparent financial stability of the dealer's business and, in general, compatability with the dealer. Since cattlemen have purchased equipment before, they should feel reasonably comfortable with this step.

Step 5

Estimate the cost/benefit of the proposed computerized management-information system. A dealer can tell the purchaser exactly what the hardware will cost; and the cattleman already should have an estimate of what the software will cost. Remember that the cattleman-buyer can take investment credit and depreciation on computer hardware just like any other piece of machinery.

The clerical cost of collecting and processing the management information should also be included. This frequently is your time or that of your spouse. Collecting and typing data into the computer is time-consuming and boring. You might even consider hiring a person to be specifically responsible for the data processing of the management information.

While determining the cost/benefit, cost of the total management information system should be projected. A Michigan State University study indicates that it may cost $500 to $600 a year to process a producer's business records through his own microcomputer. Again, an outside consultant can be useful.

Estimating the dollar benefit of having a computerized management-information system is difficult for most cattlemen to do. Today's high costs of production and high interest rates do not leave much margin for management errors. Just preventing one management error a year may well pay for the microcomputer system. As could the ability to experiment with a decision on paper before implementation.

Step 6

The final step is to make the decision whether to set up a computerized management-information system. You should consider talking to other cattlemen that already own microcomputers. Many states are offering educational seminars for ranchers and farmers to learn more about how microcomputers can enhance a producer's decision-making process. The final decision rests with you the individual. There is no blank recommendation that will fit all situations. Microcomputers can, however, be an effective management tool.

Microcomputers are becoming a more common management tool for cattlemen. Innovative producers are purchasing microcomputers to enhance their personal management-information systems. This article summarizes six recommended steps that you should go through in making the decision to purchase a microcomputer. If these six steps are followed, you will have a higher probability for a successful experience with your first microcomputer.

REFERENCES

Engler, Verlyn, E. A. Unger and Bryan Schurle. 1981. The potential for microcomputer use in agriculture. Contribution 81-412-A. Department of Agricultural Economics, Kansas State Univ.

Nott, Sherrill. 1979. Feasibility of farm accounting on microcomputers. Agricultural Economics Report No. 336. Michigan State Univ.

Successful Farming. 1982. Successful Farming farm computer news. A special survey summary. Successful Farming.

1981. A survey of on-farm computer use in Alberta. Alberta Farm Management Branch, Olds, Alberta.

48
DIRECT DELIVERY OF MARKET INFORMATION THROUGH RANCHER-OWNED MICORCOMPUTERS: A RESEARCH REPORT

Harlan G. Hughes, Robert Price,
Doug Jose

Ranchers needs for marketing information have changed dramatically since the early 1970s. Increasing price variability, rapid inflation, higher interest rates, and closer ties to world supply-and-demand conditions for agricultural commodities have resulted in increased needs for short, intermediate, and long-run marketing information. Also, ranchers continually have fewer market outlets available so that they must do a better job of marketing their product. The net result is that many ranchers are unable to adequately evaluate marketing alternatives and, thus, are often unable to make good marketing decisions.

In late June 1981, Cooperative Agreement Number 12-05-300-522 was signed between the USDA Extension Service and the Colorado State University Cooperative Extension Service on behalf of the Western Livestock Marketing Information Project to give ranchers decision assistance. The agreement was to conduct a pilot study concerning the feasibility of direct electronic delivery of marketing and management information to farm and ranch families.

Ranchers base marketing decisions on information from both internal and external sources. Accounting records, herd performance records, and budgets are examples of internal information used. Market news, outlook reports, price forecasts, weather forecasts, and research reports are examples of external information. Internal and external information are required for almost all short, intermediate, and long-run marketing decisions.

Needs for short-run market information commonly relate to selling decisions. There are sometimes substantial risks associated with selling agricultural commodities today rather than waiting a few days, or vice versa. Short-run decisions are relatively simple to analyze in a budgeting sense as the costs are readily predictable. The difficult element is the probability of price increases and decreases.

Typically, university and government outlook specialists have not provided short-run market information. It has generally been left to the commodity brokerage firms and other private organizations to provide short-run market in-

formation. These sources tend to discount the risk and uncertainty aspects.

Intermediate-run needs for market information relate to
such decisions as purchasing of stocker and feeder cattle,
crop selections, fertilizer application, feed choice, and
other decisions that do not result in immediate revenue.
These decisions are generally more complex as the information needed to evaluate possible outcomes is more compli-
cated and has more chance of error. University and government outlook specialists generally have been most active in
providing intermediate-run information.

Examples of long-run market-information needs include
land purchases, irrigation development, machinery selection,
cattle herd expansion, and the construction of livestock
production units. These decisions, although not made as
frequently as the previous types, require significant information to allow for success of a farm business. Although
farm management economists have devoted much time and effort
to investment analysis, outlook specialists in the university and government realm generally have concentrated very
little on this long-run arena.

Because of variability in agricultural prices and production, as well as high financing requirements, producers
may risk bankruptcy before profits from an investment can be
realized. Long-run market information can also be useful in
assessing the amount of risk that a specific producer can
afford when making investment decisions.

A comprehensive marketing-information system, used properly, could play a major role in stabilizing or increasing
net ranch income during the 1980s. In the coming years,
ranchers are going to need more marketing information, delivered faster, and available in an easy-to-use form. Computers can and should play a major role in such a marketing
-information system and the associated educational needs.
The rapid development of electronic technology also presents
an exceptional opportunity for the Cooperative Extension
Service to assume an even greater role in the delivery of
timely market information.

DELIVERY OF MARKET INFORMATION IN THE WEST

The problem of delivering timely market information in
the western U.S. is compounded by the vast geographical dis-
persion of producers. The extension specialists and county
agents must travel extensively to accommodate the needs of
farmers and ranchers. Most newspapers carry very little, if
any, current market data. Farm magazines are major sources
of intermediate-run market information, but timeliness of
that information does not meet the standards necessary for
decision making in today's economic environment.

The Western Livestock Marketing Information Project
(WLMIP) was created over 25 years ago in recognition of
the void that existed in the delivery of timely market in-

formation to livestock producers in the West. The proven
record of WLMIP as a major source of useful intermediate-run
market information for the region has been well documented
(WLMIP, 1977; Bolen, 1949). However, the changing complexi-
ties of the livestock market, combined with the rapid growth
in computer technology, present the need and opportunity for
WLMIP to expand services. These opportunities include direct delivery of market information at the producer level,
as well as increased service to professional economists and
others in the West. It also presents the opportunity for
WLMIP to go from almost exclusive emphasis on intermediate-
run market information to expanding short-run information.

AGNET--AGRICULTURAL COMPUTER SYSTEM

AGNET is a time-sharing computer network headquartered
in Lincoln, Nebraska. There are over 2500 subscribers to
the network with a total yearly connecttime of over 75,000
hours. This averages out to 8.5 users per hour concurrently
on a 24-hour 7-day-a-week basis. AGNET is being utilized
for problem-solving and information networking. The system
is very "user-friendly" and is designed for use by people
with no computer background. Ranchers are allowed to subscribe to the AGNET System by paying variable costs associated with operating the system.

PREVIOUS STUDIES

In 1979 a survey was sent to state extension service
administrators inquiring about the priority of marketing extension programs. (Watkins and Hoobler, 1980). Thirty-
seven of the 44 state administrators returning the survey
placed extension marketing programs in the range of "important" to "of highest importance." The following summary
statement was taken from the report:
  - "It is recommended that each state Cooperative
    Extension Service administration, in cooperation
    with their marketing specialists and representa-
    tive clientele groups, examine the results of
    this national study, analyze their state's spe-
    cific needs, determine where a cooperative ef-
    fort is needed with other states, and develop
    plans for renewing and/or initiating programs to
    effectively manage the problems."
Brown and Collins (1978), University of Missouri, con-
ducted a national study in 1977 on the information needs of
large commercial farms. Their study revealed that:
  - Commercial family farmers and ranchers perceive
    marketing information as their number one need.
  - Extension and universities were rated the most
    important source of production technology, but
    only of minor importance as a source of market-
    ing information.

- Farmers, agribusiness, extension, and the agricultural media all expressed the belief that marketing information is critical now and will continue to be critical in the future. They also agreed that present sources of market information are inadequate.

A joint USDA/NASULGC study (1968) committee recommended in 1968 "that extension increase its emphasis on marketing and farm-business management while reducing the percentage of effort in husbandry and production." The study goes on to say, "Extension should gradually shift towards giving more in-depth training to producers and to wholesaling information through supply firms."

Most extension marketing-program-appraisal studies generally include recommendations for experimentation with the latest electronic and computer innovations. For example, New York dairymen in a 1977 telephone survey felt that extension could improve its effectiveness by placing more emphasis on the use of the computer as an educational tool. (Ainsle et al., 1977).

In spite of the emphasis placed more than 10 years ago on changing extension priorities, little progress has been made to implement these program shifts. This is mainly due to extension administrators' reluctance to changing priorities of their extension programs. We are hearing the same priority requests coming from producers today as we did a decade ago. This paper reports on one pilot project that attempted to respond to some of these priority requests.

PILOT PROJECT

The state of Wyoming piloted a basic electronic market information system on the AGNET computer network during 1978-79 (Skelton, 1980). Four objectives of the pilot system were:

1. To collect price information of interest to Wyoming producers.
2. To provide county extension offices with the ability to retrieve market information that allowed them to put together today's, yesterday's, last week's, last month's, or last year's markets of interest for use by their producers.
3. To provide simple, down-to-earth interpretations of what market prices and associated outlook mean to Wyoming producers.
4. To provide price forecasts for extension personnel to use with producers in planning.

The Western Livestock Marketing Information Project piloted some initial work in computerized market-information delivery in 1980. Major livestock reports (Cattle on Feed, Hogs and Pigs, etc.) were placed on AGNET, complete with analysis and interpretation. WLMIP was instrumental in substantially increasing listings of producers with hay for sale and making these listings available to areas hardest hit by the drought of 1980. In addition, WLMIP served as a clearing house for drought conditions in many areas of the western plains region. This information was collected and transmitted throughout the region and forwarded to the office of the Secretary of Agriculture in Washington, D.C.

OBJECTIVES OF THIS STUDY

In an effort to provide an evaluation of electronic delivery of market information, a cooperative agreement was signed between the Colorado State University Extension Service on behalf of WLMIP and the USDA Extension Service. Subsequent cooperation was obtained from the University of Wyoming and the University of Nebraska. Four of the six objectives of the pilot study reported in this report are:

1. Research and develop mechanisms for direct producer access to AGNET via farmer-owned microcomputers and computer terminals.
2. Add current livestock market news information on AGNET for retrieval by producers and others.
3. Improve the documentation of marketing information and other pertinent information on AGNET and make it available to producers.
4. Evaluate the effectiveness and efficiency of this new delivery system in providing useful information to farmers.

RESULTS

Objective 1: Research and develop mechanisms for direct producer access to AGNET by farmer-owned microcomputers and computer terminals.

The technology of communication between computers of different brands and types is an involved science of its own. Different hardware requires different communication protocols and procedures. Much additional work is needed in this area that is receiving a lot of interest at the current time.

The importance of networking between microcomputers and mainframe computers housing networks such as AGNET is becoming increasingly obvious. As an example, of the 12 producers that participated in the pilot study, 9 accessed AGNET through the use of microcomputers. The other 3 used "dumb terminals" for communication. It is the authors' opinion that microcomputers will become more the norm in producer hardware than dumb terminals. The reason is that the microcomputer can also be used to solve on-the-farm types of production and marketing problems, handle production and accounting records, and handle other applications.

Current technology is readily available to enable a microcomputer to operate as a dumb terminal in communicating with AGNET. Generally, all that is required is a modem (telephone coupler) and software for the microcomputer, which is generally included with the hardware coupler. Such packages for microcomputers generally run in the price range of $500 or less. However, the technology involved in making a microcomputer into an "intelligent terminal" with AGNET is more complex. A high degree of interest in this type of software appears evident throughout the western region.

The important of operating a microcomputer in an intelligent mode with AGNET arises from the tremendous potential savings in telephone costs. A large part of the user's time during any terminal session is now spent typing in the needed information to respond to the AGNET questions. If such files could be developed on the microcomputer before the telephone call is actually made to AGNET, much of the telephone cost could be eliminated.

The authors have experienced substantial savings in telephone costs (50% or more) when using microcomputers as intelligent terminals. The capability to access AGNET as a central warehouse for information, download the information to the microcomputer, hang up the telephone, and work with the information that has been accessed results in even more cost savings.

In summary, it is a relatively easy procedure to turn a microcomputer into a dumb terminal for communicating with AGNET. It becomes a little more difficult to operate in the intelligent-terminal mode, but software has been developed in this study that makes this possible for most brands of microcomputers. The idea of interfacing farmer-owned microcomputers and a regional computer, such as AGNET, could be one of the most significant thrusts in extension service activities for computer applications to agriculture in the coming years.

Objective 2: Add current market-news information on AGNET for retrieval by producers and others.

This objective has been pursued heavily since the beginning of the pilot project. The system has been expanded so that 17 different market-price files are going onto AGNET daily. In addition, weekly and monthly analyses are going onto AGNET. During the six-month study period, 19,873 market-price files were retrieved by all AGNET users.

In addition to providing information for its current market value, most of the information included in the price files is captured by the computer and put into historical data files. By building such a data bank, files are in place for retrieval by the user in various programs for management and marketing decisions. Most of the captured information is already available for use in various retrieval and charting programs. However, retrieval programs for AGNET market information are still under development.

In an effort outside the scope of this study, the Foreign Agriculture Service (FAS) of USDA has begun a test using AGNET markets for distribution for much of their information. The response from users of the FAS information has been very enthusiastic, and several new users have subscribed to AGNET just to receive the FAS information.

The addition of market news and other information on AGNET will be a continuing process. Feedback from participating county agents and producers during this pilot study resulted in several files being added. Additional feedback on the final end-users' evaluations points to the need for even more types of files.

Market information that is currently being placed on AGNET is almost exclusively done by volunteer labor. Therefore, relatively little money is allocated to staffing explicitly for placing market information on AGNET. Several staff hours weekly are being devoted from numerous offices to place the information on AGNET.

By relying so heavily on manual labor to provide the information to the computer network, costs are magnified and the chances for errors arise. USDA Extension Service is working with Agricultural Marketing Service for direct electronic transfers of market information from AMS to the AGNET computer. If such a system could be put in place, cost savings for staff time would be tremendous and the timeliness of the availability of the information could be much improved.

Objective 3: Improve the documentation of marketing information and other pertinent information on AGNET and make it available to producers.

One of the developments coming out of this pilot study has been an AGNET Market Information Users Guide. The guide is intended to be just that, a guide to help the new user know what type of market information is available and how to access that information. In addition to documenting the market information, management decision tools are also referenced in the guide with a brief explanation of how to access and use those tools.

AGNET is a very "user-friendly" computer system. The major part of any documentation needed by the user is accessible directly from the computer with the use of "HELP" commands built into the system. Such HELP commands are unique to AGNET. Consequently, most of the needed documentation and aids are available at any time during a terminal session and preclude the necessity of having a manual available for reference while the user is online.

Objective 4: Evaluate the effectiveness and efficiency of this new delivery system in providing useful information to farmers.

The study tested two methods of market-information delivery. The first method was actual direct delivery to farmer-owned microcomputers or computer terminals located on the farm or ranch. The second method tested was "wholesaling" market information through county agents or trained agricultural professionals. These agricultural professionals used marketing bulletin boards in the county

agent's office or in the financial institution center and
then used frequent mailings of market information from these
offices to selected producers in their area. This course of
delivery was used mainly to acquaint producers with the type
of information that could be obtained from the computer network and to test their responses to see if the information
delivered was useful. The authors were also interested in
seeing whether, after receiving the information in this manner, producers would be more interested in obtaining their
own computer hardware for direct delivery of the information.

EVALUATION OF USERS DIRECTLY ACCESSING INFORMATION

An evaluation form was sent to the producers who were
directly accessing the information from their own hardware.
Evaluation forms were returned from 12 direct-access users.
It should be noted that this group is a representative sub-
set of farmers and not all AGNET users.

Users were asked to evaluate six general types of market information they could access from the computer. The
results of that evaluation are listed in the following
table:

Evaluation By Direct Users

Information	Very useful	Slight-ly useful	Not useful	No response	Total
Futures prices	3	4	1	4	12
Cash prices	5	3	1	3	12
Commentary & interpretation	6	3	0	3	12
News releases	2	5	1	4	12
Retrieval programs	3	2	1	6	12
Conferences	1	4	0	7	12

The files that contain commentary and interpretation of
factors influencing the market were very well received by
the users who were accessing them directly. Various comments received on this question included: "Good insights."
"Comments really helped to get a feel for the market."
"More of this type of information needed."

The files on various cash prices were also very well
received by the users who were directly accessing the information. AGNET is very unusual in that several files contain
localized information for various areas within a state that
is not available anywhere else in a condensed, summarized
form. Comments included such things as: "We need more
local prices on the system." "Used these the most." "Often
AGNET is the only source of this information." "Excellent."

The files on futures prices were not perceived by the end users to be as useful as the two previously discussed categories. One of the main reasons for this is that AGNET only offers each day's open and close of the futures. Producers who are active in the futures markets find that they need more current quotes, which they obtain from their farm radios or from their brokers. Also, not many producers use the futures market. Some comments on the futures included: "Would be better if we had a detailed report on weekly futures price movement." "Information didn't fit our area completely as there were no sugar futures." "Out of date by the time the producers really need this information." "Useful if picked off daily."

The retrieval programs were not used as heavily as the authors hoped they might be. One of the main reasons was that perhaps the users did not feel they were sufficiently versed in the correct technical aspects to use the program. Typical comments for this information included: "Did not use." "What are these?" "We need a lot more information on how to run these programs." "I liked these very much and accessed them regularly."

NEWSRELEASE items were also not rated very highly by the end users. This was not too surprising as the NEWSRELEASE program on AGNET is generally considered to be more consumer oriented, although there is much good useful information for livestock and grain producers. Typical comments included: "Very few used." "Checked only on occasional basis." "Some good, some bad." "Especially liked the ones on economic issues." "Some were excellent."

The lowest-rated information source by the end users was the electronic CONFERENCES. Again, this was not too surprising. CONFERENCES are of more use to people other than farmers. It is the responsibility of an individual AGNET user to link with the electronic CONFERENCES. Although the authors had sent out the procedure for doing this via U.S. mail, it is doubtful that many of the users took the time to go through the procedure to link up to the CONFERENCES. Typical responses included: "Did not use." "So what?" "Helped sometimes." "Need more information on how to use." "Not enough conferences sales or prices."

The users were asked to evaluate the timeliness of information delivered by AGNET. The response broke down as follows: very timely--5; average timeliness--5; too late to be useful--1; no response--1.

The users, who were all paying their own computer and telephone costs, had a very high expectation of when the information should be available on the computer. Many times the information for a given day would not be available until the following morning because of the manual transfer of the information onto the system. Once again, this points out the high desirability of automatic linkages with the AMS teletype system so that the information can be available much more quickly. Typical responses to the timeliness question included: "Most information was available from

other sources at lower costs like newspapers and radio; however, this service shines in the fact that information is
available on demand." "Many times it is hard to check information everyday. Why not put on a program that records
daily futures-prices information and then on Friday evening
we could pull them off for our records."

Users were asked to report costs. Very few had kept
records of their costs, but those who did report indicated
that $50 to $75 a month was a normal combined telephone and
computer cost for accessing the AGNET information. A good
share of the users responded that they did take advantage of
nonprime-time telephone and computer costs by calling early
in the morning or late in the evening.

Only three users indicated any problems from trying to
access the information on AGNET. They also indicated that a
workshop on operating technique would have been helpful.
Most indicated that they felt it was quite easy to use the
system. However, six respondents indicated that a workshop
on how to apply the information being received from AGNET
would be extremely useful.

Users were asked to give suggestions for improving
AGNET delivery of market information and whether they felt
it was worthwhile to continue providing information across
the system. Most of the respondents who indicated that additional information would be desirable were looking for
more localized cash prices and more commentary with specific
projections for what the markets might do in the future.
The overwhelming response was that the direct delivery of
market information was extremely worthwhile and that the
project should be continued. Only two users indicated that
they did not intend to continue accessing AGNET information
regularly.

Although very few respondents put a dollar value on the
information received, the majority indicated that the cost-
benefit ratios for accessing the information were highly
favorable.

SUMMARY

The need for better information to be used by agricultural producers in making agricultural marketing decisions
has been well documented. The thrust of this study has been
to evaluate the feasibility of direct electronic delivery of
this needed market information. The development of mechanisms for direct producer access to AGNET via farmer-owned
microcomputers and computer terminals was one of the main
objectives.

The best evaluation of this project lies in the large
increase in retrievals of market information. During the
time period of the study there was over a three-fold increase in the number of times that AGNET was accessed for
market information.

Of the cooperators in this study, nearly three-fourths accessed AGNET through the use of microcomputers, while the remainder used dumb terminals for communication. It was found that operating a microcomputer in an intelligent mode with AGNET becomes increasingly desirable due to the tremendous potential in telephone savings. Savings of 50% or more resulted from using microcomputers in this manner.

It is relatively easy to turn a microcomputer into a dumb terminal for communicating with AGNET. It becomes much more difficult, however, to operate in the intelligent-terminal mode. This study uncovered software that makes this possible for most brands of microcomputers. The idea of interfacing farmer-owned microcomputers and a regional computer such as AGNET could be one of the most significant thrusts in extension service activities for agricultural computer applications in the future.

The pilot study was successful in providing current market news information to AGNET for retrieval by producers and others. A wide variety of new files has been made available in the MARKETS section on AGNET. Many of these files were the direct result of this pilot study. These files and others will continue to be available on AGNET.

Feedback from the final end user indicated the need for several more types of files, particularly of a regional type. The timeliness of the market information provided on AGNET was a concern to the producers involved in the study. Although the majority of the participants felt that the material was very helpful to them, they also expressed a desire for more timely information. This end-user evaluation points toward a critical need for direct electronic transfers of market information from AMS to the AGNET computer.

In summary, this pilot study provided much needed background information on the electronic delivery of market information. This study found that there is, indeed, a demand for the direct electronic delivery of marketing and management information to farm producers. There exits a distinct opportunity for the extension service to assume an even greater role in the delivery of this timely market information. In addition, the study provided the documentation of the need for increased development of computer applications to agriculture in information networking and evaluation of marketing alternatives. This information provides a base from which the extension service can evaluate and plan their activities in the computer arena for the future.

REFERENCES

Ainsle, et al. 1977. An evaluation of cooperative extension dairy programs. Specialist Report. Cornell University.

Bolen, Kenneth R. 1979. Economic information needs of
farmers. Report of ESCS and SEA/Extension Study.

Brown, Thomas R. and Arthur Collins. 1978. Large commercial family farms information needs and sources. A Report of the National Extension Study Committee.

Skelton, Irvin. 1980. Wyoming agricultural extension service accomplishment report for FY-1980.

Watkins, Ed and Sharon Hoobler. 1980. Report of ECOP Sub-
committee on agriculture forestry, and related industries extension marketing program and priorities survey. SEA/Extension.

USDA. 1968. A people and a spirit. Report of the joint
USDA/NASULGC study committee on cooperative extension.
Colorado State University.

WLMIP. 1977. Evaluation of the western livestock marketing
information project. Report of WLMIP Technical Advisory Committee Survey of Users.

Part 10

THE FUTURE OF
THE DAIRY INDUSTRY

49

A GLIMPSE AT DAIRYING
IN THE YEAR 2000

Robert E. Walton

It would be presumptuous of me to claim to have solid insights into what will be happening in the dairy industry in the year 2000. However, that is only 17 years away now, so it is closer than we think. In addition, I believe we all have a responsibility to look ahead a few years as part of our own personal and professional planning. Thus, it is in that context that I agreed to commit to the printed word and to you here today my far out (some may think way out!) thoughts on dairying. I must admit it has been fun working on this assignment. I invite you to join in the spirit of looking ahead with me.

LOOKING BACK

A good starting point for looking ahead is to look back at history:
- We have developed very precise and accurate genetic evaluation tools, particularly for sires widely used in artificial insemination, and have utilized this information more effectively than ever before in history. As a consequence, in recent years we have achieved a rate of genetic improvement in the U.S. of about 1% per year in milk production, compared to less than 1/4% per year improvement 20 years ago.
- In terms of production per cow, we have had increases ranging from 10% to 100% in different parts of the world. For example, we produce as much milk in the U.S. today with 11 million cows as we did in 1947 with 27 million cows. The top individual records now stand at 55,660 pounds of milk (Beecher Arlinda Ellen) and 2,230 pounds of butterfat (Breezewood Patsy Bar Pontiac). The top herd average is currently 27,239 pounds of milk and 1,030 pounds of butterfat in the Lekker herd in California.
- We have seen great emphasis on total pounds of production per cow and less emphasis on percent-

age butterfat. The past 5 years, however, have seen renewed emphasis on percentage butterfat and protein, both in the U.S. and in Europe. Today there are many bulls available that are plus (+) in breeding for percentage butterfat as well as for Predicted Difference Milk.

- We have seen dramatic shifts to Holstein cattle at the expense of other breeds. There are signs of renewed interest, however, in several of the high fat or high solids breeds as milk composition and efficiency of production in specific environments gets renewed interest. Some countries with red cattle are showing a growing interest in the Red and White Holstein.
- The superiority of U.S. cattle has been widely recognized by other countries as evidenced by the major trade in exporting semen and livestock all over the world from the U.S. This is particularly well documented for the Holstein breed, not only in terms of production per lactation but also other very important management traits such as rate of milk letdown per minute and for udder and teat structure designed for modern milking systems as well as for minimal damage or injury under commercial management systems.
- We have seen dramatic progress in the validity and accuracy of our evaluation systems for functional conformation or type. Consequently, there is renewed interest in the emphasis placed on these important factors by breeders and commercial producers alike. All major AI companies and herd associations are now using the linear evaluation system in their type evaluation programs.
- We have seen great emphasis on loose or group housing of cows and less emphasis on individual stanchion or tie stall housing.
- We have seen dramatic increases in herd sizes, so that many herds in the U.S. are operating with more than 1,000 cows per herd. Management problems tend to increase geometrically (i.e., faster) as herd size increases.
- Many computer applications to the routine management of dairy herds are now being developed and implemented.
- We have many new innovations in milking equipment and cow-handling systems utilizing new inventions of the electronic age.
- We have learned again how to produce more milk more efficiently, but we have forgotten again the critical importance of developing adequate markets for this output. Dairying perhaps does better than other segments of agriculture in

this regard, but the nature of dairying requires
daily disposal of most of our product to avoid
major disruptions. The meat industry has similar problems, as does fresh produce, but most
grains can be stored at reasonable costs without
major deterioration until needed or until markets improved.

LOOKING AHEAD

Looking ahead 17 years to the year 2,000, I foresee:
- In the field of breeding and genetics, a contin-
uation of the application of basic animal-breed-
ing principles to the practical field of dairy
cattle breeding and genetic improvement. The
basics will remain the same, i.e., special matings to produce super young bulls and progeny
testing those to sort out the best. But there
will be refinements in evaluation procedures and
perhaps some additional help from the physiologists and biochemists in better determining the
genetic makeup of the key animals in the population. I do not anticipate, however, that we
will be changing the genetic material in cattle
by chemical, radiation, or other laboratory
means during my lifetime. Nature's genetic system is beautiful, powerful, and nearly fool-
proof. Wide genetic variation exists in our
populations and will continue to be generated by
the genetic mechanism so that there is ample
room for continued genetic progress and improvement for several hundred years. We need to use
it wisely, however.

I do anticipate some important, if not major,
developments in this field, however, including
help from the physiologists and biochemists in
identifying the specific chemical nature of the
genetic material in cattle genes and identifying
early in their life those animals that carry the
most desirable genes so as to lessen some of the
heavy costs of progeny testing. I expect that
superovulation, in vitro fertilization, frozen
sexed embryos, sexed sperm, and perhaps even
cloning using nuclear transfer will be part of
our routine kit of tools in the breeding and improvement of our herds.
- In individual cow or herd performance, I anticipate:
1. Top individual records approaching 70,000
   pounds of milk and 3,000 pounds of butterfat.

2. Herd averages in the top herds approaching 35,000 pounds of milk and 1,400 pounds of butterfat.
3. Increased emphasis on the protein component of milk and probably upon total solids.
4. Increased emphasis on the important conformation points of the cow--the stress points as her production increases:
   Udder--teat size and location, median suspensory ligament, balance, and rate of milk letdown.
   Foot and leg structure.
   Pelvic area structure.
   Calving ease performance.

- In milking systems, there will be: (1) constantly improving technology that is better for the cow and her product and will be completely automated to match the uniformly perfect udders that will have been bred, and (2) the milking system will be integrated into a complete animal handling system on minicomputers that will: identify the cow, determine her individual nutritional needs, check for estrus and pregnancy, detect potential health problems, and record important daily information about the cow in the computer memory such as milk production, milk composition, feed consumed (energy equivalent), temperature, and estrus, breeding, pregnancy data, etc.
- In terms of dairy housing, we will still have many of the current systems and buildings, simply because of investment and cost, but the new systems will be very modern in line with new ideas on milking systems and animal handling concepts. The "cow psychologist" will have made contributions to our thinking regarding animal behavior and response. Although the cow will have been "genetically engineered" to better fit our systems and our needs, we will at the same time have made great progress in adapting our systems to her needs so as to more efficiently utilize the cow's potential as a producer of human food. Many of our current production systems ignore or are counter to the cow's biological needs. By the year 2000, as an economic necessity, our best dairy operations will have to be much more in tune with the cow's biological needs.

Herd sizes will undoubtedly increase on the average, just as a matter of economy of scale and efficient use of expensive supporting equipment and facilities. But I firmly believe that our rush to increase size will be strongly tempered by our cognizance of the high economic return for catering to the individual needs of these highly productive animals. Her value and replacement costs, as well as her productive potential, will make it even more worthwhile to assure her comfort, her maximum production, and her lessened risk of injury or disease as a matter of sound economics. I am not predicting (in any sense) a return to individual box stalls and individual pampering, rather I foresee improved environments within large-scale, free-stall systems that give the cow a better shot at doing her thing best for her owner.

- Energy will be an even more critical factor for the entire world in the year 2000. This will have an impact on the dairy industry in reducing the energy available to feed, house, and handle the cow and her products, and also in capturing more of the energy now lost as by-products. Heat removed from cooling the milk will be routinely recycled elsewhere in the enterprise. Methane and other fuel sources from manure and other animal wastes will be utilized for heating homes. Whey from cheesemaking will become a useful by-product, perhaps as a source of alcohol, wine, or fertilizer, rather than dumped in the sewers.

- While I must totally disqualify myself as an economist or an estate planner, I would feel remiss in this presentation in not recognizing the major impact these changes we've mentioned will have on the structure of ownership and sources of capital for the big dairy enterprises in the coming century. They will be big businesses with all that implies in terms of capital employed, sources of credit, management of resources, and continuity of management and ownership from generation to generation.

Some will likely become major corporations with shares of stock sold publicly. I believe the majority will continue to be family-type operations, but because of their size and the capital involved, they will become family corporations with stock closely held so as to assure continuity over time. They will have become too big and too costly to be routinely phased out by one generation and started up again by the next generation. Without the family-type corporation, taxes and inheritance laws would prevent their smooth transfer and operation. Nevertheless, there will always be successful dairy enterprises initiated and developed from scratch by daring entrepreneurs. That is the American way and I hope it will always be so.

LONG-RANGE PLANS

Looking ahead to the year 2000, we need to making long-range plans for marketing our product. We all know the problem of surpluses, and we must avoid that trap. The world needs and should be using more milk to improve diets everywhere. But it is best to start at home in trying to increase consumption. A look at the current (1981) per capita consumption of milk in different countries might be interesting.

Country	Milk production (metric ton)	Consumption/capita (lb)			
		Fluid milk	Butter	Cheese	Nonfat dry milk
Denmark	5,037	348	24	21	13
France	32,500	183	21	40	23
Netherlands	12,160	300	14	30	33
New Zealand	6,690	352	29	19	7
United States	54,628	139	4	17	3
West Germany	24,858	143	16	16	10

We have, however, doubled our consumption of cheese in the U.S. since 1960 and stabilized our fluid milk and butter consumption. Our dairy farmers have become much more market oriented, and in major market areas such as California where aggressive advertising and promotion has been done, there have been dramatic increases in consumption of fluid milk and other dairy products. There can be no doubt that the consumer will respond to positive promotion for dairy products.

I would suggest, as a favor to ourselves as producers, and to consumers for health reasons, that we give serious consideration as an industry to the commitment of advertising greater consumption of fluid milk. This will require a major long-range commitment, not only to promotion but also to the systems and manner in which milk is made available to the public.

You know and I know that there is no profit in producing more milk if there is not a market for that milk. We can do something about increasing that market and increasing our profits in the years ahead.

In the year 2000, the dairy enterprise will be a very sophisticated unit with input-output ratios closely monitored in terms of energy and resources used and food produced. High capital investment and limited energy resources will force this to be the case even more so than today.

As an example of the savings in cost, energy, and resources, consider the fact that in the U.S. we are producing more milk today with 11 million dairy cows than 35 years ago with 27 million dairy cows. The savings in dollars, feed, equipment, fuel, buildings, and labor for both raising and producing from those 16 million cows that are no longer required is an enormous figure. Increased production per animal is a major energy saver and emphasizes again how efficient the farmer is today in relation to other industries.

There is growing international dependency within the dairy industry involving utilization of genetic material. This international dependency also will come more and more to involve energy and feed sources, exchange of management ideas, and orderly milk marketing within and between countries.

In summary, I see the dairy industry in the year 2000 as a very healthy and basic industry; continuing to develop and incorporate innovative new concepts of management and technology; adapting to the economic realities of that era; enjoying the benefits of having allocated adequate resources to marketing as well as production; and setting new records of performance and efficiency on a cow, herd, and national basis.

NAMES AND ADDRESSES
OF THE LECTURERS AND STAFF

GEORGE AHLSCHWEDE
Sheep and Goat Specialist
Texas Agricultural Extension Service
Route 2, Box 950
San Angelo, TX 76901
--Sheep Specialist-Tour Coordinator

JOE B. ARMSTRONG
Associate Professor and
Extension Horse Specialist
Animal Science Department
New Mexico State University
Las Cruces, NM 88003
--Horse Geneticist

ROY. L. AX
Assistant Professor
Department of Dairy Science
University of Wisconsin
Madison, WI 53706
--Dairy Physiologist

HAROLD KENNETH BAKER
Meat and Livestock Commission
Box 44, Bletchley
Milton Keynes
MK2-2EF ENGLAND
--Beef Cattle Specialist & Geneticist

R. L. BAKER
Visiting Professor of Animal
    Breeding & Genetics
Animal Science Department
University of Illinois
Urbana, IL 61801
--New Zealand Geneticist

R. A. BELLOWS
Location Research Leader
USDA-ARS
Route 1, Box 2021
Miles City, MT 59301
--Beef Cattle Physiologist

W. T. BERRY, JR.
Executive Vice-President
National Cattlemen's Association
P. O. Box 3469
Englewood, CO 80155
--Beef Cattle Organization Leader

HENRY C. BESUDEN
Vinewood Farm
Route 2
Winchester, KY 40391
--All-Time Great (Sheepman)

RONALD BLACKWELL
Executive Secretary-General Manager
American Quarter Horse Association
Amarillo, TX 79168
--Horse Organization Leader

BILL BORROR
Tehama Angus Ranch
Route 1, Box 359
Gerber, CA 96035
--Cattle Breeder

MELVIN BRADLEY
Professor of Animal Science and
    State Extension Specialist
University of Missouri
Columbia, MO 65211
--Horse Specialist

B. C. BREIDENSTEIN
Director
Research and Nutrition Information
National Livestock and Meatboard
444 North Michigan Avenue
Chicago, IL 60611
--Meat Scientist

JENKS SWANN BRITT, D.V.M.
Veterinarian/Dairyman
Logan County Animal Clinic/J&W Dairy
Route 1
Russellville, KY 42276
--Veterinarian

HERB BROWN
Research Associate
Lilly Research Laboratories
P. O. Box 708
Greenfield, IN 46140
--Animal Scientist

O. D. BUTLER
Associate Deputy Chancellor
    for Agriculture
Texas A&M University
College Station, TX 77843
--Animal Scientist

EVERT K. BYINGTON
Range Scientist
Winrock International
Route 3
Morrilton, AR 72110
--Range Scientist

B. P. CARDON
Dean
College of Agriculture
University of Arizona
Tucson, AZ 85721
--Animal Scientist

ARTHUR CHRISTENSEN
Manager
Christensen Ranch
P. O. Box 186
Dillon, MT 59725
--Sheep Producer

ROBERT L. COOK
Wildlife Biologist
Shelton Land and Ranch Company
P. O. Box 1107
Kerrville, TX 78028
--Wildlife Management Specialist

CARL E. COPPOCK
Professor
Animal Science Department
Texas A&M University
College Station, TX 77843
--Dairy Nutritionist

DICK CROW
Publisher
Western Livestock Journal
Crow Publications Inc.
P. O. Drawer 17F
Denver, CO 80217
--Journalist

STANLEY E. CURTIS
Professor of Animal Science
University of Illinois
Urbana, IL 61801
--Animal Scientist

A. JOHN DE BOER
Agricultural Economist
Winrock International
Route 3
Morrilton, AR 72110
--Agricultural Economist

WAYNE L. DECKER
Professor
Department of Atmospheric Science
University of Missouri
Columbia, MO 65211
--Meteorologist and Agri Weather Specialist

R. O. DRUMMOND
Laboratory Director
U.S. Livestock Insects Laboratory
P. O. Box 232
Kerrville, TX 78028
--Entomologist

ED DUREN
Extension Livestock Specialist
P. O. Box 29
Soda Springs, ID 83276
--Animal Scientist

WILLIAM EATON
Clear Dawn Angus Farm
R. R. 1
Huntsville, IL 62344
--Registered Cattle Breeder

WILLIAM D. FARR
Farr Farms Company
Box 878
Greeley, CO 80632
--All-Time Great (Cattleman)

H. A. FITZHUGH
Animal Scientist
Winrock International
Route 3
Morrilton, AR 72110
--Animal Scientist

MIGUEL A. GALINA
Professor
Universidad Nacional Autonoma de Mexico
A.P. 25, Cuautitlan Izcalli
Edo de Mexico, MEXICO
--Animal Scientist

HENRY GARDINER
Route
Ashland, KS 67831
--Cattleman

DONALD R. GILL
Extension Animal Nutritionist
Oklahoma State University
005 Animal Science
Stillwater, OK 74078
--Nutritionist

HUDSON A. GLIMP
Blue Meadows Farm, Inc.
Route 2, Box 407
Danville, KY 40422
--Sheep Producer

MARTIN H. GONZALEZ
President
ECO TERRA SA de CV
Fernando de Borja 208
Chihuahua, Chihuahua
MEXICO
--Range Scientist

TEMPLE GRANDIN
Livestock Handling Consultant
Department of Animal Science
University of Illinois
Urbana, IL 61801
--Livestock Facilities Specialist

SAMUEL B. GUSS, D.V.M.
Professor Emeritus
Veterinary Science Extension
Pennsylvania State University
2410 Shingletown Road
State College, PA 16801
--Veterinarian

JAMES C. HEIRD
Assistant Professor
Department of Animal Science
Texas Tech University
P. O. Box 4169
Lubbock, TX 79409
--Horse Specialist

A. L. HOERMAN
Extension Livestock Specialist
Texas A&M University Extension Center
P. O. Drawer 1849
Uvalde, TX 78801
--Beef Tour Coordinator

DOUGLAS HOUSEHOLDER
Extension Horse Specialist
Texas A&M University
College Station, TX 77843
--Tour and Clinic Moderator

HARLAN G. HUGHES
Agricultural Economist
University of Wyoming
Laramie, WY 82071
--Agricultural Economist

CLARENCE V. HULET
Research Leader
U.S. Sheep Experiment Station
USDA, ARS
Dubois, ID 83423
--Sheep Physiologist

HENRYK A. JASIOROWSKI
Director, Cattle Breeding Research
Warsaw Agricultural University
AGGW-AR 02-528 Warszawa
Rakowiecka Str. 26/30
POLAND
--Animal Scientist

DONALD M. KINSMAN
Professor, Animal Industries Department
University of Connecticut
Storrs, CT 06268
--Meat Scientist

JACK L. KREIDER
Associate Professor
Horse Program
Texas A&M University
College Station, TX 77843
--Physiologist

JAMES W. LAUDERDALE
Senior Scientist
The Upjohn Company
Performance Enhancement Research
Kalamazoo, MI 49001
--Physiologist

ROBERT A. LONG
Professor
Department of Animal Science
Texas Tech University
P. O. Box 4169
Lubbock, TX 79409
--Animal Scientist

CRAIG LUDWIG
Director of TPR
American Hereford Association
715 Hereford Drive
P. O. Box 4059
Kansas City, MO 64101
--Cattle Organization Leader

JAMES P. MC CALL
Director of Horse Program
Stallion Station
Louisiana Tech University
P. O. Box 10198
Ruston, LA 71272
--Horse Specialist

WILLIAM C. MC MULLEN, D.V.M.
Large Animal Medicine & Surgery
Texas A&M University
College Station, TX 77843
--Veterinarian

JOHN W. MC NEILL
Beef Cattle Specialist
Texas Agricultural Extension Service
6500 Amarillo Blvd., West
Amarillo, TX 79106
--Beef Cattle Specialist

DOYLE G. MEADOWS
Manager
Robinwood Farm
2822 East 2nd Street
Edmond, OK 73034
--Horse Specialist

JOHN L. MERRILL
XXX Ranch
Route 1, Box 54
Crowley, TX 76036
--Range Scientist and Rancher

BRET K. MIDDLETON
Animal Science Department
Iowa State University
Ames, Iowa 50011
--Computer Cow Game Coordinator

J. D. MORROW
Executive Vice-President
International Brangus Breeders Association
9500 Tioga Drive
San Antonio, TX 78230
--Cattle Organization Leader

HARRY C. MUSSMAN, D.V.M.
Administrator
APHIS/USDA
Washington, D.C. 20250
--Veterinarian

CHARLES W. NICHOLS
Manager
Davidson & Sons Cattle Company
Route 2, Box 15-0
Arnett, OK 73832
--Cattleman

J. DAVID NICHOLS
Anita, IA 50020
--Registered Cattle Breeder

MICHAEL J. NOLAN
Executive Secretary
Health and Regulatory Committee
American Horse Council, Inc.
1700 K Street, N.W., Suite 300
Washington, D.C. 20006
--Horse Specialist

JAY O'BRIEN
P. O. Box 9598
Amarillo, TX 79105
--Cattleman

JERRY O'SHEA
Principal Research Officer
The Agricultural Institute
Moorepark, Fermoy Company
Cork, IRELAND
--Dairy Equipment and Mastitis Expert

RICHARD O. PARKER
Assistant to the President
Agriservices Foundation
648 West Sierra Avenue
P. O. Box 429
Clovis, CA 93612
--Physiologist

BRENT PERRY, D.V.M.
President
Rio Vista International, Inc.
Route 9, Box 242
San Antonio, TX 78227
--Veterinarian

GUSTAV PERSON, JR.
Extension Agent
Guadalupe County
Ag Building
Seguin, TX 78155
--Horse Tour Coordinator

L. S. POPE
Dean
College of Agriculture
New Mexico State University
Las Cruces, NM 88003
--Animal Scientist

DOUGLAS PRESLEY
Extension Agent
Bexar County
Room 310
203 W. Nueva
San Antonio, TX 78207
--Horse Clinic Coordinator

NED S. RAUN
Vice-President, Programs
Winrock International
Route 3
Morrilton, AR 72110
--Animal Scientist

PATRICK O. REARDON
Assistant General Manager
Chaparrosa Ranch
P. O. Box 489
La Pryor, TX 78872
--Range Scientist

RUBY RINGSDORF
President
American Agri Women
28781 Bodenhamer Road
Eugene, OR 97402
--Farm Organization Leader

DON G. ROLLINS, D.V.M.
Technical Veterinary Advisor
Mid-America Dairymen, Inc.
800 West Tampa
Springfield, MO 65805
--Veterinarian

BUDDY ROULSTON
Professional Horse Trainer
Brenham, TX 77833
--Horse Trainer

MANUEL E. RUIZ
Head
Programa de Produccion Animal
CATIE
Turrialba, COSTA RICA
--Nutritionist

CHARLES G. SCRUGGS
Vice-President and Editor
Progressive Farmer
P. O. Box 2581
Birmingham, AL 35202
--Journalist

RICHARD S. SECHRIST
Executive Secretary
National Dairy Herd Improvement Association
3021 East Dublin-Granville Road
Columbus, OH 43229
--Dairy Organization Leader

MAURICE SHELTON
Professor
Texas Agricultural Experiment Station
Route 2, Box 950
San Angelo, TX 76901
--Sheep Specialist

PAT SHEPHERD
Manager
South Plains Feed Yard
Drawer C
Hale Center, TX 79041
--Feedlot Manager

JOHN STEWART-SMITH
President
Beefbooster Cattle Ltd.
P. O. Box 396
Cochrane, Alberta
CANADA TOL OWO
--Cattle Breeder

GEORGE STONE
President
National Farmers Union
12025 East 45th Avenue
Denver, CO 80251
--Farm Organization Leader

JACK D. STOUT
Associate Professor
Extension Dairy Specialist
Oklahoma State University
Stillwater, OK 74074
--Dairy Specialist

MRS. BAZY TANKERSLEY
Al-Marah Arabians
4101 North Bear Canyon Road
Tucson, AZ 85715
--All-Time Great (Horsewoman)

MAURICE TELLEEN
Editor and Publisher
The Draft Horse Journal
Box 670
Waverly, IA 50677
--Horseman and Journalist

THOMAS R. THEDFORD, D.V.M.
Extension Veterinarian
Oklahoma State University
Stillwater, OK 74078
--Veterinarian

TOPPER THORPE
General Manager
Cattle-Fax
5420 South Quebec Street
Englewood, CO 80155
--Agricultural Economist

ALLEN D. TILLMAN
Rockefeller Foundation, Emeritus
523 West Harned Place
Stillwater, OK 74074
--Animal Scientist

JAMES N. TRAPP
Associate Professor
Agricultural Economics Department
Oklahoma State University
Stillwater, OK 74078
--Agricultural Economist

ROBERT WALTON
President
American Breeders Service
P. O. Box 459
DeForest, WI 53532
--All-Time Great (Dairyman)

RODGER L. WASSON
Executive Director
American Sheep Producers Council, Inc.
200 Clayton Street
Denver, CO 80206
--Sheep Association Leader

DOYLE WAYBRIGHT
Mason Dixon Farms
RD 2
Gettsburg, PA 17325
--Dairyman

GARY W. WEBB
Stallion Manager
Winmunn Quarter Horses, Inc.
Route 1, Box 460
Brenham, TX 77833
--Horse Specialist

RICHARD O. WHEELER
President
Winrock International
Route 3
Morrilton, AR 72110
--Agricultural Economist

DICK WHETSELL
Vice-Chairman
Board of Directors
Oklahoma Land & Cattle Company
P. O. Box 1389
Pawhuska, OK 74056
--Range Scientist

R. GENE WHITE, D.V.M.
Coordinator of the Regional College
of Veterinary Medicine
University of Nebraska
Lincoln, NE 68583
--Veterinarian

RICHARD L. WILLHAM
Professor of Animal Science
Iowa State University
Kildee Hall
Ames, IA 50011
--Animal Breeding Specialist

DON WILLIAMS, D.V.M.
Henry C. Hitch Feedlot, Inc.
Box 1442
Guymon, OK 73942
--Feedlot Manager and Veterinarian

JAMES N. WILTBANK
Professor
Department of Animal Husbandry
Brigham Young University
Provo, UT 84602
--Beef Cattle Physiologist

CHRIS G. WOELFEL
Dairy Specialist
Texas Agricultural Extension Service
218 Kleberg Center
College Station, TX 77843
--Dairy Tour Coordinator

JAMES A. YAZMAN
Animal Scientist
Winrock International
Route 3
Morrilton, AR 72110
--Dairy Goat Specialist and Nutritionist

Other Winrock International Studies
Published by Westview Press

Hair Sheep of West Africa and the Americas, edited by H.
A. Fitzhugh and G. Eric Bradford

Future Dimensions of World Food and Population, edited by
Richard G. Woods

Other Books of Interest from Westview Press

Carcase Evaluation in Livestock Breeding, Production and
Marketing, A. J. Kempster, A. Cuthbertson, and G. Harrington

Energy Impacts Upon Future Livestock Production, Gerald
M. Ward

Science, Agriculture, and the Politics of Research, Lawrence
Busch and William B. Lacy

Developing Strategies for Rangeland Management, National
Research Council

Proceedings of the XIV International Grassland Conference,
edited by J. Allan Smith and Virgil W. Hays

Animal Agriculture: Research to Meet Human Needs in the 21st
Century, edited by Wilson G. Pond, Robert A. Merkel, Lon D.
McGilliard, and V. James Rhodes

Other Books of Interest
from Winrock International[*]

Ruminant Products: More Than Meat and Milk, R. E. McDowell

The Role of Ruminants in Support of Man, H. A. Fitzhugh,
H. J. Hodgson, O. J. Scoville, Thanh D. Nguyen, and
T. C. Byerly

Potential of the World's Forages for Ruminant Animal Production, Second Edition, edited by R. Dennis Child and Evert
K. Byington

Research on Crop-Animal Systems, edited by H. A. Fitzhugh,
R. D. Hart, R. A. Moreno, P. O. Osuji, M. E. Ruiz, and
L. Singh

Bibliography on Crop-Animal Systems, H. A. Fitzhugh and
R. Hart

[*]Available directly from Winrock International, Petit Jean Mountain,
Morrilton, Arkansas 72110